SCHAUM'S OUTLINE OF

Theory and Problems of
Immunology

GEORGE PINCHUK

Department of Biological Sciences
Mississippi State University

Schaum's Outline Series

McGRAW-HILL

New York   Chicago   San Francisco   Lisbon   London   Madrid   Mexico City
Milan   New Delhi   San Juan   Seoul   Singapore   Sydney   Toronto

GEORGE V. PINCHUK was born in Ukraine. He obtained his M.D. at Kyiv Medical University and his Candidate of Medical Sciences (Ph.D., Medical Sciences) degree at the Institute of Medical Genetics, Moscow. Dr. Pinchuk moved to the USA in 1990 and worked as a postdoctoral researcher at the University of Southern California, the Fred Hutchinson Cancer Research Center, and the Virginia Mason Research Center. His published research is the area of molecular immunology, with the focus on the human antibody repertoire. Dr. Pinchuk taught immunology at Seattle Pacific University and the University of Washington. Currently, he is on the faculty of Mississippi State University, where he is responsible for a program in immunology and where he teaches an immunology course.

Schaum's Outline of Theory and Problems of
IMMUNOLOGY

Copyright © 2002 by The McGraw-Hill Companies, Inc. All rights reserved. Printed in the United States of America. Except as permitted under the Copyright Act of 1976, no part of this publication may be reproduced or distributed in any form or by any means, or stored in a data base or retrieval system, without the prior written permission of the publisher.

4 5 6 7 8 9 10 11 12 13 14 15 16 17 18 19 20 CUS CUS 0 9 8 7

ISBN 0-07-137366-7

Sponsoring Editor: Barbara Gilson
Production Supervisor: Clare Stanley
Editing Liaison: Maureen B. Walker
Project Supervision: Keyword Publishing Services Ltd.

Library of Congress Cataloging-in-Publication Data applied for.

McGraw-Hill
A Division of The McGraw-Hill Companies

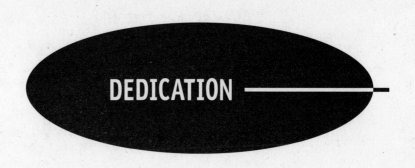

DEDICATION

*To the memory of my father, Vadim Pinchuk,
a wonderful man and an outstanding scientist*

PREFACE

This book is an attempt to provide students who study immunology in colleges and universities, as well as everyone who wants to refresh, deepen, and systematize their knowledge of immunology, with an outline of an up-to-date immunology course. Its focus is on cellular and molecular mechanisms of immune responses. The author considers this book to be an attempt to supplement such excellent comprehensive immunology textbooks as *Cellular and Molecular Immunology* (by A.K. Abbas, A.H. Lichtman, and J.S. Pober, W.B. Saunders Co., Philadelphia *et al.*, Fourth edition, 2000), *Kuby Immunology* (by R.A. Goldsby, T.J. Kindt, and B.A. Osborne, W.H. Freeman and Co., New York, Fourth edition, 2000), and others. Throughout the book, the author used a question-and-answer format (which is so characteristic for *Schaum's Outlines*), based on real questions that real students – both undergraduate and graduate – ask, and real answers that real immunology professors, including the author himself, try to give.

I am grateful to a large number of people who helped me during the conception and the writing of this book. I will mention but a few. John Welborn, my colleague and friend in the Department of Biological Sciences of the College of Arts and Sciences at Mississippi State University, Mississippi State, MS, was first to suggest that I might write the book, and graciously served as a go-between the publishers and me at the stage of its conception. Glenn Mott, the Editor of the McGraw-Hill Professional Book Group, New York, NY; Jennifer Chong and Maureen Walker, of the same company provided me with invaluable guidance and advice during the course of the writing. Tracy Sanchez Ambrose of TypeMaster Inc., Choudrant, LA, proofread and edited a number of chapters and gave me a tremendous encouragement during the writing, making me believe that I can write and supporting me with her steadfast friendship and humor. Jeff Rudis helped me with a competent and sound advice in word processing and file management. Maureen Allen and her associates of Keyword Publishing Services, Barking, UK, were extremely collegial and efficient during the final preparation of the manuscript for printing. My colleagues at the Department of Biological Sciences and, of course, my students deserve my warmest acknowledgment for being patient and forgiving at the times when I was somewhat overwhelmed with the work on this book at the expense of my responsibilities at school. Last but in no way least, I am very much indebted and thankful to my family members – to my wife Lesya, daughter Maryana, and to my mother, Ludmila Pinchuk, for bearing with me and being there for me always.

Feci quod potui, faciunt meliora potentes.

GEORGE V. PINCHUK

CONTENTS

CHAPTER 1   **Overview of Immunity and the Immune System**   **1**
Introduction   1
Discussion   2
     General Features of Immune Responses   2
     Clonal Selection Hypothesis   7

CHAPTER 2   **Cells, Tissues, and Organs of the Immune System**   **13**
Introduction   13
Discussion   14
     Lymphocytes   14
     Macrophages   17
     Dendritic Cells   20
     Other Nonlymphoid Cells that Participate in
         Immune Reactions   22
     Lymphoid Tissues and Organs   23

CHAPTER 3   **Antibodies and Antigens**   **32**
Introduction   32
Discussion   33
     An Overview of Antibody Structure   33
     Antigen–Antibody Interaction   40
     Antibody Heterogeneity and Hybridoma
         Technology   43
     Laboratory Uses of Antibodies   47

CHAPTER 4   **Maturation of B Lymphocytes and Expression of Immunoglobulin Genes**   **55**
Introduction   55
Discussion   56
     Generation of the Primary Antibody Repertoire   56
     Antibody Gene Mutations and the Generation of
         Secondary Antibody Repertoire   68

CHAPTER 5   **The Major Histocompatibility Complex**   **77**
Introduction   77
Discussion   78

CHAPTER 6	**Antigen Processing and Presentation**	**94**
	Introduction	94
	Discussion	95
CHAPTER 7	**T-Lymphocyte Antigen Recognition and Activation**	**108**
	Introduction	108
	Discussion	109
CHAPTER 8	**B-Lymphocyte Activation and Antibody Production**	**128**
	Introduction	128
	Discussion	129
CHAPTER 9	**Immunologic Tolerance**	**143**
	Introduction	143
	Discussion	144
CHAPTER 10	**Cytokines**	**158**
	Introduction	158
	Discussion	158
	A. Common Properties of Cytokines	158
	B. Cytokines that Mediate and Regulate Innate Immunity	163
	C. Cytokines that Mediate and Regulate Adaptive Immunity	170
	D. Cytokines that Stimulate Hemopoiesis	177
CHAPTER 11	**Innate Immunity**	**182**
	Introduction	182
	Discussion	183
CHAPTER 12	**Effector Mechanisms of Cell-Mediated Immunity**	**199**
	Introduction	199
	Discussion	200
CHAPTER 13	**Effector Mechanisms of Humoral Immunity**	**212**
	Introduction	212
	Discussion	213
CHAPTER 14	**Immunity to Microbes**	**237**
	Introduction	237
	Discussion	238

Contents

CHAPTER 15	**Transplantation Immunology**	**253**
	Introduction	253
	Discussion	254
CHAPTER 16	**Immunity to Tumors**	**270**
	Introduction	270
	Discussion	271
CHAPTER 17	**Autoimmunity and Autoimmune Diseases**	**282**
	Introduction	282
	Discussion	283
CHAPTER 18	**Immunodeficiencies**	**293**
	Introduction	293
	Discussion	294
INDEX		**311**

CHAPTER 1

Overview of Immunity and the Immune System

Introduction

Immunology is a science that studies **immunity**. Historically, immunity has been understood as a defense against, or resistance to, contagious (infectious) diseases. It has become apparent, however, that the mechanisms that confer protection against the above diseases also operate when a body mounts a reaction against some innocuous substances. Such a reaction is triggered when certain substances that are not made in the body ("foreign" substances) invade the body from outside. The mechanisms of immunity can protect against diseases that might be caused by the foreign agents but, on the other hand, these same mechanisms can themselves injure the body and cause disease. Therefore, immunity was redefined as a reaction against foreign substances, including – but not limited to – infectious microorganisms. This reaction may or may not be protective. In some instances, it is aimed at altered (e.g., malignantly transformed) self substances, or even to unaltered self substances. This reaction is quite complex, involves many different cells, molecules, and genes (collectively termed **the immune system**), and is aimed essentially at maintaining the genetic integrity of an individual, protecting it from the invasion of substances that can bear the imprint of a foreign genetic code. The response of the immune system to the introduction of foreign substances is called the **immune response**.

Immunity is a part of a complex system of defense reactions of the body. These defense reactions can be innate or acquired. **Innate (or natural) immunity** refers to the work of mechanisms that pre-exist the invasion of foreign substances. These include physical barriers like the skin and mucosal surfaces;

chemical substances (mostly proteins) that neutralize microorganisms and other foreign particles; and specialized cells that engulf and digest foreign particles. The mechanisms of innate immunity are **non-specific**, i.e., they do not discriminate between different kinds of foreign substances. Also, the innate immunity is **non-adaptive**, i.e., the nature or quality of the reaction to a foreign substance does not change when the organism encounters this substance repeatedly. **Acquired immunity** refers to a reaction that is caused by the invasion of a certain foreign substance. The elements of this reaction pre-exist the invasion of the foreign substance, but the reaction itself is generated strictly in response to a certain foreign agent (which is called an **antigen**) and changes its magnitude as well as quality with each successive encounter of the same antigen. The acquired immunity is highly **specific**, i.e., the system discriminates between various antigens, responding with a unique reaction to every particular antigen. The acquired (or specific) immunity is highly **adaptive**, i.e., the nature or quality of the reaction to an antigen changes after the encounter with this antigen, and especially when the organism encounters the same antigen repeatedly. The ability of the immune system to "remember" an encounter with an antigen and to develop a qualitatively better response to it is called **the immune memory**. This feature is a paramount property of specific immunity.

In the subsequent sections, we will dissect the particular mechanisms of immunity and characterize the elements of the immune system and the properties of immune responses.

Discussion

GENERAL FEATURES OF IMMUNE RESPONSES

1.1 What is the purpose of the immune reaction?

Essentially, it is to rid the organism of foreign antigens. From birth until death, an organism is surrounded by a host of microorganisms, many of which are dangerous. Using antigen receptors, the immune system continuously screens myriads of substances in the body, discerning among them and mounting an attack against those that are foreign. The end result of a successful immune attack is the destruction of foreign substances and particles, including microbial cells, viruses, various toxins and also tumors. Once destroyed by the immune system, foreign substances or particles or their remains are cleared from the body.

1.2 What is an antigen?

Immunologists use the term "antigen" in two senses. Originally, antigens were understood as substances that can trigger, or **gen**erate, immune responses. The other definition of antigen is a substance that the immune system can specifically recognize with the help of antigen receptors expressed on lymphocytes or secreted by them (see below).

CHAPTER 1 Immunity and the Immune System

1.3 Can all substances be antigens?

No. In order to be an antigen, a substance must be of enough complexity to bear an imprint of potential "foreignness." For example, a protein can be an antigen, because it has a complex structure determined by the sequence of its amino acids. The latter, in turn, is determined by an individual's genetic code. A different individual may have (and usually does have) a different genetic code for the same protein. On the other hand, simple inorganic chemicals like water or salt, or simple organic molecules like glucose, cannot be antigens because it does not matter in which biological individual they have been synthesized: their structure will be the same anyway.

1.4 Can substances other than proteins be antigens?

Yes. Most antigens are proteins, but polysaccharides, certain lipids, and nucleic acids also can trigger immune reactions. Besides, some relatively simple organic chemicals and chemical groups can be specifically recognized by the immune system, although they cannot trigger immune reactions. Such substances are called **haptens**. The immune response specific to a hapten can be triggered if the hapten is chemically coupled with a protein. The latter in this case will be called a **carrier**.

1.5 How does the immune system react against antigens?

For this purpose, the immune system uses molecules called **antibodies**, and cells called **lymphocytes**. Antibodies are protein molecules synthesized by a class of lymphocytes called **B lymphocytes** (or **B cells**). Antibody molecules recognize antigens through physical contact. They can be either expressed on the surface of B cells, or secreted. The other class of lymphocytes, **T lymphocytes** (or **T cells**) expresses molecules that also can recognize antigens through physical contact. These molecules are somewhat similar to antibodies, yet of a different structure and, unlike antibodies, they are never secreted. The molecules that are made by both classes of lymphocytes (T and B) and that are able to recognize antigens are called **antigen receptors**. Antibodies are **B-cell antigen receptors**, and antibody-like molecules expressed on T lymphocytes are **T-cell antigen receptors** (**TCRs**). The antigen receptor is what determines the specificity of any given lymphocyte. Only lymphocytes can make antigen receptors and, therefore, recognize antigens. However, some other cell types can be, and often are, involved in the immune response, although they cannot specifically discern between antigens. These antigen-nonspecific cells aid lymphocytes during specific immune responses and are called **accessory cells**. (See a more detailed discussion of lymphocytes and accessory cells in Chapter 2.)

1.6 Do all organisms have lymphocytes and antibodies?

The specific immune system, or the system that mediates adaptive immunity, is a feature of higher vertebrates. Lymphocytes and their specific antigen receptors appear in jawed fishes and are more diverse and efficient in amphibians, reptiles, birds, and mammals. More primitive organisms have nonspecific immunity, however. Molecules and cells, resembling those that are parts of the nonspecific immune system in higher organisms, are operative in insects, worms, and even sponges.

1.7 How did immunologists learn about T and B lymphocytes?

From observations and experiments with components of immune reactions. Immunity can be **active** or **passive**. Active immunity refers to the immune reaction that develops in an organism after the introduction of an antigen (**immunization**). An organism that is not immunized but receives blood cells or serum from an actively immunized individual acquires passive immunity. From observations on animals acquiring passive immunity with a transfer of either serum or cells, immunologists learned that immunity could be **humoral** or **cellular** (or cell-mediated). The former is conferred by substances dissolved in serum and other body fluids (Latin *humori*). Today we know that these soluble substances are antibodies and that they are produced by B lymphocytes. Cells, more precisely, lymphocytes and accessory cells with the necessary participation of T lymphocytes, confer cellular immunity. T lymphocytes play a major role in the recognition of antigens and their elimination, but they do not produce antibodies (Fig. 1-1).

1.8 Why does the immune system need both T and B lymphocytes?

These two classes of lymphocytes are designed to take care of two different classes of antigens. T cells are designed primarily to fight foreign substances that are hidden within the organism's cells (intracellular). Among these substances are viruses and intracellular bacteria. Proteins made by these intracellular parasites are displayed on the membranes of the infected cells. The TCR (see Chapter 7) is built so that it can recognize parts of these proteins (**peptides**) in conjunction with certain structures expressed on host's cell membranes and called **major histocompatibility complex (MHC) molecules**. (We will discuss them in detail in Chapters 5 and 6.) **T cells** can, therefore, **exclusively recognize entities attached to the membranes of the host's own cells**. This pattern of recognition helps them to "concentrate their attention" exclusively on the organism's own cells, screening them for signs of **infection by viruses or other intracellular parasites, as well as of malignant transformation**. On the other hand, **B cells** and antibodies that they make (Chapter 3) are designed primarily to fight foreign antigens that are located **in the extracellular space**. Among these substances are extracellular microorganisms, toxins, and extraneous chemicals. Unlike TCRs, antibodies recognize antigens in their native form, which does not require the antigen's attachment to cellular membranes and conjunction with MHC. Thus, B cells are very efficient weapons against "loose" extracellular microbes. They can reach them with the help of secreted antibodies that can float almost everywhere in the body. The T cells,

CHAPTER 1 Immunity and the Immune System   5

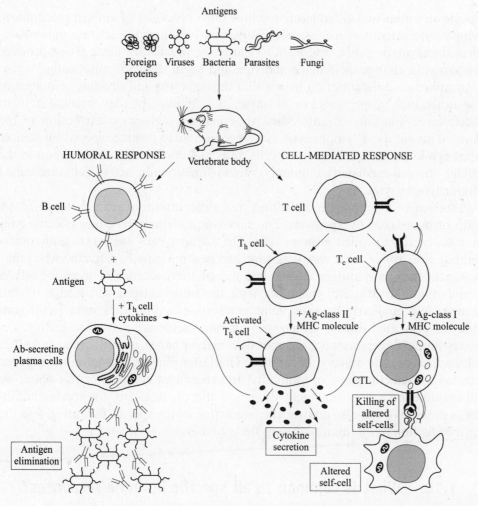

Fig. 1-1. Humoral and cellular immunity as two "arms" of the specific immunity. Antigens that invade higher vertebrates trigger humoral and cellular (or cell-mediated) specific immune responses. The former involve B cells that may differentiate into antibody(Ab)-secreting plasma cells; the latter involve T cells that may differentiate into T_h or T_c. While antibodies recognize antigens in their native form, T cells recognize antigenic peptides that are incorporated into self MHC molecules ("altered self").

however, not only recognize intracellular antigens, but also control and regulate the function of B cells in most immune responses.

1.9 How exactly does the immune system deal with antigens?

A complex series of processes, collectively called the **immune response**, follows the contact of antibody or TCR with antigen. All specific immune responses undergo **phases**. Initially, the **antigen recognition** must occur, which means that an antibody, expressed on the surface of a B lymphocyte, or a TCR, expressed on the surface of a T lymphocyte, must bind it with a certain affinity. Only those antibodies or TCRs that are specific, i.e., complementary, to an antigen can bind it. Processes of **lym-**

phocyte activation and differentiation follow these processes of antigen recognition. Lymphocyte activation means, essentially, that lymphocytes activate many complicated enzymatic processes, begin to transcribe previously silent genes, produce new proteins, change their shape and size, and begin to divide mitotically.

Lymphocyte differentiation means that the activated and dividing cells acquire new functional properties. For example, B lymphocytes may become able to secrete large amounts of antibodies; thus making a major contribution in the humoral immunity. T lymphocytes may become able to produce special substances called **cytokines** and activate other cells, thus making a major contribution in the cellular (or cell-mediated) immunity. These functionally active cells are called **effector lymphocytes**.

Processes of lymphocyte activation and differentiation are accompanied by death of many cells by **apoptosis**. The surviving cells may be either effector lymphocytes or the so-called **memory cells**. The memory cells are functionally quiescent but able to live for a very long time and become rapidly activated when they encounter the same antigen again. Because of the existence of memory cells a second, third, fourth, etc. encounter with the same antigen will lead to faster, stronger, and qualitatively better immune responses. Antibodies and effector lymphocytes generated during the primary immune response as well as during secondary immune responses act together with macrophages, granulocytes, and other cells and, eventually destroy the antigen. This latter phase of the immune response is called the **effector phase** of immunity. In subsequent chapters of this book, we will examine the recognition, activation, and effector phases of the specific immunity in detail. We will also analyze the interaction of the specific (adaptive) and the nonspecific (innate) immunity during these phases.

1.10. What is common to all specific immune responses?

There are several features that pertain to all specific immune responses.

- **Specificity**. Each response is uniquely specific to a particular antigen. In fact, antigen receptors of lymphocytes are able to recognize parts of complex antigenic molecules. The part of an antigen that an antigen receptor uniquely recognizes is called **antigenic determinant** or **epitope**.
- **Diversity**. All immune responses involve lymphocytes whose antigen specificity is already determined. The array of antigenic specificities of lymphocytes that exist at any given moment of time is tremendous (approximately one billion or more). It has been proven (see below) that this enormous diversity of specificities exists independently of exposure to antigens, and is being created by molecular mechanisms intrinsic to T and B lymphocytes. The total number of antigenic specificities created by these mechanisms is called the lymphocyte **repertoire**. As we will discuss in later chapters of this book, the size of normal immune repertoire is huge; it includes billions of different antibody and TCR specificities.
- **Memory**. Immunological memory is the ability to "remember" a previous encounter with the antigen, and to develop a faster, stronger, and qualitatively better response to the antigen when it is encountered again. Such responses are

CHAPTER 1 Immunity and the Immune System 7

called **secondary** (or second-set) or **recall** immune responses. As we stated in the previous section, these responses are faster, stronger, and qualitatively better than primary responses due to the fact that memory cells mediate them.

- **Specialization**. Immune responses to different antigens may involve different molecular and cellular mechanisms for the sake of maximizing the efficiency of these responses. For example, antiviral responses are most efficient when T lymphocytes are involved; responses to extracellular bacteria work best when B cells produce antibodies of certain classes; responses to parasites must involve B cells, T cells, and nonlymphoid cells called eosinophils; etc.
- **Self-limitation**. Normally, all immune responses wane with time after antigen stimulation. One reason for that is the successful elimination of the antigen that caused the response. The other reason is the existence of negative feedback mechanisms, which will be discussed later.
- **The ability to discriminate between self and nonself**. The immune system is said to "tolerate" self-antigens. The latter are substances that are produced by the organism that is the host of the immune system; these same substances can behave as foreign antigens when exposed to an immune system of a genetically different individual. Because of tolerance of the self, the host normally is not harmed by its own immune system. Cellular and molecular mechanisms of self-tolerance are being intensively studied and will be discussed later.

CLONAL SELECTION HYPOTHESIS

1.11 Are antigen receptors made before the immune response commences?

Yes. The molecular mechanisms that create antigen receptors operate independently of antigen exposure. If a laboratory rodent is raised under germ-free conditions, its lymphocytes will still have antigen receptors specific to various microbial and viral antigens. The hypothesis that antibodies are made before the antigen invasion and independently of this invasion was first advanced by Paul Ehrlich in the early 1900s. Later, when immunologists discovered a tremendous variety of antigens, Ehrlich's hypothesis became unpopular and was challenged by an "instructionist" theory, proposed by K. Landsteiner; it stated that immune cells make nonspecific molecules that *become* specific antigen receptors only after antigens "shape" or "mold" them. This theory implied that antigens served as "templates" for antigen receptors. The instructionist theory was proven wrong by N.-K. Jerne, who showed (in the late 1950s to early 1960s) that laboratory mice produced antibodies to antigens they had never encountered.

1.12 Can lymphocytes change the specificity of their antigen receptors during their lifetime?

As a general rule, this does not happen. The specificity of one unique antigen receptor, expressed by one given lymphocyte, is not changed throughout the lymphocyte's life. (An exception from this rule is the so-called receptor editing, a phenomenon that we will discuss later.) Moreover, daughter lymphocytes resulting from a parental lymphocyte's mitotic division also do not change the specificity of the antigen receptors that they inherit. In 1957, Burnet postulated that cells of the specific immune system develop as **clones**. A cell that makes a receptor specific to certain antigen originates from a separate precursor, and can make genetically identical progeny (clone). While all cells of any given clone have identical receptors, each clone differs from any other clone by the specificity of its antigen receptor. According to Burnet, the entire diversity of lymphocyte clones pre-exists antigen encounter. Further, Burnet hypothesized that in the absence of antigen, lymphocyte clones do not live long. The encounter of a lymphocyte clone with its specific antigen, however, selectively rescues this particular clone from death, sending a signal that stimulates the viability and expansion of this particular clone (Fig. 1-2). This **clonal selection hypothesis** later received tremendous experimental support, and it is currently considered that this hypothesis more or less adequately explains the work of immune system *in vivo*.

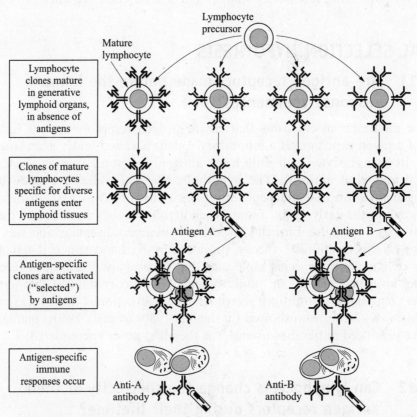

Fig. 1-2. The clonal selection hypothesis. Each antigen (A or B) selects a pre-existing clone of specific lymphocytes and stimulates the proliferation of that clone. The diagram shows only B cells, but the same principle applies to T cells.

CHAPTER 1 Immunity and the Immune System

1.13 How do we know that the clonal selection hypothesis is largely correct?

In the late 1960s to early 1970s, immunologists learned how to culture lymphocytes at limiting dilution, i.e., in such a way that one lymphocyte is placed in a miniature well of a tissue culture tray. Under these conditions, single lymphocytes may divide and produce antibodies. If the cultured lymphocytes were taken from an animal that had been immunized with several different antigens, antibodies to these different antigens were detected in different cultures, and never in one. In other words, individual lymphocytes and their clonal progeny always produce antibody of one specificity. Later, the same was shown to be true for T lymphocytes and their antigen receptors. Further, the binding of an antigen to an individual lymphocyte can be visualized by labeling. Using this approach, experimenters showed that one lymphocyte could be bound by one antigen, but never by many. Also, antigens can be radioactively labeled, and when such antigens bind their specific lymphocytes, the latter are killed by the radiation. Still, the animal that received the injection of the radioactive antigen will be perfectly able to respond to a very wide array of other antigens, indicating that only the clone specific to the radioactively labeled antigen was killed.

Finally, proteins that make up antigen receptors have been examined in detail. The amino acid sequence of these proteins, as well as the sequence of nucleotides in the genes that code for these proteins, has been solved. These studies show that individual antigen receptors have unique combining sites (the parts of their molecules that combine directly with the antigen). Any two different lymphocyte clones will always have two distinct antigen receptors, each with its own amino acid sequence at the combining site. Taken together, this entire evidence supports Burnet's hypothesis very strongly.

1.14 If antigen encounter stimulates proliferation of lymphocyte clones, will an immunization or infection lead to a massive increase in the number of blood cells?

No. We have to realize that at any given moment of time a vertebrate organism has at its disposal at least 10^9 different lymphocyte clones, each with its own specificity. The invasion of any individual antigen will cause proliferation of the clone that is specific to it, exclusively. (Some antigens, for example, large protein and polysaccharide antigens, have more than one **epitope**, i.e., more than one site that can be bound by antibodies or TCR; in this case, several different clones may react to the antigen, each of them to a separate epitope.) The nonspecific clones will not be affected, however. Therefore, no matter how strong the proliferative reaction is, the overall increase in the number of cells in the blood will be negligible. Suppose two clones react to two distinct epitopes of an antigen and each clone expands 10,000-fold. The overall increase in the number of cells will be 20,000, which is still a negligible increase (0.002%) over the number of all lymphocyte clones (~1 billion).

1.15 What is the main practical significance of the clonal selection hypothesis?

Perhaps it is the development of **hybridoma technology**. Based on the principal postulate of the clonal selection hypothesis – that lymphocytes exist in clones, each clone possessing its own antigen specificity – G. Kohler and C. Milstein in 1975 discovered a way to "immortalize" clones with particular, useful antigenic specificity through their fusion with myeloma cells. The resulting "hybridomas" can be used (and are being extensively used) for production of useful **monoclonal antibodies**, as well as for studies on antibody genes. The impact of hybridoma technology on basic science, medicine, pharmaceutical industry, agriculture, etc., turned out to be so big that in 1984 Kohler, Milstein, and Jerne were awarded Nobel Prize. We will discuss the details of hybridoma technology in Chapter 3.

Questions

REVIEW QUESTIONS

1. A "common sense"-based belief about immunology is that it is a science that studies how body defends itself against harmful microbes. To what extent is this belief true?
2. Simple molecules like salt and water cannot be antigens. Why?
3. Due to genetic defects, a person X has no antibodies and a person Y has no T lymphocytes. Which person is more likely to succumb to: (a) influenza; (b) diphtheria; (c) malignant myeloma?
4. Why is one of the basic features of the immune system called "memory?" What is common between immune memory and memory conferred by the activity of neurons? What is different?
5. Both antibodies and TCRs are antigen receptors and, as such, they can discern between different antigens. In later chapters, we will show that MHC molecules also show selectivity in their interaction with antigenic peptides. Based on the material of this chapter, how can you prove that MHC molecules are not antigen receptors?
6. In patients with infectious mononucleosis – a disease caused by the so-called Epstein–Barr virus (EBV) – lymphocyte counts in patients increase manifold over those in healthy individuals. Does this mean that the immune system reacts against the virus?
7. Laboratory mice can be grown in sterile conditions, so that no contact with microbial antigens is possible. Design an experiment that would use these mice and address the question, whether the "instructionist" or the "selectionist" theory of antibody diversity is true.
8. Would it be correct to say that hybridoma technology served as a decisive proof in favor of the clonal selection hypothesis?

CHAPTER 1 Immunity and the Immune System

MATCHING

Directions: Match each item in Column A with the one in Column B to which it is most closely associated. Each item in Column B can be used only once.

Column A

1. Antibody
2. TCR
3. Macrophage
4. Carrier
5. Innate immunity
6. Cytokine
7. Peptide
8. MHC
9. Tolerance of the self
10. Clonal selection

Column B

A. A type of accessory cell
B. A substance secreted by T lymphocytes
C. Discrimination between self and foreign antigens
D. Molecules that present peptides to T lymphocytes
E. Part of a protein molecule
F. Lymphocyte and its progeny activated by specific antigen
G. B cell antigen receptor
H. The entity that interacts with peptides and MHC
I. Nonspecific defense
J. Haptens are attached to it

 Answers to the Questions

REVIEW QUESTIONS

1. Defense against microbes is a major physiological function of the immune system; however, it is not the only function of the immune system and not the only form in which immunity can manifest itself. Many antigens are not microbial, and not all immune responses against antigens confer protection against these antigens (some are even injurious). Yet, immunology studies all antigens and all immune responses. Therefore, the "common sense"-based notion about immunology is only partially correct.

2. Because they are the same regardless of where they were synthesized. More complex molecules like proteins, polysaccharides, glycolipids, and nucleic acids may be different in genetically nonidentical organisms and thus be self to one organism and foreign to another.

3. Person X will be more susceptible to diphtheria infection because its causative agent is an extracellular bacterium. Person Y will be more prone to influenza (a viral infection whose causative agent is an intracellular virus) and to myeloma (a malignant tumor that is normally under surveillance of T lymphocytes).

4. Although exact cellular and molecular mechanisms of memory mediated by the nervous and the immune systems are different, there is the following deep similarity: both

imply that an organism may learn from previous encounter with an object and, due to such learning, develop a response that is qualitatively different from the response to this object when it is encountered for the first time.

5. It has been mentioned in this chapter that accessory cells express MHC molecules. Clonally distributed antigen receptors, on the other hand, are expressed exclusively on lymphocytes.

6. No. Specific immune responses are clonal. The observed massive increase in lymphocyte counts witnesses not a specific immune reaction, but a polyclonal activation of lymphocytes by the virus.

7. Such an experiment must include quantitation of antibody-producing cells in animals kept under sterile conditions (experimental group) and under normal conditions (control group). Cells that produce antibodies to common environmental microbial antigens should be quantified. If the results show that animals kept under sterile conditions still produce antibodies to the above antigens, it should favor the "selectionist" and disfavor the "instructionist" hypothesis.

8. Yes, because it allowed investigators to capture clones of lymphocytes immortalized by fusion with myeloma cells. These clones were shown to be expanded by immunization of the animal, in full agreement with the clonal selection hypothesis.

MATCHING

1, G; 2, H; 3, A; 4, J; 5, I; 6, B; 7, E; 8, D; 9, C; 10, F

CHAPTER 2

Cells, Tissues, and Organs of the Immune System

Introduction

Like many other functions of the body, the immune function is performed with the help of specialized cells. As we have already stated in Chapter 1, these cells are T and B lymphocytes operating together with accessory cells. Higher vertebrate organisms contain many billions of lymphocytes and accessory cells. These cells can be found throughout the body; they form compact or diffuse agglomerates, in which they are organized in such a way that the performance of their specific immune function is greatly facilitated. Thus, they can be collectively called a tissue, and by convention are usually called **lymphoid tissue**. Compact agglomerates of the lymphoid tissue are collectively called the **lymphoid organs**.

In this chapter, we will discuss the "natural history" and properties of T cells, B cells, accessory cells and some other cells involved in immunity. We will briefly analyze structural and functional organization of the lymphoid tissue and the lymphoid organs, trying to show, how exactly this organization facilitates the immune function of lymphocytes and accessory cells. Note that more information on the details of the structure and function of lymphoid organs will be presented in latter chapters, during the description of the dynamics of immune responses.

Discussion

LYMPHOCYTES

2.1 How do lymphocytes look?

Morphologically, both T and B lymphocytes are small cells whose diameter (8–10 mm unless activated) is comparable with that of large bacteria. They are spherical in shape and have a relatively large nucleus surrounded by a thin rim of cytoplasm. Organelles in resting lymphocytes are developed rather poorly. Electron microscopy reveals a number of microvilli on their surfaces.

This "boring," bland morphological pattern says virtually nothing about the complexity of the function that lymphocytes perform. Methods based on immunochemistry and molecular biology show that the surface of a lymphocyte contains thousands of different kinds of molecules, many of which serve as important intermediaries during antigen recognition, cellular activation, differentiation, and attack against antigens.

2.2 Where are lymphocytes located?

Lymphocytes mature from their **precursors** in the so-called **primary**, or **generative**, **lymphoid organs**. These are the **bone marrow**, the **bursa of Fabricius** (in birds), and the **thymus**. After maturing, lymphocytes circulate in the peripheral **blood**, although they do not function there. The recognition of antigens, as well as the complex events that follow this recognition, happen mostly in the so-called **secondary lymphoid organs**. These are anatomically defined organs like the **spleen**, **lymph nodes**, **tonsils**, **appendix**, and **Peyer's patches** (accumulations of lymphocytes in the small intestine). In addition, lymphocytes accumulate as diffuse conglomerates in all tissues except the central nervous system.

2.3. Why are the primary lymphoid organs called "generative?"

In these organs, lymphocytes are "generated;" they mature from their precursors. In the 1950s to 1960s, it was firmly established that all "blood cells," including lymphocytes, originate from a common precursor cell called a "**stem cell**." The process of B lymphocyte maturation from stem cells takes place in the bone marrow.

In birds, B lymphocyte precursors develop in the bone marrow and then in a special organ called the "bursa of Fabricius." Only those avian lymphocytes that pass through the bursa during embryogenesis become cells that produce antibodies. There are no known homologues of the bursa in mammals. It is thought that the entire process of B lymphocyte maturation in mammals can take place in the bone marrow. Therefore, the abbreviation "B" may point to the bursa of Fabricius or to the bone marrow.

CHAPTER 2 Organs of the Immune System

T lymphocyte precursors must undergo a stage of their maturation that takes place entirely in the organ called the thymus. The latter is an encapsulated glandular organ located in the anterior mediastinum (behind the chest bone). The thymus is relatively large in late fetuses and newborn infants and then undergoes an involution. In a newborn or in a young infant it occupies a sizeable space behind the chest bone; in an adult human, the thymus is invisible, hidden inside the layers of fat tissue. This organ is classified as "generative" because, at least in early life, it is entirely responsible for the finalization of T-lymphocyte maturation. The above process occurs through rigorous selection of the T-cell precursors in the thymus (which will be discussed in more detail in Chapter 9).

2.4 Are lymphocytes, as cell populations, homogenous?

No. Both T and B lymphocytes are functionally heterogeneous and consist of more than one subpopulation (or cell subset). T lymphocytes can be subdivided into two major functionally distinct subsets: **helper T cells** (T_h) and **cytolytic T cells** (T_c). The former population consists of cells that do not attack antigens directly but, rather, differentiate into producers of biologically active substances called **cytokines**. Through these cytokines, as well as through direct contact, T_h promote or enhance the function of other cells (notably, B lymphocytes and macrophages). T_c attack and destroy foreign cells directly by attaching to them and releasing cytotoxic granules that break the integrity of the target cell's membrane and also destroy its DNA. T_h and T_c express TCRs that have the same structure. Other cell-surface molecules are different in these two subsets, however. In particular, most T_h express a molecule called **CD4**, while most T_c express a somewhat similar but structurally different accessory molecule called **CD8**. All T cells that express CD8 recognize antigenic peptides in conjunction with MHC molecules that belong to the so-called "**Class I**" of the MHC molecules. The T cells that express CD4 recognize the antigenic peptides in conjunction with the so-called "MHC **Class II**" molecules (more about this in Chapters 5 and 6).

B lymphocytes also consist of two functionally distinct subpopulations or subsets. A subset called B-1 lymphocytes produces polyreactive antibodies, i.e., antibodies that can bind a more or less wide variety of antigens. B-2 lymphocytes produce antibodies that are usually monoreactive. The paramount feature of B-1 lymphocytes is that they transcribe the gene that codes for a protein called CD5. This transcription may or may not result in the translation and the surface expression of CD5. B-1 lymphocytes are very abundant in fetal life, and in the adult they tend to accumulate in certain body compartments, e.g., the peritoneal cavity and the omentum. The percentage of B-1 cells is very small in adult mice, and in adult humans may vary from a few percent to 25–30% of all B lymphocytes. It is thought that B-1 cells develop from precursors that are distinct from those of B-2 lymphocytes and constitute a self-sustaining (or self-replenishing) population of cells. B-1 and B-2 cells profoundly differ in their responses to stimulation of certain enzymatic (signal transduction) pathways. The exact function of B-1 lymphocytes remains unknown.

2.5 What are "large granular lymphocytes?"

Strictly speaking, they are not lymphocytes. These cells resemble lymphocytes morphologically, although they are larger and contain preformed granules. (As we will detail later, T_c develop granules only when they "prepare" to kill their target, and they do not store these granules.) Importantly, however, the "large granular lymphocytes" do not make specific antigen receptors. A more appropriate name for these cells is "**natural killer (NK) cells**," and their function will be discussed in Chapter 13.

2.6 What are "activated lymphocytes," and what is the difference between them and "resting" lymphocytes?

The term "lymphocyte activation" refers to the sequence of events that follows the engagement of the cell's antigen receptor. For full activation, a number of molecules called **accessory molecules** must also be bound (we will discuss the role of these molecules later.) The lymphocyte activation may be caused by the antigen specific for a given lymphocyte clone, or by agents that mimic antigens but are not clone-specific. (Such agents are called "polyclonal activators.") When a lymphocyte is activated by either of the above, it becomes larger – its diameter grows from 8–10 to 10–12 μm, and it changes its shape from round to hand mirror-like. The large hand mirror-shaped activated lymphocytes are often called **lymphoblasts**.

Lymphocyte activation always includes **transduction of signals** from outside, delivered via the antigen receptor and accessory molecules, to the inside of the cell through signal transduction pathways. The signal transduction culminates in the **activation of gene transcription**, and is often accompanied by the **expression of new surface proteins**. We will discuss signal transduction pathways and the regulation of gene transcription in the cells of the immune system in Chapters 7–9.

The most important difference between resting lymphocytes and activated lymphoblasts is that the latter are able to divide mitotically, and differentiate, i.e., acquire new properties that are necessary for dealing with antigens (see Section 1.9). As we have already mentioned in Section 1.9, differentiated lymphocytes may become effector or memory cells. For example, activated B lymphocytes can divide and differentiate into memory B lymphocytes or **plasma cells** – short-living, egg-shaped cells that acquire the ability to secrete very large quantities of antibodies. Activated T lymphocytes can divide and differentiate into memory T lymphocytes or effector T cells. Among the latter, T_h are the cells that produce cytokines, and T_c are the cells that release cytotoxic granules upon attachment to target cells. (We will discuss functions of effector and memory lymphocytes in more detail in subsequent chapters.)

2.7 Is lymphocyte activation always followed by differentiation?

No. As we have already mentioned in Section 1.9, some activated lymphocytes do not differentiate but instead, die by **apoptosis**. Recently, it has been discovered that

CHAPTER 2 Organs of the Immune System

apoptosis is the fate of many – maybe even the majority – of activated T and B lymphocytes. Apoptosis is a process that leads to cell death without "spilling the cell's guts," i.e., allows the lymphocyte to die with its outer membrane relatively intact. The details of the apoptotic process are described in cell biology textbooks. Activated lymphocytes die because of a monotonous, repeated stimulation by antigen and/or concentrations of cytokines that are too high. This phenomenon is called **activation-induced cell death**, and will be discussed in more detail in Chapter 9.

2.8 Do resting lymphocytes also die by apoptosis, and what protects them against it?

Yes, resting (not activated) lymphocytes die by apoptosis unless they are rescued from it either by their specific antigen or by a polyclonal activator. This process is called **programmed cell death**. The life span of a resting lymphocyte is commonly three to four days; however, at least some resting murine lymphocytes can live for a few weeks or even months without being contacted by antigen. Generally, the lymphocyte life span and the conditions that prevent lymphocytes from undergoing programmed cell death are subject of an ongoing research.

2.9 What are accessory cells?

Those are nonlymphoid cells that do not make specific antigen receptors but participate in the immune reactions together with lymphocytes. The two principal kinds of accessory cells are **macrophages** (mononuclear phagocytes) and **dendritic cells**. The dendritic cells were discovered and described some 80 years after the macrophages and, perhaps because of this historical reason, most immunology textbooks describe the macrophages first. It should be noted, however, that the role of dendritic cells for the proper function of the immune system is in no way less significant than the role of macrophages. In addition, several other populations of nonlymphoid cells can play the role of accessory cells at certain circumstances.

MACROPHAGES

2.10 How do macrophages look, and where are they located?

Macrophages are the product of differentiation of bone marrow-derived cells called **monocytes**. A typical monocyte is just slightly larger than a typical resting lymphocyte and is spherical or egg-shaped. In contrast to lymphocytes, monocytes and macrophages have a much better developed cytoplasm rich in organelles, especially lysosomes, vacuoles, and a well-developed cytoskeleton that allows them to grow amoeba-like pseudopodia. Monocytes are able to circulate in the blood and lymph. They become macrophages after they move from the blood into tissues. Macrophages can be found in many organs of the body and in all types of the connective tissue. Unlike lymphocytes, monocytes and macrophages can firmly adhere to plastic surfaces, a property which is used in laboratory manipulations with these cells.

2.11 What is the function of macrophages?

Macrophages actually perform several different functions, participating both in nonspecific and specific immunity. Functions that are primarily related to innate (nonspecific) immunity, such as phagocytosis and secretion of various biologically active substances, are described in Chapter 11. Here, we will just mention that **phagocytosis** is a process by which a cell engulfs foreign particles and breaks them down using the cell's digestive enzymes. The mechanism that allows macrophages to discern foreign particles from self is not understood. An important point should be made, though, that macrophages do not discern *between different* foreign particles. Thus, unlike antigen-specific lymphocytes, a macrophage can engulf a wide variety of microbes. To engulf a particle, the macrophage must first form pseudopodia, and then attach, and surround the particle with them. After the particle is surrounded, it is internalized by the macrophage and fused with lysosomes. The latter contain enzymes that break down (hydrolyze) the macromolecules of the particle into smaller molecules. For example, proteins are reduced to peptides.

The variety of **biologically active substances** produced by macrophages includes bactericidal substances like reactive oxygen species (ROI) and cytokines that recruit inflammatory cells from blood to tissues. (The mechanisms of inflammation are described in detail in physiology and pathology textbooks, and in some detail in Chapter 11.) Some cytokines produced by macrophages, notably molecules called **interleukins**, play an important role in the induction of fever (see Chapter 10). Other biologically active substances produced by macrophages act as growth factors for connective tissue cells and are important for tissue repair (wound healing).

Macrophages also perform such functions as antigen presentation and opsonization. **Antigen presentation** is a function of macrophages and other accessory cells directly related to specific immunity. As we have already mentioned, T lymphocytes cannot recognize antigens in their native form, but can recognize peptides derived from protein antigens after the latter have been processed and appropriately presented. Macrophages and other accessory cells possess an intracellular machinery that enables them to digest proteins, breaking them down into peptides, and to "traffic" the derived peptides to the sites where they are joined to self-MHC molecules (see Chapters 5 and 6). The MHC molecules with antigenic peptides that are incorporated into the MHC peptide-binding region or domain are expressed on the macrophage's membrane and "shown" (or presented) to the T cells. The mechanisms of processing and trafficking the antigenic peptides and of their assembly with the MHC molecules are tightly regulated. Antigen processing and presentation allows not only the T-cell recognition to occur but also allows the

CHAPTER 2 Organs of the Immune System

two principal populations of T cells – $CD4^+$ or $CD8^+$ – to be recruited in the immune response differentially, depending on whether endogenous or exogenous antigens are being presented. We will discuss the mechanisms of antigen presentation in later chapters.

In addition to the above-mentioned functions, macrophages can bind substances called **opsonins**. The latter are molecules that attach to the surface of a microbe and can be recognized by special receptors expressed on macrophages. The recognition of opsonins increases the efficiency of phagocytosis, production of ROI and other macrophage activities. Opsonins include some classes of antibodies and fragments of proteins that belong to complement system (see Chapters 11 and 13).

2.12 What is "macrophage activation?" How do "activated" macrophages differ from nonactivated?

Macrophages are called activated when they actively perform at least one of their known functions. Thus, a macrophage activated in terms of one function may not be activated as far as another function is concerned. Unlike activated lymphocytes, activated macrophages may be morphologically identical to their resting counterparts. Macrophages become activated due to signals from other cells (most commonly, T lymphocytes) delivered through the secretion of cytokines.

2.13 Who discovered macrophages, and why was this discovery important?

In the 1880s, Eli (Ilya) Metchnikoff observed that certain cells of a jellyfish accumulate near the site of entrance of a plant thorn. If the thorn is small enough, these cells can engulf it and break it down into invisible components. Metchnikoff and his colleagues later described these "phagocytes" in other organisms and showed that these cells were present in the blood and connective tissue. Since these cells could attack not only plant thorns but also bacteria, Metchnikoff postulated that they are important for immune reactions.

The discovery of macrophages provoked one of the longest and most dramatic debates among early immunologists. In the late 1880s, von Boehring and Kitasato discovered serum antibodies, and in the 1900s Ehrlich formulated what was later called a "**humoral theory of immunity**." According to that theory, antibodies were the principal devices of immunological defense. Metchnikoff never agreed with this view, arguing that cells, and not soluble substances, are principally responsible for immune reactions. Because of the great authority of Ehrlich, Metchnikoff's views were at first dismissed. However, Wright, Landsteiner, and others found evidence supporting these views and advanced a "**cellular theory of immunity**." Later, it became apparent that both cells (lymphocytes and accessory cells, including macrophages) and humoral factors (antibodies and cytokines) are equally important for immunity.

2.14 Are "histiocytes," "Kupffer's cells," "osteoclasts," "microglial cells," and "alveolar macrophages" macrophages?

Yes. The above-listed terms, often used in special literature, refer to the preferred localization rather than to special kinds of cells. "Histiocytes" (from Greek "histos" – tissue) are macrophages found in loose connective tissue. "Kupffer's cells" are macrophages that migrate to the liver. "Osteoclasts" are macrophages that are located in bone islets. "Microglial cells" (or "microglia") are macrophages that form the connective tissue surrounding the cells of the central nervous system. "Alveolar macrophages" are macrophages that can be found in airways. All these types of cells perform the functions of macrophages described above.

2.15 What is the "reticuloendothelial system?"

It does not exist. Histologists who worked in the late 19th and early 20th century (notably, Aschoff) thought that macrophages and endothelial cells (i.e., epithelial cells that line the inside of the blood vessel walls) are equally capable of phagocytosis. This notion, based on the observation that both cell types can internalize certain dyes, led to the suggestion that macrophages and endothelial cells have a common biological function and should be united into a "system." Later, however, it has been shown that endothelial cells do not phagocytose; their ability to imbibe dye particles is based on a completely different cellular mechanism. Because of the functional difference between macrophages and endothelial cells, the term "reticuloendothelial system" should be avoided.

DENDRITIC CELLS

2.16 What are dendritic cells, and what is their significance?

Dendritic cells (DC) are bone marrow-derived cells that are an extremely important kind of accessory cell. One feature of dendritic cells that makes them outstanding is their unusual ability to process and present antigens in the most efficient way, thus triggering a very strong immune response. DC are the most efficient presenters of exogenous peptides to T lymphocytes, because they express the highest number of the Class II MHC molecules per cell. DC are also potent producers of cytokines, regulators of lymphocyte functions, and powerful mediators of immunological tolerance (see later chapters).

2.17 How do DC look, and where are they located?

DC were first identified as a distinct cell type based on their peculiar morphology. They are rather small, although slightly larger than lymphocytes (diameter 10–12 μm), and they show membranous or spine-like projections. Microscopically, dendritic cells often seem to be surrounded by a "veil." Langerhans first described

CHAPTER 2 Organs of the Immune System

such cells in the skin. He noticed that the projections that these cells make resemble projections made by neurons, called "dendrites." It was believed at first that these cells, called **Langerhans cells**, are characteristic only for the skin. Later, cells resembling Langerhans cells were found in the blood, thymus, and peripheral lymphoid organs, as well as in nonlymphoid organs (liver, pancreas, peritoneum, etc.). In the peripheral blood, DC are rare; only about one in 1,000 blood cells can be identified as a DC. The frequency of DC is greater in the peripheral lymphoid organs. It has been hypothesized that DC develop from their progenitors in bone marrow and then migrate into the skin. There, they acquire their morphology and function, and further migrate into the blood and peripheral lymphoid organs.

2.18 What are "interdigitating DC" and "follicular DC?"

The classification of DC into "interdigitating" and "follicular" is perhaps wrong. Virtually all cells that originate from bone marrow progenitors and possess the morphology and functions of DC can be called "interdigitating," because they tend to penetrate ("interdigitate") the interstitium of the peripheral lymphoid organs. Therefore, there is hardly a need for the term "interdigitating DC," which might imply that these cells are a special subset of DC. As for "follicular DC," most immunologists believe that those are not bone marrow-derived and, therefore, they cannot be placed into the same category as dendritic cells. The reason they are called "dendritic" is because their morphology is similar to conventional ("interdigitating") DC. The function of follicular dendritic cells (FDC) is also different from that of dendritic cells, and will be discussed in Chapter 4.

2.19 Are DC a homogenous population?

No. There are at least **two subsets** of DC that represent two subsequent stages of their **maturation**. **Immature** DC are the cells that just moved from bone marrow into the skin and from there to other tissues. Langerhans cells are morphologic correlates of immature DC. They are capable of **capturing antigens** with extremely high efficiency. Yet, they are not able to present the processed antigens to T lymphocytes and induce immune responses. **Mature** DC develop from the immature cells within approximately 24 hours after capturing antigens. These mature DC express very large numbers of MHC molecules and so-called accessory molecules (see below). This makes these cells able to present the processed antigens to T lymphocytes efficiently, causing strong immune responses. The antigens captured by immature dendritic cells promote their development into mature DC.

2.20 What are "myeloid" and "lymphoid" DC?

DC can be also classified into **myeloid DC** and **lymphoid DC** according to the type of progenitors that give rise to them during bone marrow stage of their maturation. These two subsets of DC differ in the cytokines that they produce and in the exact role they play in immune responses.

OTHER NONLYMPHOID CELLS THAT PARTICIPATE IN IMMUNE REACTIONS

2.21 What are granulocytes, and what is their role in immune reactions?

Granulocytes are bone marrow-derived white blood cells (leukocytes) that are distinct from both lymphocytes and monocytes. Their characteristic feature is the presence of abundant cytoplasmic granules (hence the name). Granulocytes participate in nonspecific defense mechanisms, but also in specific immunity when stimulated by cytokines. Since granulocytes participate in inflammation, they are sometimes called "inflammatory cells" or "inflammatory leukocytes." Many granulocytes are capable of phagocytosis. We will discuss some functions of granulocytes in Chapter 11 and elsewhere.

2.22 What are neutrophils, basophils, and eosinophils?

They are kinds of granulocytes. These names, coined by old histologists, originate from the tendency of granulocytes to be stained by different dyes according to the acidity or alkalinity of their cytoplasm. If a granulocyte has an acidic cytoplasm (pH less than 7), it will be readily stained by basic dyes (basophil). If a granulocyte has a basic cytoplasm (pH more than 7), it will absorb acidic dyes like eosin (eosinophil). If a granulocyte has a neutral cytoplasm (pH approximately 7), it will be equally well stained with basic and acidic dyes (neutrophil). The functions of neutrophils, basophils and eosinophils differ from each other. We will discuss them later in the chapters that describe different immune responses.

2.23 Are granulocytes accessory cells?

Formally speaking, no. These cells do not process and present antigens, and thus do not play the necessary "third party" role for T-cell antigen recognition the way macrophages or dendritic cells do. Granulocytes participate mostly in the effector phase of immune reactions – in particular, in the immune inflammation characteristic for the delayed-type or immediate hypersensitivity reactions (to be discussed later).

2.24 Are there any other types of accessory cells besides macrophages and dendritic cells?

B lymphocytes have properties of accessory cells because, in addition to their role as producers of antibodies, they can also process antigens and present them to T lymphocytes. We will discuss the antigen-presenting role of B lymphocytes in Chapter 8. **Endothelial cells** and some other cell types are thought to perform some of the accessory cells' functions when Class II MHC molecules are induced on their surface in the presence of cytokines (see Chapter 10).

CHAPTER 2 Organs of the Immune System

LYMPHOID TISSUES AND ORGANS

2.25 What is the bone marrow, and what processes related to the immune function take place there?

Bone marrow is a soft, tender, sponge-like substance found inside most bones of young individuals and in flat bones (vertebrae, chest bone, pelvic bone, etc.) of mature individuals. Bone marrow is a fine meshwork of cells that belong to the connective tissue. These include reticular cells, fat cells, and maturing precursors, or **progenitors,** of red blood cells, white blood cells – lymphocytes, monocytes, dendritic cells, and granulocytes – and platelets. Bone marrow is the principal site of the maturation of immune cells from their precursors. In addition, bone marrow is a site of active antibody production because in addition to maturing progenitors of blood cells, it contains numerous **plasma cells** that secrete large quantities of antibodies.

2.26 What precursors of immune cells are present in bone marrow, and what are their properties?

The earliest precursor of immune cells is the same as the precursor of other "blood cells," namely the **stem cell**. This is a cell type capable, essentially, only of proliferation, thus maintaining a more or less stable cellular pool. Because of signals that are being sent to proliferating stem cells, some of them become the so-called **committed precursors**. We do not fully understand what signals are required to turn some of the bone marrow stem cells into committed precursors, and why only some of the stem cells react to them. Bone marrow reticular cells and other cells, as well as a variety of the so-called hemopoietic cytokines (see Chapter 10) are involved in this signaling. The committed precursors continue to mature and eventually diverge so that their progeny develops into a more immediate precursor of only one certain "blood cell" type.

2.27 What is the thymus and what immune (and other) cells can be found there?

The **thymus** is a glandular, encapsulated organ located in the upper mediastinum (behind the upper portion of the chest bone). It is the principal site of T lymphocyte maturation in all higher vertebrates. The removal of the thymus soon after birth ("neonatal thymoectomy") was shown in mice and other laboratory animals to lead to profound deficiencies of the immune response and to an almost complete lack of T lymphocytes. In newborn and very young animals and humans, the thymus is relatively large and occupies most of the higher mediastinum. It is, for example, a substantial obstacle to surgeons operating on the opened heart of newborn infants. With age, the thymus undergoes an involution. In an adult human, the thymus is almost completely replaced by fat tissue. It is not clear what takes care of the maturation of new T lymphocytes in adults. Either the small part of the thymus that remains after its involution is sufficient, or some extrathymic sites of T-cell maturation take over the thymus with age.

The thymus consists of two layers: the external layer, which is called the **cortex**, and the internal (softer or "mushier") layer that is called the **medulla** (Fig. 2-1). The cortex and the medulla have a somewhat different histological organization, but both are thought to be important for T-cell maturation and selection. Immature precursors of T cells are brought into the thymus (initially to the cortex) via the afferent blood vessels. Altogether, the thymus in a newborn human can "house" many hundred million of the arriving precursors. As will be detailed later, the majority of these "new arrivals" dies without ever making it to mature T cells and are engulfed and digested by the macrophages, which are very abundant in the thymus (especially in the medulla). Besides T lymphocytes and macrophages, the thymus contains numerous dendritic cells and epithelial cells, which are thought to play an important role in the thymic selection. Nonlymphoid thymic cells also produce thymic **hormones**. The exact role of these substances is not known.

2.28 What are lymph nodes?

Lymph nodes are aggregates of lymphoid tissue whose size can vary from fractions of millimeter to several centimeters in diameter. These aggregates are located along the way of **lymph** via large and small lymphatic vessels. Lymph is, essentially, the interstitial fluid mixed with blood plasma that oozes through tiny blood capillaries into the lymphatic vessels. The movement of the lymph through the lymphatic vessels is facilitated by the heartbeat, because the largest lymphatic

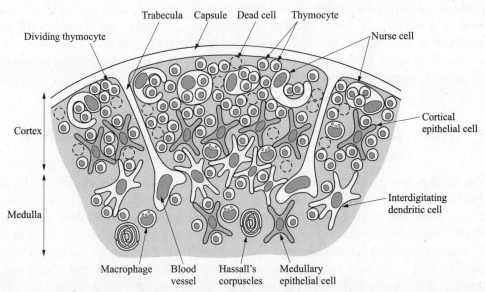

Fig. 2-1. Thymus (a schematic diagram). Shown are several lobules (compartments) separated by trabecules (strands of connective tissue); cortex (the outer layer densely populated by cells), medulla (the inner, less populated layer), maturing thymocytes, and nonlymphoid thymic cells. Nurse cells are a special kind of thymic epithelial cells. Hassal's corpuscles are concentric layers of degenerating epithelial cells.

CHAPTER 2 Organs of the Immune System

collector (the so-called thoracic duct) drains into very large veins, where negative pressure during the relaxation of the cardiac muscle sucks the lymph in. On its way, lymph percolates through the lymph nodes. In addition to the lymphatic vessels, the lymph nodes receive blood that flows into them through afferent blood vessels. The various antigens collected from various sites of the body and borne by lymph, and lymphocytes, brought into the lymph nodes by the flow of blood through the afferent blood vessels, meet inside the lymph nodes. This property of lymph nodes makes them the arena of antigen recognition and lymphocyte activation.

2.29 What is the structure of lymph nodes, and what events take place there?

Lymph nodes, like the thymus, consist of the cortex and the medulla (Fig. 2-2). The cortex is clearly divided into two structural and functional "zones." One of these "zones" consists of lymphoid **follicles** – rounded, spherical aggregates that are

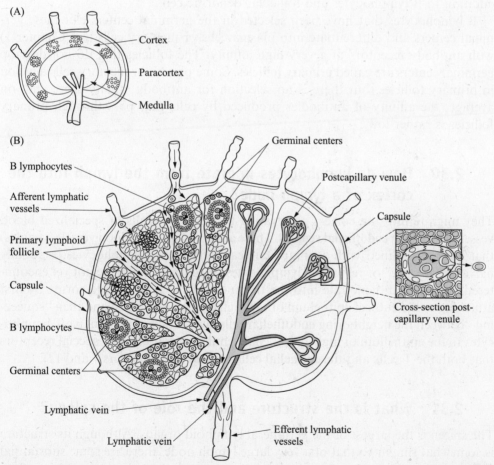

Fig. 2-2. Lymph nodes (a schematic diagram). (A) The three layers of a lymph node; (B) the arrangement of the cells and the vasculature.

made almost exclusively of B lymphocytes. The other "zone" is the so-called **parafollicular cortex** that consists mostly of T lymphocytes (of which the majority is CD4-expressing T_h cells), some accessory cells, and rare B cells. Among the accessory cells, one can find macrophages and dendritic cells, which are relatively numerous in lymph nodes. The current working hypothesis, supported by considerable experimental evidence, states that lymph-borne antigens are processed and presented to T cells by accessory (especially dendritic) cells in the parafollicular cortex. Also there, in the parafollicular spaces, the activated T cells interact with the B lymphocytes specific to the same antigens to which the T cells are specific (this process being called the **cognate interaction**). The B cells are activated due to such interaction. Some of the activated B lymphocytes subsequently move to follicles.

Soon after immunization, the central area of some (not all) of the follicles develops into what is called the **germinal centers**. These are the sites where activated B lymphocytes clonally expand, mutate their antibody genes (see later), and are selected for the affinity of their antibodies to the specific antigen. The latter is exposed to the B cells by FDC, which form an abundant mesh in the internal part of the germinal center. Some evidence suggests that a separate class of dendritic cells, called "germinal center dendritic cells," is present in the germinal centers, in addition to B lymphocytes and follicular dendritic cells.

B lymphocytes that have been selected in the germinal centers exit these germinal centers and differentiate into plasma cells or memory cells (see Chapter 1) with antibody receptors of a very high affinity. The follicles that do not develop germinal centers are called **primary follicles**. Some plasma or memory cells develop in primary follicles, but there is no selection for antibody affinity there and, on average, the affinity of antibodies produced by cells that pass through primary follicles is rather low.

2.30 How do lymphocytes migrate from the lymph into the cortex of a lymph node?

They migrate into the cortex, penetrating through the walls of specialized blood vessels called **high endothelial venules**. These are postcapillary blood vessels (venules) that have walls lined with a unique cubical epithelium. Lymphocytes, especially naive (or "virgin," or resting) T lymphocytes (the T cells that have not yet encountered antigen) can penetrate into the thickness of the lymph node cortex by first attaching to the cubical epithelium of the high endothelial veins, and then "squeezing" between the neighboring endothelial cells. The attachment of naive T lymphocytes to the epithelium of high endothelial venules is facilitated by special receptors that both the T cells and the epithelial cells express (see Chapters 7 and 12).

2.31 What is the structure and the role of the spleen?

The spleen is the largest of the peripheral lymphoid organs. Although its structure is somewhat similar to that of a very large lymph node, there are some substantial differences both in the anatomical organization of the spleen, as compared to lymph nodes, and in its exact role in immunity.

CHAPTER 2 Organs of the Immune System

The spleen (Fig. 2-3) consists of the so-called **white pulp** and the so-called **red pulp**. The former is, essentially, a large collection of lymphocytes that surround the branching afferent blood vessels. The "cuffs" of lymphocytes – predominantly T cells, as well as some B lymphocytes – that are intimately attached to small branches (arterioles) of the splenic artery, are called "**periarteriolar lymphoid sheaths**" (PALS). The other components of the white pulp are follicles, aggregates that resemble the above-described follicles of the lymph nodes. The follicles contain predominantly B lymphocytes and are usually attached to PALS. The PALS and the follicles together are surrounded by a thin rim of lymphocytes and macrophages called the **marginal zone** (a structure that is unique to the spleen). Many of the follicles develop germinal centers during immune responses. The red pulp is, essentially, the collection of red blood cells, accessory cells (both macrophages and dendritic cells), and rare T or B lymphocytes. Interestingly, there are no high endothelial venules in the spleen.

A substantial functional difference between the spleen and the lymph node is that in the latter naive lymphocytes recognize, and become activated by, antigens borne by lymph. The spleen, however, is the site where *blood*-borne antigens are recognized. In practical terms, when an animal or person is immunized

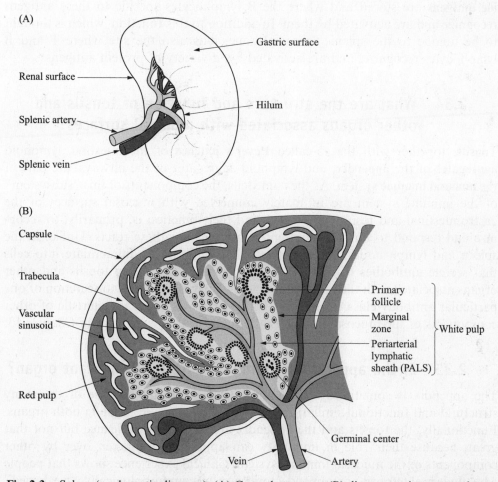

Fig. 2-3. Spleen (a schematic diagram). (A) General structure; (B) diagrammatic cross-section.

intradermally or subcutaneously, the recognition of antigen(s) and the activation of specific lymphocyte clones will take place predominantly in the regional lymph nodes. If an animal or person is injected intravenously, though, or if external antigens enter the bloodstream directly because of a large opened wound, the arena of the antigen recognition and the lymphocyte activation will be the spleen.

2.32 What is the function of the red pulp?

The red pulp can be viewed as part of the nonspecific body defense systems. The macrophages that are present there in large numbers efficiently engulf and digest a variety of foreign particles, as well as "aged" (senescent) red blood cells of the host. Because of this latter property, old histologists sometimes called the spleen "a cemetery of red blood cells." Also, a variety of toxic products dissolved in blood are detoxicated in the spleen's red pulp.

2.33 Does the marginal zone have a unique function?

Presumably, yes. It is thought that the marginal zone is the site where **polysaccharide antigens** are stored and where the B lymphocytes specific to these antigens recognize and are activated by them. In addition to this function, which is thought to be unique to the splenic marginal zone, it is also the site where T and B lymphocytes recognize, and are activated by, a variety of protein antigens.

2.34 What are the structure and function of tonsils and other organs associated with mucosal surfaces?

Tonsils, together with the so-called **Peyer's patches** of the intestine, lymphoid aggregates of the **appendix**, and lymphoid aggregates of the airways, are part of the **mucosal immune system**. As the name tells, the components of this "subsystem" of the immune system are intimately connected with mucosal surfaces of the gastrointestinal and respiratory tracts, and their function is, primarily, to recognize and respond to antigens that enter the body via these tracts. Unlike in the spleen and lymph nodes, where activated B lymphocytes differentiate into cells that secrete antibodies of all classes (see later), the B cells in tonsils and other organs associated with the mucosal immune system have a bias to secretion of one particular antibody class called IgA. This property is also characteristic of other components of the mucosal immune system and will be explained in Chapter 13.

2.35 Is the appendix an immunologically significant organ?

The appendix is sometimes called "the intestinal tonsil" because of the many structural and functional similarities between the lymphoid tissue in both organs. Functionally, the tonsils and the appendix are of some significance but not that great, because their role in immunity can apparently be taken over by other components of the mucosal immune system. Clinical experience shows that people who underwent tonsil- or appendectomy are not immunologically "weakened"

CHAPTER 2 Organs of the Immune System

compared with those who retain the tonsils or appendix. This perhaps is an example of the immune system's plasticity, its ability to function normally even when some of its components are defunct or missing.

2.36 What are ectopic lymphoid tissues?

These are the sites where lymphoid aggregates are not normally present but can develop during a disease. For example, during rheumatoid arthritis, lymphocytes accumulate and organize into lymphoid structures on the synovial surfaces of the cavities in small joints (mostly the joints between finger and toe phalanxes). In some cases of rheumatoid arthritis, even true follicles with germinal centers can develop in these synovial lymphoid aggregates. The reason for the formation of these ectopic lymphoid aggregates, and the antigens to which lymphocytes in these aggregates are responding, are unknown. It is thought, however, that the build-up of the ectopic lymphocyte tissue plays a substantial role in the pathogenesis of rheumatoid arthritis and some other diseases. We will return to this issue in Chapter 20.

Questions

REVIEW QUESTIONS

1. If a volume of whole blood is carefully layered on a volume of a solution of Ficoll (a polysaccharide) with density 1.077 mg/ml, and allowed to sediment, lymphocytes and monocytes will not reach the bottom of the tube but instead will float at the border between the two layers. Imagine that you collected the floating cells in a pipette. How would you separate lymphocytes from monocytes?
2. What diseases are likely to develop in chickens that underwent bursaectomy (a surgical removal of the bursa of Fabricius)?
3. Children born without a thymus (as a result of the genetic defect called DiGeorge syndrome) are prone to various viral and fungal infections; however, if they survive their first year, their chances to overcome these infections increase. Offer a hypothesis explaining this.
4. As we will discuss in detail in Chapters 5 and 6, MHC Class I molecules are expressed on all nucleated cells of the body, while MHC Class II molecules are expressed only on some cells. Knowing this, which subset of T cells – T_h or T_c – would you name as a more important one for the defense against viruses?
5. Plasma cells can undergo malignant transformation and become myelomas – tumors that grow as rapidly expanding and metastasizing clones. What difference would you expect to find if you compare the serum of a myeloma patient and the serum of a healthy individual?
6. It has been mentioned in Chapter 1 that secondary immune responses are faster, stronger, and qualitatively better than primary immune responses. Explain how memory cells contribute into each of these three differences.

7. In what way are programmed cell death and activation-induced cell death similar, and in what way are they different?
8. A grant proposal offers to employ a new method for characterization of activated macrophages. This method involves labeling of particles that macrophages ingest. According to the author of the proposal, his method will be useful in obtaining a pure population of activated macrophages. How would you comment on this proposal?
9. How would you prove the statement made in this chapter that FDC, in spite of their morphology, are not really DC?
10. As you learned from this chapter, lymph-borne antigens are trafficked to lymph nodes, and blood-borne antigens to the spleen. Does this mean that there are no antigens in the thymus?
11. What type of immunodeficiency (impairment of immune response) would you expect in a person whose spleen was removed?
12. If a laboratory mouse is kept in sterile conditions throughout life, how would its secondary lymphoid organs differ from a normal mouse?

MATCHING

Directions: Match each term in Column A with the one in Column B to which it is the most closely associated. Each item in Column B can be used only once.

Column A

1. T_c
2. B-2
3. Langerhans cells
4. Microglia
5. Basophils
6. Committed precursors
7. Involution
8. Parafollicular cortex
9. Periarteriolar lymphoid sheaths
10. Alveolar macrophages

Column B

A. Dendritic cells of the skin
B. Degradation of the thymus with age
C. Part of the mucosal immune system
D. T cell-rich zone of lymph nodes
E. A subset of B lymphocytes
F. Part of the splenic white pulp
G. A subset of T lymphocytes
H. A subset of granulocytes
I. Macrophages of the central nervous system
J. Are found in bone marrow

Answers to Questions

REVIEW QUESTIONS

1. Place the collected cells on a plastic surface and incubate, and then separate those cells that did not adhere from those that did. Non-adherent cells will be mostly lymphocytes, and adherent cells mostly macrophages.
2. Since the bursa of Fabricius is crucially important for development of B cells and not T cells, one should expect diseases caused by extracellular bacteria (see Chapter 1).

CHAPTER 2 Organs of the Immune System

3. One hypothesis might be that the thymus is not the only place where T cells mature. The significance of the thymus may be high during fetal life and the first year of postnatal life, but later these putative "extrathymic sites" may take over. This hypothesis would be in line with the fact that in spite of the apparent involution of the thymus with age, new T lymphocytes are being produced throughout life.
4. T_c must be more important because they interact with MHC Class I-expressing cells and thus would cover all cells that might be infected with viruses.
5. In myeloma patients, the amount of antibodies in the serum must be much higher; besides, since myelomas grow as clones, antibodies secreted by one particular clone must predominate.
6. Secondary responses are faster because memory cells have already been formed by the moment a secondary response commences (therefore no need for lag period). They are stronger because memory cells have already been expanded after the first encounter with antigen (therefore more cells will become effector cells after the next encounter). They are qualitatively different because memory cells are products of lymphocyte selection (therefore, e.g., antibodies will be of higher affinity).
7. Similar in that they both occur by apoptosis; different in that programmed cell death is due to a lack of growth factors (cytokines) or other rescuing signals while the activation-induced cell death is due to a repeated stimulation by excessive concentrations of these factors.
8. The method offered by the author hardly fits the goal of the proposal because phagocytosis is not the only manifestation of macrophage activation. A macrophage may be called activated because it produces ROI or performs some other function.
9. They are not bone marrow-derived, they do not migrate from the bone marrow to the peripheral lymphoid organs through the skin, and they do not express large quantities of MHC Class II antigens.
10. No, it rather means that there are virtually no *foreign* antigens in the thymus. There is, however, a wide variety of self antigens there.
11. Since only the spleen has marginal zone where polysaccharide antigens are accumulated and recognized, the expected deficiency would be insufficient response to these antigens. Bacterial capsules often consist of polysaccharides, hence, the patient will probably suffer from chronic bacterial infections.
12. There will be no germinal centers in the mouse kept under sterile conditions, because these develop only in response to antigenic stimulation.

MATCHING

1, G; 2, E; 3, A; 4, I; 5, H; 6, J; 7, B; 8, D; 9, F; 10, C

Antibodies and Antigens

Introduction

The words "**antibody**" and "**antigen**" have been perhaps so widely used by popular literature that in the minds of many lay people the science of immunology is as closely associated with these words as it is with resistance to infectious diseases. Very broad audiences know that antibodies are molecules capable of "fighting" foreign (and potentially harmful) substances. Antibodies elicited by vaccination protect the vaccinated individuals from diseases. On the other hand, it is perhaps much less known to the broad public that antibodies are but *one of the three classes of molecules able to discern between antigens* (the other two being TCR and MHC). Antibodies occupy a very special place among these three classes, however, because they: (1) have the broadest range of antigen specificities; (2) can possess a bigger affinity or strength of the binding to antigens; and (3) are best studied, both for historical reasons (known by their biological effects since the 1880s and by their chemistry since the 1930s), and for technical reasons (are relatively easy to be purified from mixtures and studied biochemically). In this chapter, we will highlight essential features of antibodies as molecules, and show how their biochemical properties help them to perform their functions.

We will emphasize on the notion that any antibody molecule always has two unequal parts, of which a smaller part actually mediates the binding of the specific epitope, while a larger part is responsible for the effector functions of the antibody. The former, being structurally unique for the given B-lymphocyte clone, is called the **variable (V-) region** of the antibody. *This part of the antibody molecule determines the specificity of a given antibody molecule*. The latter, which is structurally conserved, is called the **constant (C-) region**. In our discussion, we will show that effector functions of antibody molecules can only be performed when the V-region of the antibody is bound by the specific epitope. Nevertheless,

CHAPTER 3 Antibodies and Antigens

these *effector functions are mediated through binding of the C-region of the antibody molecules* to various entities that include components of **complement**, or cell-surface receptors (e.g., receptors expressed by macrophages or inflammatory leukocytes).

Another avenue of our discussion will be the unique three-dimensional secondary, tertiary, and quaternary conformation of antibody molecules. An important feature of the above configuration is the ability of an 110-amino acid stretch of the antibody chain to fold into a sphere, or globule, that always retains peculiar physicochemical parameters. This globule is called an **antibody (or immunoglobulin) domain**. Domains are repeated, homologous units that make up not only antibodies, but also some other molecules (TCRs, some accessory molecules, etc.). In this chapter, we will examine domains of antibody molecules and show how the unique molecular structure of the antibody V-region domains optimizes the antigen binding and tremendously broadens the range of antibody specificities. Further, we will analyze binding of antibodies to antigenic epitopes, and will outline some quantitative aspects of antigen–antibody interaction.

Antibodies are extremely versatile weapons of the immune system, but they are also useful tools for science, as well as for laboratory and clinical medicine, agriculture, and biotechnology. In this chapter, we will discuss not only the structure and functions of antibodies, but also the laboratory use of antibodies. This discussion will be preceded by a short outline of the principles and some applications of **hybridoma technology** – the current paradigm of obtaining homogenous antibodies of predefined specificity.

Discussion

AN OVERVIEW OF ANTIBODY STRUCTURE

3.1 Why are the terms "antibody" and "immunoglobulin" being used interchangeably?

By their chemical nature, antibodies are protein molecules that belong to what older biochemists called "**globulins**." When serum proteins move in the electrical field during gel electrophoresis, they segregate into classes according to their size. Smaller molecules that move faster were called albumins, and larger molecules that move slower, globulins. The latter were divided into three classes designated by the Greek letters, alpha, beta, and gamma. In the late 1930s to early 1940s, M. Heidelberger, A. Tiselius, and E. Kabat showed that molecules that were able to bind antigens (antibodies) segregated with the gamma-globulin class of serum proteins. The contemporary term "immunoglobulins" reflects the fact that antibodies are part of the immune system, as well as the fact that they are globulins.

3.2 Are the secreted antibodies found exclusively in the serum, and are the cell surface-expressed antibodies exclusively on B lymphocytes?

No. Although the liquid portion of the blood is the richest source of antibodies, other biological fluids contain antibodies as well. Antibodies have been detected in, and purified from, saliva, tears, cerebrospinal fluid, and other secretions. Interestingly, mucosal secretions have a bias for antibodies that belong to the class of immunoglobulins called IgA (see later). Although only B lymphocytes make antibodies, these molecules can be captured and carried by many types of nonlymphoid cells that have the so-called Fc receptors. These receptors can bind the antibody constant region, leaving the variable region (and the opportunity for the antibody to bind antigens) intact. The Fc receptors belong to the immunoglobulin superfamily and are present on macrophages, mature dendritic cells, granulocytes, NK cells, mast cells, and some other cell types. The ability of antibody constant regions to interact with Fc receptors is a very important feature that influences many of the effector functions of antibodies (see Chapter 13).

3.3 How did immunologists determine that all antibody molecules have two separate regions, one of which participates in the antigen binding while the other does not?

This was determined through experiments with the so-called **limited proteolysis** of antibody molecules. This enzymatic treatment cleaves protein molecules into distinct parts. If papain was used as the proteolytic enzyme, the antibody molecules were reproducibly cleaved into three parts. Of these, two were identical to each other and retained the ability to bind antigen. The third fragment did not bind antigen and tended to crystallize. Potter and Edelman, who carried out these experiments in the late 1950s to early 1960s, called the first two fragments **Fab** (fragment, antigen-binding), and the third fragment **Fc** (fragment, crystalline) (Fig. 3-1). If another enzyme, pepsin, was used in the limited proteolysis system, the results were somewhat different: the Fc fragment was unstable and degraded quickly, while the two Fab fragments appeared to be bonded to each other. Potter and Edelman called these "doublets" of Fab fragments **F(ab′)$_2$**. It was from these experiments that the spatial separation of the antigen-binding region (Fab) and the antigen-nonbinding region (Fc) became obvious. Besides, the fact that one antibody molecule appeared to have two identical Fab portions prompted Potter and Edelman to suggest that antibody molecules consist of polypeptide chains that are identical "mirror images" of each other when positioned along the longitudinal axis of symmetry.

3.4 What is the basic structure of antibody molecules?

In spite of their ultimate heterogeneity, all antibody molecules have the same basic, principal structure. Each and every antibody molecule consists of four

CHAPTER 3 Antibodies and Antigens

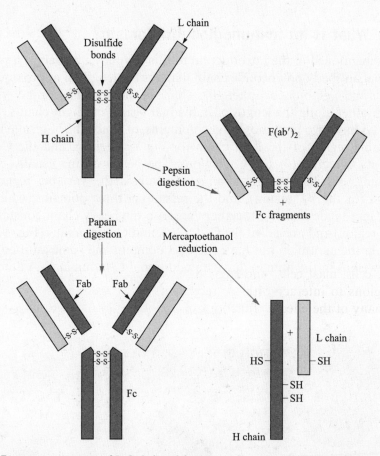

Fig. 3-1. Prototype structure of IgG deduced from experiments with the use of limited proteolysis.

polypeptide chains. Of these, two chains are identical to each other and have a molecular weight of approximately 55 to 70 kDa; they are called antibody **heavy chains**. The other two chains are also identical to each other and have a molecular weight of approximately 24 kDa; they are called antibody **light chains**. Each light chain is attached to a heavy chain, and the two heavy chains are attached to each other through covalent –S–S– -bonds. Each heavy and each light chain has its own variable and its own constant region. This macromolecule consisting of four polypeptide chains is called a monomeric antibody molecule because, as we will discuss later, antibodies of some isotypes form multimeric aggregates, in which several monomeric molecules are joined together. In the two identical heavy and the two identical light chains, the variable and the constant regions are identical. However, the structure of the heavy chain variable and constant regions is different from the structure of the light chain variable and constant regions. The entity that can physically bind an epitope, called the antibody variable region, is actually the heavy chain variable region and the light chain variable region that are juxtaposed (brought close together) in space. Thus, each monomeric antibody molecule has two *identical* antibody variable regions. The specificity of any antibody to its epitope is, therefore, unique, but the "valency" of the interaction of a monomeric antibody with its specific epitope is two, i.e., if two *identical* epitopes are available, one specific monomeric antibody molecule can bind them both simultaneously.

3.5 What is an immunoglobulin domain?

As already mentioned in the introduction to this chapter, a **domain** is a portion of an individual antibody polypeptide chain that has the length of approximately 110 amino acids and a globule- or sphere-like shape of the fold. The antibody domains follow each other along the length of individual heavy and light chains. The light chains always have two immunoglobulin domains, of which the N-terminal domain is the variable region of the light chain (or the V_L region), and the C-terminal domain is its C-region (or the C_L region). The heavy chains usually consist of four domains, of which the N-terminal domain is the variable region of the heavy chain (or the V_H region), and the remaining three domains, which follow each other in a tandem array, together comprise the heavy chain constant region (or the C_H region) and are called the C_H1, C_H2, and C_H3 domains (Fig. 3-2). Some antibodies in the mouse do not have the C_H3 domain, and some antibodies in the mouse, as well as in the human, have an additional C-terminal C_H4 domain. The

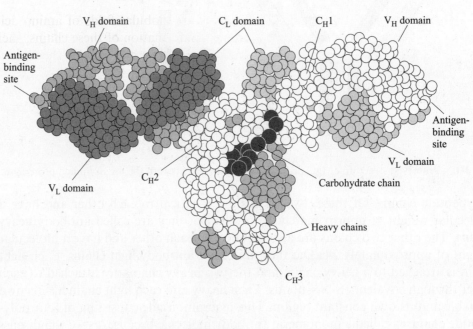

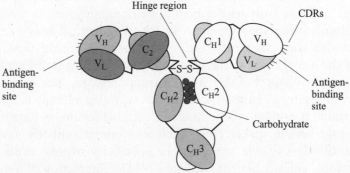

Fig. 3-2. Immunoglobulin domains. A three-dimensional model; the lower illustration is a schematic diagram.

CHAPTER 3 Antibodies and Antigens

portion of an antibody molecule between the C_H1 and C_H2 domains is very flexible and forms the so-called **hinge**, the region where the V_H and the C_H1 domains together with the V_L and C_L domains that are attached to them form an angle, allowing the variable regions to deviate in space from the longitudinal axis (Fig. 3-2). *The exact shape of the hinge region and the exact size of the angle that it forms may vary in antibodies of different isotypes* (see later). The interaction with antigenic epitopes is always mediated through portions of the V_H and V_L domains. The interaction with complement and Fc receptors, which is very important for the proper performance of the antibody's effector functions, is mediated through the C_H2 domains. The ability of the C_H2 domain to interact with complement and Fc receptors, again, depends on the particular antibody isotype.

3.6 How are the V_H and V_L domains built, and what enables them to interact with antigen epitopes?

Like all immunoglobulin domains, V-domains are globular folds of amino acid chains. The globule (or sphere) is the tertiary conformation of these chains; their secondary structure can be either alpha-helical, or of the "beta-pleated sheet" kind (see biochemistry and molecular biology manuals for details). What makes a particular V-domain able to interact with its specific antigen epitope is *the unique sequence of its amino acids*. Within a B lymphocyte clone, all cells make antibody molecules that have an almost identical amino acid sequence in the V-domains ("almost" because of the somatic mutations that will be discussed later). No two different clones, however, produce antibody molecules where the sequence of amino acids in the V-domains would be identical between the clones. Because of the unique sequence of amino acids within its boundaries, any given V-domain displays a unique complementarity to its specific antigen. Early studies suggested that antigen epitopes interact with small sites underneath the spherical surface of V-domains, so that the complementary interaction and bonding occurs in formations that resemble "grooves." Later studies, however, showed that in many cases an antigen epitope and the surface of the V-domain that is specific to this particular epitope interact without forming "grooves." In other words, *the antigen-binding site* of the antibody molecule is actually "planar," i.e., it *is a part of the V-domain's concave spherical surface*. Such "planar" interaction is an adaptation that allows antibody V-domains to better accommodate antigen epitopes in their native form.

3.7 Is the entire V-domain able to interact, and does it actually interact with its specific epitope?

No. Within any given V-domain, there are subregions that form noncovalent bonds with specific epitopes, and subregions that do not. The amino acid sequence in the first kind of V-domain subregions varies from one clonospecific domain to another so extensively that sometimes no two amino acids are found in the same position in two different domains. In the second kind of V-domain subregions, the extent of variability in the amino acid sequence is smaller. Certain positions often, or almost always, contain the same amino acid in different V-domains produced by different B lymphocyte clones.

Initially, the first kind of "subregions" within V-domains were called **hypervariable regions** because of their extremely extensive amino acid variability from clone to clone. Later, immunologists began to call these "subregions" **complementarity-determining regions (CDR)**, because it became apparent that it is these "subregions" who actually interact with complementary (specific) antigen epitopes.

3.8 If the function of CDR is to bind antigen, what is the function of the rest of the V-domain?

Actually, it is just to make sure that the domain retains its three-dimensional structure. For this reason, these "other" subregions of V-domains were called **framework regions (FR)**. The "conservative" amino acids are always there in the same position within the FR for a very valid reason: they form covalent –S–S– bonds, or other chemical bonds, with other amino acids in the FR, thus preventing the unique globular construction of the V-domain from falling apart.

3.9 How many individual CDR and individual FR exist within a V-region, and how are they positioned there?

In a typical V_H domain, there are three FRs and three CDRs. If one imagines the amino acid chain that folds into an immunoglobulin domain unwind, the relation between the FRs and the CDRs will be as follows: the most N-terminal part of the chain will be FR1, followed by CDR1, then by FR2, then CDR2, then FR3, and, finally, CDR3. The latter is the most C-terminal part of the chain that forms the V-domain. In a typical V_L domain, there are four FRs and three CDRs. Notably, the length of individual FRs and CDRs varies in V-domains coded by different V-region gene segments (see Chapter 4). As a result of the folding, the CDRs are brought together and become the outermost, "exposed" part of the domain, while the FRs become, for the most part, "hidden" inside the globule. Thus, the CDRs acquire the possibility to interact with antigen epitopes and form bonds with complementary, specific epitopes.

3.10 What are antibody isotypes?

These are kinds, or types, of the *constant* regions of the antibody heavy and light chains that dictate the particular shape of antibody molecules and their *effector functions*. Unlike different antibody specificities that are determined by different amino acid sequences in the V-region, *different isotypes are determined by the amino acid sequences within C-regions*. For example, there are five major "kinds" or **classes** of human antibody heavy chains, called α, γ, δ, ε, and μ. They are determined by five different kinds of amino acid sequence in the C-region of these chains.

CHAPTER 3 Antibodies and Antigens

Depending on the "kind" of the heavy chain that they have, the whole antibody molecules are said to belong to the following classes: IgA, IgG, IgD, IgE, and IgM (the Latin letter in the name of the class corresponding to the Greek letter of the "kind" of the heavy chain). Further, within the "kind" of antibody heavy chains called γ, there are several small variations, and the chains showing these variations (but still belonging to the gamma "kind") are called γ 1, γ 2, γ 3, and γ 4. (In the mouse, the same isotypes are called γ 1, γ 2a, γ 2b, and γ 3.)

Antibody light chains, depending on the amino acid sequence in their constant regions, also can be of different "kinds." The two of these, known in all higher vertebrates, are called κ and λ. The term isotype can refer to the "kind" of the heavy or light chain of the antibody molecule; for example, one can say that α, γ, μ, etc., are antibody heavy chain isotypes, and kappa and lambda are antibody light chain isotypes. Alternatively, the term isotype can refer to the "kind" of whole antibody molecule; for example, one can say that the isotype of this particular antibody is "IgM, κ," while the isotype of another antibody is "IgG1, λ." As we will discuss in the next chapter, a separate antibody constant region gene determines each of the known antibody heavy and light chain isotypes.

One crucially important thing to remember about isotypes is that they have nothing to do with antibody specificity. Antibodies that belong to the same isotype may have a tremendous diversity of specificities. Conversely, the antibody that has a particular specificity can exist as more than one isotype; for example, the antibody that is specific to one particular epitope of a bacterial antigen can be made initially as IgM and then become IgG. Its specificity will not change, but its isotype will. This phenomenon, called the **isotype switch**, will be discussed in more detail in Chapter 4. The other important thing to remember about isotypes is that different heavy chain isotypes enable antibodies to perform this or that effector function better. For example, antibodies of the IgM isotype are the best at binding complement (see Chapter 13); antibodies of the IgG1 and IgG3 isotypes are good at it, although not as good as IgM; antibodies of the IgG2 isotype are mediocre; and antibodies of the IgG4, IgD, IgA, and IgE isotypes do not bind complement at all. We will discuss the effector functions of antibodies later, as we move to the dynamics of immune responses.

3.11 What are the different "shapes" of antibody molecules conferred by their isotype?

When expressed on the B lymphocyte membrane, an antibody molecule has a rather uniform "shape" that is not dependent on the isotype. When secreted, antibody molecules may have a "shape" that is conferred by its isotype. For example, IgM and IgA antibodies, but not IgG and IgE antibodies, have the so-called **tail pieces** – short nonglobular sequences of amino acids on the C-terminus of the heavy chain. These tail pieces have the ability to be bound by –S–S– bonds and by a special polypeptide called the **J-chain**. This allows IgM and IgG to exist as stable multimeric molecular aggregates. Usually, secreted IgM exist as pentamers, and secreted IgA as dimers or trimers.

3.12 Are membrane-expressed antibodies and secreted antibodies structurally similar or different?

They are very similar, although not identical. The membrane form has an extra C-region segment that consists of approximately 26 uncharged, hydrophobic amino acid residues followed by a varying number of charged (usually basic) amino acids. This segment is coded by a separate exon (see Chapter 4). The hydrophobic stretch of amino acids allows the membrane form of antibody molecules to be embedded in the phospholipid bilayer of the lymphocyte membranes, while the C-terminal hydrophilic segment serves as a cytoplasmic portion of the transmembrane molecule. Interestingly, in IgM and IgD, the cytoplasmic portion of the molecule is extremely short; it consists of just three amino acid residues.

3.13 Can all antibody isotypes be secreted?

All except IgD. The latter is always embedded in the B-lymphocyte membrane. No effector function associated in this isotype is known.

3.14 What is the function of light chain isotypes?

It is not clear so far. There are no known differences in the performance of effector functions by the immunoglobulins that contain kappa chains versus the immunoglobulins that contain lambda chains.

3.15 Besides binding complement and Fc receptors on macrophages and inflammatory leucocytes, what other effector functions can antibodies perform?

IgE antibodies are able to bind a special kind of Fc receptors, called Fc-εR1, expressed on the surfaces of mast cells and basophils. These cells contain special granules that can be released when the Fc-εR1 is cross-linked. The released granules cause a series of effects collectively known as allergic reaction or the immediate-type hypersensitivity. This special effector function of IgE antibodies will be thoroughly discussed in Chapter 13. IgA antibodies play a crucial role in the work of the mucosal compartment of the immune system; their function will be discussed in Chapter 13.

ANTIGEN–ANTIBODY INTERACTION

3.16 What is the nature of epitopes that can be recognized by antibodies?

Epitope is, by definition, anything that can be bound by the specific antigen receptor (antibody or TCR). Epitopes that are recognized and bound by antibody V-regions – mainly by their CDR, to some extent also by parts of their FR – may

CHAPTER 3 Antibodies and Antigens

vary in their exact chemical nature. They can be stretches of amino acids, sugar groups, nucleotides or their subcomponents, or parts of organic molecules bound by, e.g., phosphodiester bonds. Importantly, the epitopes recognized by antibodies do not need special processing and can be recognized in their native form. As already mentioned, this is one cardinal difference between them and the epitopes recognized by TCR.

3.17 Do all epitopes retain their natural three-dimensional shape when recognized by antibodies?

No. Antibody V-regions can also be specific to epitopes that have lost their natural three-dimensional shape because of the protein denaturation. Such epitopes are called **linear epitopes** or determinants. The epitopes that are complementary to their specific antibodies only in their unique three-dimensional form are called **conformational epitopes** or determinants. In many cases, antibodies specific to conformational epitopes cease to recognize them if the protein is denatured. Conversely, if an antibody is generated against a linear epitope of a denatured protein, it may not recognize the same region of the protein molecule in its native form. Some antibodies, however, can cross-react between linear and conformational epitopes that correspond to the same region of the protein antigen molecule. When protein molecules are subjected to chemical modification, e.g., proteolysis, new epitopes can appear; these are called **neoantigenic determinants**.

3.18 What is antibody affinity?

It is, essentially, the "*tightness of the fit*" between the epitope and the antibody V-region. Two antibodies produced by two different B-cell clones, or even by one B-cell clone but at two different stages of its differentiation and selection in the germinal centers (see Chapter 1 and later chapters), may have V-regions similar enough to bind to the same epitope, but differing in the strength, or tightness, of the binding. The latter may change if even one amino acid in the V-region (especially in the CDR) is replaced. Affinity is a very important parameter of the antigen–antibody interaction and a useful characteristic of an antibody preparation or reagent.

3.19 How is antibody affinity measured?

The best way to have an idea about the antibody affinity is to measure its **dissociation constant** (K_d). The interaction between the epitope and the corresponding part of the antibody V-region ("paratope") is noncovalent and reversible; hence, one can imagine it as a two-way reaction that is in the state of dynamic equilibrium, similar to reversible chemical reactions, ionic dissociation, etc. (see basic chemistry manuals for details). If the concentration of the antibody is limiting and constant, and the concentration of the antigen varies, then the exact molar concentration of the antigen that saturates one-half of the "paratopes" in the system will depend solely on the quality of the epitope–"paratope" fit. One would need more antigen to half-saturate a "sloppy" antibody and less antigen to half-saturate a "tightly fit" antibody. This molar concentration of the antigen that is enough to saturate one-half of the "paratopes" of the given antibody in a system where the antibody concentration does not vary is called the antibody's K_d. Obviously, the bigger the K_d, the worse the "tightness of the fit," or, in other words, the bigger the K_d, the lower the affinity. Conversely, the smaller the K_d, the better the "tightness of the fit," or the smaller the K_d, the higher the affinity. A common way to express the K_d is to use molar concentrations in an exponential form with a negative power. For example, the two antibodies may have affinities 10^{-8} M and 10^{-11} M. Apparently, the K_d of the first antibody is higher (10^{-8} is bigger than 10^{-11}); hence, the affinity of the first antibody is lower.

3.20 How can one determine the K_d?

There are a number of ways to do that. Traditionally, antibody K_d was measured through the experimental procedure called **equilibrium dialysis**. The principle of this procedure is as follows: Antigens and antibodies are placed on the two sides across a semi-permeable membrane, so that, if they are not bound to each other, they can diffuse through it but, if they are bound, they cannot. If one varies the concentrations of the free antigen and knows the concentration of the free antibody that is kept constant on one side of the membrane, one can directly deduce the molar concentration of the antigen that binds one-half of the antibody molecules in the system. More recently, assays based on the liquid phase inhibition of the antigen–antibody interaction in the solid phase became widely used for the experimental determination of the K_d.

3.21 What is antibody avidity?

It is an overall strength of the binding between the antigen and **all** of the individual antibody molecule's binding sites. As already mentioned, the affinity is the "fitness" of a particular V-region to its specific epitope. Antibodies, however, have at least two identical V-regions (see Section 3.14), and multimeric antibodies can have more than two; for example, a pentameric IgM molecule has ten identical V-regions (see Section 3.12). If two or more identical epitopes are available, each of them can be bound by the V-region of the corresponding specificity; the strength of the binding will then be determined by the affinity of the V-region *and* by the number of the V-regions involved. In this regard, the affinity of individual V-regions may be low, but if many of them are involved, the overall strength of the binding, or the avidity, will be sufficiently high. For example, individual V-regions of IgM antibodies often have a low affinity for their epitopes (for the reasons discussed later); yet, since ten individual V-regions may be involved, the avidity of the IgM molecules to the antigen that can present the multiple identical epitopes can be high. The avidity is not a sum but, rather, is a product of multiplication of individual affinities of a multimeric antibody.

CHAPTER 3 Antibodies and Antigens

ANTIBODY HETEROGENEITY AND HYBRIDOMA TECHNOLOGY

3.22 If one major obstacle hampering the progress in antibody research could be named, what would it be?

The tremendous heterogeneity of antibodies was one major obstacle. Although all antibodies belong to the gamma-globulin class of serum proteins, they are in fact a very complex mixture of molecules that differ in size, shape, charge, subunit composition, and other structural features. Moreover, serum and other biological fluids contain many different antibodies that originate from different B lymphocyte clones (**polyclonal**), each having a distinct specificity for antigen. Even among the serum antibodies specific to one epitope, different individual molecules may have different affinities of the binding. The advent of hybridoma technology in 1975 was the decisive factor that solved many of the methodological problems related to the antibody heterogeneity.

3.23 Did immunologists always work with heterogeneous antibodies before hybridomas were invented?

No. Beginning from the 1950s, antibodies produced by myelomas began to serve as a valuable object of research. Myelomas (or plasmacytomas) are malignant tumors that originate from terminally differentiated effector B lymphocytes called plasma cells. As malignant tumors, these cells form clones that grow continuously and produce **monoclonal** antibodies. Myeloma-derived monoclonal antibodies were isolated in large quantities from biological fluids of the patients, as well as from the fluids of myeloma cells cultured *in vitro*. In most cases, it is impossible to determine the antigen specificity of these antibodies; therefore, they generally are not instrumental in the studies of antigen–antibody interactions. Nevertheless, they served as a valuable tool for structural studies. These studies, performed in the late 1960s and early 1970s, were extremely important for the subsequent characterization of the content of the human and murine antibody V-region gene loci. (We will discuss the genetic determination of the antibody V-genes in detail in Chapter 4.)

3.24 What are hybridomas, and why are they so advantageous for antibody research?

Hybridomas are cell lines that stem from hybrids between normal lymphocytes and malignant tumors of the lymphoid origin. As G. Kohler and C. Milstein first demonstrated in 1975, such hybrids can be made *in vitro*, and culturing them in the presence of chemicals that select against the nonhybridized parent tumor cells can ensure their continuous replication. Monoclonal antibody-producing hybridomas are products of the **fusion** between B lymphocytes and myeloma cells. These fused cells inherit from the parental myeloma the ability to grow continuously and be cloned in culture, and from the normal lymphocyte, the ability to produce antibodies. Importantly, the antibodies made by hybridomas are exactly the same that are made by parental lymphocytes, so their specificity can be *pre-determined* by immunizing the donor of lymphocytes with an antigen of interest. Since stable cloned hybridomas can be maintained in culture for an indefinite period of time, the homogenous antibodies that they secrete can be obtained in unlimited quantities. Such antibodies were successfully used for the analysis and preparative purification of various antigens, as well as for extensive studies of the structure and genetic coding of immunoglobulin molecules.

3.25 How does the mentioned selection of hybrids work?

If a number of myeloma cells is mixed with a number of normal lymphocytes and treated with a fusion-promoting agent (most often, polyethylene glycol), only a small fraction of the cells will form hybrids. This makes the selection against the parental myeloma a must, because without such a selection the parental myeloma cells will quickly overgrow the hybrids. Usually, the following approach helps select against the parental myeloma. The tumors used in the hybridoma technology are mutants that lack the ability to make certain enzymes because of a genetic defect. One of such enzymes, called hypoxanthine-guanine phosphoribosyltransferase (HGPRT), controls a "back-up" or "salvage" pathway of nucleotide synthesis. This salvage pathway is used by cells when the "normal," conventional pathway of nucleotide synthesis is blocked by a chemical called aminopterin. When wild-type myelomas are placed in a medium that contains aminopterin, they still can divide, because the HGPRT that they make allows them to use the salvage pathway and synthesize new DNA. $HGPRT^-$ mutants, however, stop growing and eventually die in the presence of aminopterin, because they cannot utilize the salvage pathway. However, if an $HGPRT^-$ mutant is fused with a normal lymphocyte, the resulting hybrid re-acquires the ability to survive in the presence of aminopterin because of the gene complementation. Therefore, unfused parental myeloma cells will die in the presence of aminopterin, and hybrids will not. Unfused parental lymphocytes have the HGPRT gene and therefore are resistant to aminopterin, but they will die anyway, simply because, like any other normal cells, they cannot be maintained in long-term cultures. Their lifespan in culture is approximately 1–2 weeks. To ensure the utilization of the salvage pathway by the hybrids, the medium where they are cultured must be supplemented with hypoxanthine and thymidine that serve as precursors for the nitrogenous bases synthesized. Thus, the complete medium that blocks the growth of unfused myeloma and ensures the growth of hybrids must contain aminopterin, hypoxanthine, and thymidine. Such a medium is commercially available and known as "**HAT**" (from "**h**ypoxanthine, **a**minopterin, and **t**hymidine").

CHAPTER 3   Antibodies and Antigens

3.26 How many of the hybrids successfully selected in the HAT medium actually produce the monoclonal antibody of interest?

It depends on the antigen used for the immunization of the donor of normal lymphocytes, and on the immunization schedule. Some antigens are powerful **immunogens**, i.e., they elicit strong immune responses and expand the clone(s) of B lymphocytes that recognize their epitope(s) to a great extent. In this case, a large fraction of the hybrids – several percent, or even the majority of the hybrids – will be found to secrete the antibody of interest. In the case where an antigen is a weak immunogen, only a few hybrid cells out of several thousand may produce the desired antibody. Repeated immunization and the use of the so-called **adjuvants** – nonspecific enhancers of immune responses – may somewhat increase the frequency of useful antibody-producing hybrids.

3.27 Can hybridomas grow *in vivo*, like myelomas?

Yes. In fact, hybridomas inoculated in mice secrete much more antibody than hybridomas grown in tissue culture. While in the culture fluids the concentration of monoclonal antibodies is usually of the order of micrograms (or even fractions of a microgram) per 1 ml, the serum and the ascitic fluid of mice with inoculated hybridomas may have a concentration of the monoclonal antibody that exceeds 1 mg/ml. To facilitate the inoculation and growth of hybridomas as ascites-producing tumors, immunologists usually prepare the animal recipients by injecting mineral oil and some chemicals into their peritoneal cavities prior to the injection of hybridoma cells.

3.28 Can hybridomas that produce human antibodies be obtained?

Yes, although it is much more difficult to make them, compared to the hybridomas that produce antibodies of the mouse origin. So far, investigators have not been successful in growing stable HGPRT⁻ mutant myelomas that originate from human plasma cells. Therefore, the only possibility to make a hybridoma that will produce human antibodies is to fuse human lymphocytes with a nonhuman (mouse, rat, or hamster) myeloma. This inter-species cell fusion often results in the loss of human chromosomes from the nascent hybrids. The yield of viable and antibody-producing hybrids is, therefore, rather low. Yet, a number of useful and stable inter-species hybrids that produce monoclonal antibodies of the human origin has been obtained. These "**heterohybridomas**" were instrumental in the characterization of the human immune responses to certain pathogens, as well as for in-depth studies of the human antibody repertoire.

3.29 Are there any alternatives to hybridoma technology that allow investigators to produce useful monoclonal antibodies?

Yes. Human, as well as rodent, B lymphocytes can be strongly activated, driven to proliferation, and malignantly transformed ("immortalized") by a virus that belongs to the herpes virus family and is called the **Epstein–Barr virus (EBV)**. The yield of stable clones that produce specific antibodies is much lower after the treatment of the B lymphocytes with the EBV than after their fusion with a myeloma. Nevertheless, the techniques based on the EBV-induced transformation allowed immunologists to obtain some interesting monoclonal antibodies of the human origin. Recent advances in molecular biology brought another alternative to the hybridoma technology called **phage display library** technique. This approach is based on the ability of some expression vectors to display functional antibodies, transcribed and translated from cloned immunoglobulin genes, on the surface of bacteriophages. Collections of bacteriophages where some individual phages contain specific antibodies can be screened much like hybridomas, although in substantially larger numbers. The phage display library approach turned out to be useful for the dissection of the human antibody repertoire. We will discuss this approach in somewhat more depth in Chapter 4, after the description of the V- and C-region antibody genes.

3.30 Other than the applications of hybridoma technology to basic science, what major practical goals did this technology allow to be accomplished?

Monoclonal antibodies produced by hybridomas are extremely valuable tools in medicine, used both for diagnostic as well as therapeutic purposes. For example, monoclonal antibodies that recognize epitopes of some hormones and fetal antigens are now routinely used for home pregnancy tests. Monoclonal antibodies detecting alleles of the blood group and histocompatibility antigens are gradually replacing alloantisera in the blood and histocompatibility typing and have become indispensable in organ transplantation. Antibodies to oncofetal antigens, such as alpha-fetoprotein (AFP) or carcinoembryonic antigen (CEA) are successfully used for the early diagnosis of cancer, especially the cancers of liver and stomach. There is no doubt that similar reagents will soon be used for the early diagnosis of the breast, prostate, lung, ovary, and other cancers. A monoclonal antibody called OKT3, specific to a T-lymphocyte antigen called CD3, is widely used in transplantation clinics to prevent or halt the reactions of transplant rejection. Several monoclonal antibodies to tumor-specific antigens were used for cancer therapy in the form of immunotoxins. The latter are antibodies conjugated with deadly toxins that can kill the cell that expresses the antigen to which these antibodies bind. Immunotoxins were shown to be a very efficient weapon against human tumors, at least in the early stages of their growth.

CHAPTER 3 Antibodies and Antigens

3.31 What are the major obstacles in the further development of these clinical applications of hybridoma technology?

One major obstacle for the therapeutic use of the products of hybridoma technology is that the vast majority of monoclonal antibodies to important human antigens are of the murine origin. When such antibodies are injected into a patient, they trigger powerful species-specific immune responses. It is, therefore, necessary to combine the infusion of these antibodies with immunosuppression, which is not always desirable for the patient. One possible way to work around this problem is to "humanize" the murine monoclonal antibodies through a genetic engineering approach. Such "humanization" includes cloning of the gene that codes the V-region of a useful mouse anti-human antibody, and ligating it with a gene that codes for a human antibody C-region. The resulting chimeric antibody molecule will not have the mouse C-region-associated species-specific epitopes that cause undesirable immune reactions in the patient.

LABORATORY USES OF ANTIBODIES

3.32 What are the main uses of antibodies in laboratories?

Antibodies are versatile tools that help investigators to achieve a variety of goals. Their versatility is based on their specificity for antigens. Because of this specificity, antibodies can be used for: (1) detection of antigens; (2) quantitation of antigens; (3) purification of antigens from complex mixtures; (4) characterization of the physicochemical properties of antigens; and (5) visualization of antigens in tissues, cells, and subcellular compartments.

3.33 What assays are employed for detection and quantitation of antigens?

Historically, a number of techniques designed for this purpose were based on the ability of antigens and antibodies to form insoluble **immune complexes**. Since the valency of antibodies is always more than one (two for monomeric IgG and IgE, four or six for IgA, and ten for IgM), many individual antigen molecules that express identical epitopes can be **cross-linked** and, within a certain diapason of antigen and antibody concentrations, a kind of insoluble crystallic lattice can form. The reaction leading to the appearance of insoluble immune complexes was called **precipitation**. It was widely used for detection of various antigens in the first half of the 20th century. A variant of the precipitation assay, called **immunodiffusion**, was especially popular in the 1950s and 1960s; it is based on the formation of visible antigen–antibody complexes in semisolid (usually agarose) gels. With the advent of **solid-phase assays**, however, the popularity of precipitation assays decreased.

3.34 What are these solid-phase assays?

There are two principal kinds of solid-phase immunoassays: a **radioimmunoassay (RIA)** and an **enzyme-linked immunosorbent assay (ELISA)**. Both are based on the ability of antigens and antibodies to interact in the solid phase, e.g., when one of the two components (the antigen or the antibody) is adsorbed onto the bottom of polystyrene trays. If present, the other component (the specific antibody or the antigen) binds to the attached component; all non-attached substances (e.g., nonspecific antibodies, other proteins) are removed by vigorous washing. The binding is visualized by attaching a label or tag to the bound component. In RIA, the tag is radioactive, and the number of radioactive counts serves as a measure of the binding intensity. In ELISA, the tag is enzymatic, and the change in the absorbance caused by the action of the enzyme on the added substrate is a measure of the binding intensity (Fig. 3-3).

3.35 How exactly are antigens quantified in solid-phase assays?

Antigens or antibodies can be quantified because standard preparations of them, purified by powerful biochemical techniques such as affinity chromatography, are available from industry or government-associated suppliers of standard reagents. If one knows the concentration of antigens or antibodies in these preparations – information usually supplied by the manufacturer – one can titer them and build a **standard curve** (a plot that shows radioactive counts or absorbance corresponding to these standard concentrations). Then one can measure the absorbance or radioactive counts in an experimental sample where the concentration of the reactant (antigen or antibody) is unknown. Finally, one can see where the experimental measurement fits in the standard curve. By applying statistical methods like **linear regression analysis**, it is possible to deduce the concentration of the reagent in the experimental sample from the standard curve with a high degree of precision.

3.36 What is a "direct" and an "indirect" solid-phase immunoassay?

A "direct" solid-phase immunoassay refers to an assay where there are only two components of the reaction (an antigen and an antibody), and one of them is labeled. An "indirect" assay refers to an assay where there are three components: an antigen, an antibody specific to it, and the so-called **"secondary" (anti-immunoglobulin) antibody**, which is labeled. Secondary antibodies are produced by immunization of certain species of animals with whole serum or partially purified serum immunoglobulins from other species of animals. For example, if one injects serum immunoglobulins of a mouse into a rabbit, the rabbit will generate rabbit-anti-mouse secondary antibodies. When labeled, these are versatile, because they can be used for detection and quantitation of any murine antibody regardless of its specificity.

CHAPTER 3 Antibodies and Antigens

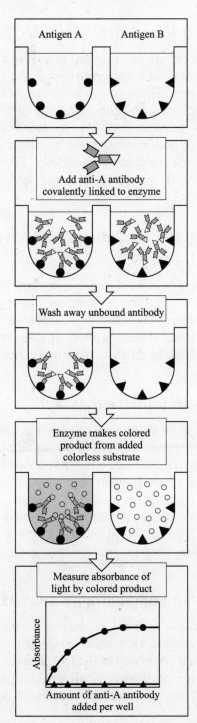

Fig. 3-3. Solid-phase binding assay. In this figure, an enzyme-linked solid-phase binding assay is shown. The same principle applies to radioimmunoassay, where the label is not an enzyme but a radioactive isotope.

3.37 What is a "sandwich" assay?

It is a popular modification of a quantitative solid-phase assay where two antibodies, specific to two different epitopes of the same antigen, are allowed to react with the antigen. The first of these antibodies is adsorbed onto the bottom of a polystyrene tray, and the antigen is added so that the solid phase-bound antibody binds it. After this, the second antibody is allowed to react with the antigen. The second antibody can be directly labeled or allowed to be bound by a labeled species-specific antibody. The antigen appears to be between the two layers of antibody like the butter between two pieces of bread in a sandwich.

3.38 What are the advantages of indirect and double-sandwich assays?

As already mentioned, indirect solid-phase immunoassays employ secondary antibodies that are versatile, and able to detect species-specific antibodies of all specificities. In addition, indirect and especially double-sandwich assays have a much higher sensitivity. The lower detection limit in these modifications of the RIA is less than 10 picograms (pg); in ELISA, the limit is between 10 and 100 pg.

3.39 What is an inhibition assay?

It is an assay where the *fixed* amount of antibody is allowed to react with *increasing* concentrations of the specific antigen in the liquid phase (test tubes). After such a reaction, the mixtures from the test tubes are tested for their ability to bind the same antigen in the solid phase (polystyrene trays). The more antigen is bound in the liquid phase, the less of the free antibody remains, and the smaller the binding in the solid phase. The decrease in this residual binding, however, depends not only on the amount of antigen, but also on its "fitness" to the antibody V-region. Therefore, inhibition assays are perfect for measuring antibody affinity.

3.40 What is an immunofluorescent assay?

It is an assay based on the labeling of antibodies with **immunofluorescent dyes**. When such an antibody binds its specific antigen on the cellular surface, the cell appears to "glow" in the dark vision field of a special **fluorescence microscope**. Antibodies carrying fluorescent dyes of different color can bind different antigens, which allows an investigator to characterize the expression of more than one antigen simultaneously.

3.41 What is flow cytometry, and why is it so popular?

It is a modification of the immunofluorescence assay wherein the cells that have been subjected to binding of fluorescent-labeled antibodies pass through very thin tubing. Such a treatment makes them "go in a single file." Laser beams that excite

CHAPTER 3 Antibodies and Antigens

the cells' fluorescence meet the cells on their way. The images of the fluorescing cells are processed and stored by a sophisticated computer interphase. Using this technique, an immunologist can visualize various populations and subpopulations of cells that express certain antigens in bigger or smaller amounts, or compare the expression of an antigen with such parameters as the size of the cells, their internal complexity, etc. Moreover, by using magnets that warp the cells' trek in a way that depends on the expression of particular antigens, an experimenter can physically separate cells that express or do not express certain antigens or express them in bigger or smaller quantities. This procedure is called **cell sorting**, and the device that can perform flow cytometry and cell separation is called the **fluorescence-activated cell sorter (FACS)**, or **flow cytometer**.

3.42 Which immunoassays are used to characterize the chemical nature of an antigen?

The two modern assays most commonly used are **immunoprecipitation** and **Western blotting**. Immunoprecipitation is a somewhat misleading term because it does not point at the essential component of the assay, namely, a radioactive labeling of the material from which a certain antigen is being precipitated. In this assay, cells are incubated in a medium that contains a radioactive isotope incorporated into an amino acid. The specific antibody is added and cross-linked, so that an insoluble precipitate forms. The precipitate is pelleted by ultracentrifuging and run on an electrophoretic gel with subsequent radioautography. The band that appears on the radiosensitive film corresponds to the protein band on the gel and can be compared with marker bands so that the molecular weight of the antigen can be established. Also, by applying the so-called reducing conditions to the electrophoresis one can investigate the tertiary structure (i.e., the subunit composition) of the proteinaceous antigen.

Western blotting is similar to immunoprecipitation in that it is also based on electrophoresis. However, in this assay, the mixture of proteins, not an isolated antigen, is subjected to the electrophoresis. The material from the electrophoretic gel is then transferred to nitrocellulose paper sheets, and the latter are placed in the solution with a specific antibody. The bound antibody is then exposed to the secondary antibody, which is labeled with radioactivity or an enzyme. The visualized protein band that appears on the paper sheet (which is called a "blot") can provide an investigator with essentially the same information as the immunoprecipitated band.

3.43 Which immunoassays are used to localize the antigen in tissues, cells, and subcellular compartments?

To this end, special modifications of immunoenzymatic staining with subsequent microscopy are used. Frozen tissue sections are stained with antibodies to the antigen(s) of interest and then developed with secondary antibodies labeled with enzymes, most commonly horseradish peroxidase or alkaline phosphatase. The reaction conditions are such that the color reaction triggered by the enzyme is not spread all over the section but develops locally. It can then be observed under a microscope. An approach that allows an investigator to locate antigens in subcellular compartments is called an **immunoelectron microscopy.** It is based on the conjugation of antibodies with electron-dense particles, such as colloidal gold.

Questions

REVIEW QUESTIONS

1. How exactly would you prove experimentally that antibodies are gamma-globulins?
2. You want to buy a preparation of antibody for your research and browse a biotech company catalog. There, you found two items: one – a whole antibody molecule and the other – an F(ab')$_2$. Which one would you buy if you were interested in reducing the possibility of nonspecific (background) binding?
3. One cell expresses two identical epitopes. Another cell expresses the same two epitopes, but they are somewhat further apart. Will the antibody molecule specific for this epitopes react only with one of the cells, or with both? Why?
4. Based on the material of this chapter, how would you object to proponents of the "instructionist" theory of antibody diversity (see Chapter 1)?
5. How would you comment on the statement, "Antibody isotype determines its effector functions?"
6. As we will detail in later chapters, IgG and IgA antibodies are characteristic of secondary immune responses and possess higher average affinity than IgM antibodies, which are used mostly in primary immune responses. Does it mean that the strength of an IgG or an IgA binding to an antigen is always bigger than the strength of an IgM binding?
7. An investigator raised an antibody against a purified protein by cutting a protein band out from an electrophoretic gel and injecting the material into a rabbit. He now plans to study the expression of the protein on cell membranes. What caution does this investigator have in mind as he plans his experiments?
8. One antiserum has an average K_d of 10^{-8} M, another of 10^{-10} M. Based on this information, can one judge which of the two antisera has a higher concentration of antibodies?
9. What exactly is the target of selection by HAT medium? Is it a positive or a negative selection?
10. What are major advantages of hybridoma technology over EBV transformation?
11. You have the following materials and equipment: antigen (a mixture of proteins), polyacrylamide, gel buffer, gel plates, an antibody specific to a protein of your interest, and a secondary (anti-immunoglobulin) antibody labeled with an enzyme. What assay would you set up to characterize the molecular weight of the protein of your interest, and what additional materials and equipment do you need for that?
12. A laboratory is interested in the detection of a chemical whose concentration in biological fluids is extremely low (within the picogram range). Another laboratory is

CHAPTER 3 Antibodies and Antigens

screening hybridoma supernatants and is interested in processing the maximal possible amount of samples within a limited time. A third laboratory is comparing affinities of different antibodies. What immunoassays would you recommend to set up in each of the three laboratories?

13. Many modern science and clinical laboratories are equipped with flow cytometers (or fluorescence-activated cell sorters). What are the main reasons for that?

MATCHING

Directions: Match each item in Column A with the one in Column B to which it is most closely associated. Each item in Column B can be used only once.

Column A
1. V-region
2. HGPRT
3. CDR
4. Tail pieces
5. Affinity
6. Avidity
7. Flow cytometry
8. ELISA
9. Immunoprecipitation
10. Colloid gold

Column B
A. Number of identical V-regions
B. Joining monomeric IgM in a multimer
C. 100% variability of amino acids
D. Immunoelectron microscopy
E. Cross-linking
F. Deficient in useful myelomas
G. Is responsible for antigen specificity
H. Is lower when the K_d is higher
I. Allows sorting of cells
J. Can be set up as a "double sandwich"

Answers to Questions

REVIEW QUESTIONS

1. Immunize a laboratory animal with a protein antigen and obtain an antiserum. Divide the antiserum into two aliquots. Run one of the aliquots on a polyacrylamide gel and observe albumin, alpha-, beta-, and gamma-globulin fractions. Allow the other aliquot to react with the excess of antigen so that most or all antibodies are removed (for example, through precipitate formation). After this, run the second aliquot on another gel and compare the picture with the first gel; gamma-globulins will disappear or show as faint bands.
2. The $F(ab')_2$. Whole antibody molecules will bind a number of cells nonspecifically, via their C-regions (because of the presence of Fc-receptors).
3. It may still react if its hinge region is flexible enough. This, in turn, depends on the antibody heavy chain isotype.

4. We now know that the antigen specificity of an antibody is determined by amino acid sequence in the V-region of this antibody. Since the latter is, in turn, determined genetically, it cannot be changed by antigens, which was the main postulate of the instructionist theory.

5. Rather, antibody heavy chain isotype (or antibody class and subclass) determines what effector functions antibodies will perform best. Light chain isotypes do not matter in this regard.

6. No, because the avidity of pentameric IgM antibodies (ten antigen-binding sites) can be higher than the avidity of IgG antibodies (two sites) or of IgA antibodies (four or six sites).

7. It might be that the investigator's protein lost its conformational epitopes during its electrophoretic purification. If this protein expresses conformational epitopes in its natural membrane-binding form, the antibody may not detect them.

8. No. The only judgment we can make knowing K_ds is about the antibodies' affinities, not about their quantity.

9. The target is an HGPRT-deficient myeloma cell. It is negative (selects against these mutant cells).

10. The yield of stable clones producing specific monoclonal antibodies is higher when hybridomas are made than when B cells are immortalized by EBV.

11. Western blotting. Additional materials – substrate for the enzyme, indicator for color reaction, nitrocellulose paper; additional equipment – apparatus for transferring proteins from gel to paper.

12. The first laboratory probably needs RIA, the second laboratory probably needs indirect ELISA, and the third laboratory probably needs inhibition ELISA.

13. The main reason is that FACS can provide unique information about antigens expressed on the cell surface, as well as about various cell populations and subsets that express certain antigens. Solid-phase immunoassays, albeit sensitive, cannot replace FACS in this regard.

MATCHING

1, G; 2, F; 3, C; 4, B; 5, H; 6, A; 7, I; 8, J; 9, E; 10, D

CHAPTER 4

Maturation of B Lymphocytes and Expression of Immunoglobulin Genes

Introduction

How are antibodies coded genetically? This seemingly simple question used to be, in fact, one of the most "burning" questions of immunology for decades. Antibodies are protein molecules and, as such, must be coded by special genes. Yet, the diversity of antibody specificities is unprecedented. Not one of the molecules produced by B-cell clones has an amino acid in its variable region that exactly repeats the amino acid in the V-region of another B-cell clone's product. Altogether, there are more than one billion clones; hence, there should exist more than one billion species of antibody protein molecules. How many distinct antibody genes might be inherited and "housed" in the zygote (and in all the cells that originate from it) if we are looking at the order of one billion or more of the distinct protein molecules? If the genome of each cell has one billion or more of the antibody genes, will there be any room left for other genes? If the genome contains a limited number of antibody genes that are further diversified by somatic mechanisms, what are these mechanisms?

The tremendous diversity of antibody specificities was difficult to explain in the terms of conventional genetics. Another difficult issue was the existence of constant and variable regions within the boundaries of the same antibody molecule. As early as in 1950s, it was clearly demonstrated that allelic variations, confined to constant regions of antibody polypeptide chains, exist and can be inherited as a trait in a classical Mendelian fashion. In other words, a given antibody polypeptide chain appeared to be controlled by a single Mendelian gene. How to reconcile this finding with the notion that the same polypeptide chain contains a V-region, which is different in different antibody-producing cells of the same organism that seems to have inherited just one gene for this chain? Again, if the diversity of V-regions is not heritable but somatic, what are the somatic mechanisms that create it, and why do they selectively affect the V-, and not the C-regions of the same protein molecules?

Although many immunologists and geneticists tried to approach this issue, two principal misconceptions hampered the progress in the area. The first of these misconceptions was the idea that one polypeptide chain is always coded by one gene. The rule, "one gene – one protein," received such a tremendous experimental support that it seemed to be something that could never be violated. The second misconception was that genes occupy their places in the chromosome like beads occupy their places in bead strings. Within the classical framework of the chromosome theory of inheritance, it was almost impossible to imagine that *genes can actively change their places*, physically *move*. Only a handful of insightful geneticists challenged these views.

It was not before the above misconceptions were dispensed with that the issues around the concept of genetic determination of antibodies were solved. The "deconstruction" of the above misconceptions began when W. Dryer and J. Bennett proposed (in 1966) that **one single antibody polypeptide chain can be coded by more than one gene**, and that the coding of the constant and the variable region within the same chain is the responsibility of two entirely different classes of genes. The following discussion will focus on the main steps in the discovery of antibody genes. We will highlight the main points concerning the content of the genetic loci that contain these genes, their selection during B lymphocyte maturation, and their involvement in immune responses.

Discussion

GENERATION OF THE PRIMARY ANTIBODY REPERTOIRE

4.1 Why, and in what context was the proposal of Dryer and Bennett so significant?

Their idea was the only logical way out of the paradox created by the discovery of **antibody allotypes**. Allotypes are products of distinct alleles, and allotypic determinants, detectable by specific antibodies, are genetic markers of antibody heavy and light chains that can be inherited as Mendelian alleles. By the 1950s, it had been clearly demonstrated that, for example, in rabbits a series of the so-called Gm allotypes has a pattern of inheritance most consistent with the existence of one single Mendelian gene coding for all rabbit antibody heavy chains of the IgG class. If this is so, how can one explain that within the pool of rabbit IgG polypeptide

CHAPTER 4 B Lymphocytes and Immuno Genes

chains – all of which are coded by one Mendelian gene – there exist many millions of different chains, each having its own specificity (or, using the terminology developed in the 1960s, its own variable region)? In an attempt to solve the paradox, Dryer and Bennett theorized that the single Mendelian gene that codes for all of the rabbit IgG antibody "polypeptides," and reveals itself in the form of allotypes, in fact codes only for a *part* of the rabbit IgG polypeptide molecule. This part, according to Dryer and Bennett, is *not responsible for antibody specificity* and is, most likely, invariable or *constant*, at least within a certain class of antibodies of a species (e.g., rabbit IgG). The other part of the rabbit IgG antibody polypeptide chain is responsible for the antibody specificity, variable, and is controlled by a genetic entity that is completely different from the Gm allotype-coding gene. This latter part of the antibody chain is likely to be controlled by a series of genes that are expressed differently in different antibody-producing clones, rather than by one single Mendelian gene. Dryer and Bennett proposed to call the single gene that codes for the first of the two mentioned parts of the antibody molecule a **C-gene**, and the genes that code for the second (variable) part **V-genes**. The authors went on to postulate that the information stored in the C-gene and in the V-genes is used for the biosynthesis of one single polypeptide because it is somehow "assembled" and decoded from one single DNA or RNA unit.

4.2 What experiments supported Dryer and Bennett's postulates? Was it difficult to test this apparently revolutionary idea right away?

No sound experimental evidence in favor (or against) these postulates was obtained until the late 1970s. The main reason for such a delay was that the experimental procedures were not as developed at that time, as was the theoretical thinking of advanced specialists in the field. Although the studies of antibodies as proteins were rapidly progressing because of the many refinements of the amino acid sequencing techniques, the studies of the antibody genes led to a much more modest progress. These latter studies relied mostly on the electrophoresis of the whole DNA or its fractions that were thought (often erroneously) to contain certain genes. Thus, the following crucial questions remained unanswered: is the C-gene really separate from the V-genes? How many of the putative V-genes exist? How exactly is the information for the C- and the V-regions combined to produce an integral antibody molecule? Why are the C- and the V- genes not expressed in antibody-nonproducing cells? The existing methodology offered no approach to these questions.

4.3 What new techniques helped to test the postulates of Dryer and Bennett and the questions that arose?

Essentially, two new methodologies: **restriction analysis** and **Southern blotting**. In the early 1970s, the enzymes called restriction endonucleases were discovered. These are of bacterial origin and able specifically to recognize short DNA sequences in eukaryotic cells, cutting the DNA strand at the site of recognition. If one treats the whole DNA extracted from a cell with a restriction endonuclease (or restrictase), one obtains a "digest" or a mixture of DNA fragments with sizes that can be defined by electrophoresis and are characteristic for this particular restrictase. A radioactively labeled complementary nucleic acid probe can selectively bind to a fragment that has a certain size. The probes can be obtained in such a way that their nucleotide sequence corresponds to the amino acid sequence in a protein of interest. The procedure of obtaining the above digests and comparing them, as well as comparing the results of the specific complementary probe hybridization, under different conditions is, essentially, the procedure of restriction analysis.

Southern blotting is a technique that allows an investigator to manipulate with restriction fragments, or other DNA fragments of certain size by transferring them from electrophoretic gel to nitrocellulose (or similar) paper, and exposing the paper sheets to complementary labeled probes. (It is similar to Western blotting described in Chapter 3, except that the material analyzed is not protein, but DNA.) It was a combination of restriction analysis and Southern blotting that paved the way to a breakthrough in the understanding of antibody genes.

4.4 What was the essence of the breakthrough brought through the use of these new techniques?

By the end of the 1970s, substantial information about the amino acid sequence of antibody V-regions was acquired. As mentioned in Chapter 3, the sequence of V-regions of antibody heavy and light chains produced by various myelomas was completely deciphered. Several laboratories, notably the laboratories of S. Tonegawa and P. Leder, set out to take advantage of that by designing V-region-specific, complementary nucleic acid probes, and using them for the antibody gene restriction analysis. These investigators used probes that were specific to the most C-terminal portion of the antibody light chain V-region. By comparing the hybridization patterns after treating different clones of lymphoid origin with restriction enzymes, they sought to verify the prediction that followed the Dryer–Bennett postulate about C- and V-genes, i.e. that, unlike C-genes, the V-genes are different in different B cell clones. Their work, however, resulted in the finding that seemed most unusual at the time.

Tonegawa called the above-mentioned probe "J," from "junctional" between the V- and the C-regions of the light chain. He noticed that the J probe hybridized to the restriction fragments whose size was only slightly different in different lymphoid tumors but *dramatically differing between B lymphocytes and any other cell type* (Fig. 4-1). The size of the restriction fragment that hybridized with the J probe was substantially smaller in B cells than in any other cells. The only way to interpret this result was to assume that the genetic entity coding for the N-terminal portion of the antibody V-region (a V-gene segment) and the genetic entity coding for the adjacent C-terminal part of the V-region (a J-gene segment) *in all cells except B lymphocytes are separated by a long stretch of DNA*; in B lymphocytes, however, this stretch is *removed* so the

CHAPTER 4 B Lymphocytes and Immuno Genes

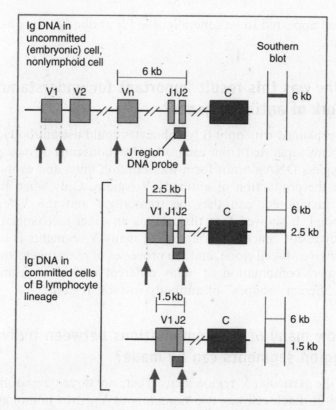

Fig. 4-1. Antibody-variable region gene rearrangements. Shown is the Southern blot analysis of DNA from a nonlymphoid cell and two different B-lymphoid cells (e.g., B-lymphoid tumors). The DNA was treated with a restriction enzyme and the resulting fragments separated by electrophoresis prior to blotting. The sites (target sequences) where the enzyme cuts are shown by arrows. The size of the fragment containing a portion of the immunoglobulin κ light chain gene called the joining (J-) segment is determined by using a radioactive probe that specifically binds to J-segment DNA. The size of J-segment-containing DNA fragments in the B cells is smaller than in the non-B cells; this indicates that the specific sites where the restriction enzyme cuts the DNA are in a different position in the genome of the B cells compared with the non-B cell. Ig, immunoglobulin. From Abbas, K., A.H. Lichtman, and J.S. Pober, *Cellular and Molecular Immunology*, fourth edition, W.B. Saunders, 2000, p. 135 (their Fig. 7-7).

two gene segments physically *join* each other, *are brought to proximity*, i.e., they are *juxtaposed*.

Based on their results, Tonegawa and others correctly predicted that there are many different antibody **V-region gene segments** in the germline DNA. These segments are organized into loci, and for the V-region of the antibody light chain, these loci include V-region segments of two kinds – V and J segments. In all cells of the body the V segments are separated from each other, as well as from the J segments, by long stretches of irrelevant DNA. In **B lymphocytes**, however, **V-region gene segments rearrange**, i.e., a V segment is brought to proximity to a J segment. Their other correct assumption was that in a given B lymphocyte, only one V segment out of many, and only one J segment out of several, are chosen for the rearrangement. Both of these assumptions were later confirmed by many other laboratories and appeared to be generally true for antibody heavy as well as light chain V-regions.

4.5 Why was this result important for understanding the work of antibody genes?

First of all, it explained why only B lymphocytes could use antibody genes. Short V-region segments separated from each other by thousands of base pairs of irrelevant, meaningless DNA do not form transcription units and cannot, therefore, be utilized for the production of antibody V-regions. Only when the individual segments are juxtaposed, can they be transcribed and the V-region can be expressed. Second, it showed that there exists an exact mechanism that creates the antibody diversity. Since only one out of many V segments is used for rearrangement in a given B-cell clone, and the processes of rearrangement are stochastic, then the mere combination of many different individual segments already creates many different "shapes" of antibody-variable regions.

4.6 How many of the combinations between individual V-region segments can be made?

It depends on the particular V-region locus. There are three types of them: V_H, V_κ, and V_λ (Fig. 4-2). Each cell has two homologous V_H, two homologous V_λ, and two homologous V_λ loci, one locus per chromosome (all three types of loci are mapped to different chromosomes). The light chain V-region loci (V_κ and V_λ) consist of two kinds of gene segments: V and J segments. The V_H loci consist of three kinds of segments. They have, in addition to the V and the J segments, an additional kind of segments called **D segments** (for "**d**iversity"). The number of V segments in the human V_H locus (V_H segments) is known to be 80 to 100 (depending on the individual). The number of human D segments is approximately 30, whilst J_H segments number six. The number of V_κ and V_λ segments, as well as the number of individual segments in the V-region loci in the mouse is not exactly known, but good estimates have been made. It is generally thought that in the human as well as in the mouse, the V_κ locus has approximately 500 individual V_κ segments and four J_κ segments. The V_λ locus in the mouse is small – it has only two V_λ and two J_λ segments. In the human, the V_λ locus is bigger and is thought to contain approximately 100 V_λ and several J_λ segments.

As one can see from these data, there exists a substantial **germline diversity** of V-region gene segments. When rearrangements begin, there is a sizeable "starting material" available, so that a variety of different V, D, and J segments can be picked in different B-lymphocyte clones. The total number of all possible combinations between the V, the D, and the J segments (termed "**combinatorial diversity**") cannot, according to simple rules of combinatorics, be greater than the product of multiplication between the three numbers. For example, the maximal number of combinations between all the different segments in the human V_H locus cannot be greater than 100 (i.e., the number of individual V_H segments) × 30 (the number of D segments) × 6 (the number of J_H segments) = 18,000 (Fig. 4-2).

CHAPTER 4 B Lymphocytes and Immuno Genes

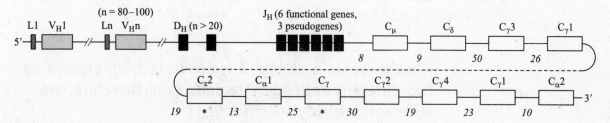

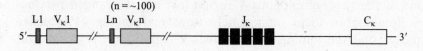

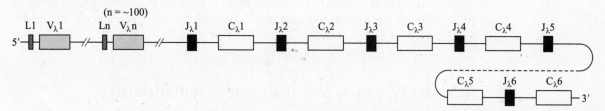

Fig. 4-2. Germline organization of antibody variable region loci. The human heavy chain, κ chain, and λ chain loci are shown, as indicated. Exons and introns are not drawn to scale. Numbers in italics refer to approximate lengths of DNA segments in kilobases. Asterisks indicate nonfunctional pseudogenes. Gene segments are indicated as follows: L, leader (or signal) sequence; V, variable; D, diversity; J, joining; C, constant.

4.7 Are there any other known mechanisms that explain how all of the antibody specificities are created?

Yes. Clearly, the above-mentioned combinations alone cannot account for all of the 10^9 antibody specificities. It is now known that during rearrangements, individual V, D, and J segments very rarely join each other precisely "side to side." They very often overlap each other to a greater or smaller degree. Alternatively, they can join in such a way that a gap between them remains; this gap is filled with random nucleotides with the help of a special enzyme called terminal deoxyribonucleotidyl transferase (TdT). Such **junctional diversity** creates a colossal number of different amino acid sequences in the V-regions, because in two different B-cell clones, even identical V segments used for rearrangement can join in a wide variety of ways. Yet another factor that further diversifies the antibody repertoire is a stochastic, **independent pairing of heavy and light immunoglobulin chains**. In one B-cell clone, only one heavy and one light chain V region can form (see also below). In a large population of B lymphocytes, however, the tremendous number of different V_H regions and the tremendous number of different V_L regions, created due to the combinatorial and the junctional diversity, are pairing at random, further increasing the overall diversity of the assembled V-regions.

4.8 What prevents a diploid B lymphocyte from expressing two different antibody V-regions (and therefore two different antibody specificities)?

Both homologous V-region loci rearrange. What prevents a B lymphocyte from expressing two different V-regions is the phenomenon of **allelic exclusion**. This means that if one of the two homologous V-region loci has rearranged productively, the other homologous locus immediately stops rearranging. (**Productive rearrangement** means that the joining of individual V-region gene segments did not result in the appearance of stop codons of frame shift; see molecular biology textbooks for details.) The exact molecular mechanism of allelic exclusion is unknown. One hypothesis claims that the protein product of the first productive rearrangement signals back to the nucleus and somehow arrests the process of rearrangement on the alternative locus. Allelic exclusion operates both on the V_H and on the V_L loci.

4.9 Do V_H and V_L loci rearrange simultaneously?

No. The V_H loci rearrange first. If one of them rearranges productively, the other stops rearranging, and one V_H-region is transcribed. The message from the rearranged V_H, D, and J_H segments (the "$V_H D J_H$ unit") is joined with the message from the Cμ gene during the mRNA splicing (see later sections of this chapter). This allows the μ chain to be translated. The nascent μ chain signals to the V_L loci, and they begin to rearrange. If both V_H loci rearranged nonproductively and no μ chain is formed in the cell, the V_L loci will not even start rearranging. This "unlucky" cell will have no antibody on its surface and will die shortly.

4.10 Do V_κ and V_λ loci rearrange simultaneously?

The current working hypothesis states that they do *begin* to rearrange simultaneously. Once one of the two homologous V_κ or V_λ loci rearranges productively, the homologous locus stops to rearrange (allelic exclusion). Moreover, if either a V_κ or a V_λ locus rearranges productively, both homologous loci coding for the V-region of the other light chain isotype are immediately shut down. This phenomenon, peculiar to the antibody light chain V-region loci, is known as **light chain isotype exclusion**. Because of this phenomenon, B lymphocytes almost always express either kappa or lambda chains, and very rarely both. (The existence of double-positive $\kappa^+\lambda^+$ lymphocytes is well documented, however; this tells us that the light chain isotype exclusion does not happen with such inevitability as the allelic exclusion on the V_H and the V_L loci.)

CHAPTER 4 B Lymphocytes and Immuno Genes

4.11 If all four V_L loci fail to rearrange productively, will the B lymphocyte still express the μ chain and acquire some new properties because of that?

It might express the μ chain, but often it does not. The isolated antibody heavy chains cannot be expressed on the cell membrane; to that end, they need to be assembled with light chains. In some precursors of the maturing B lymphocytes (see later), molecules called **surrogate light chains** exist and allow the μ chains to be expressed on the membrane when joined to them. Yet, even if a B-cell precursor expresses the heavy chain together with the surrogate light chain, it does not "graduate" into the status of a B lymphocyte and dies shortly. The surrogate light chains have no V-regions, and thus the assembly of a heavy chain and a surrogate light chain is not a functional antibody capable of interacting with antigens.

4.12 How many rearrangements are productive, and how many nonproductive?

The analysis of rearranged antibody V-genes from genomic DNA, which contains both productively and nonproductively rearranged sequences, showed that the ratio between the former and the latter is, roughly, 40% to 60%. Thus, the probability that at least one of the V_H loci is rearranged productively is 0.4, and the probability that at least one of the two V_H loci and at least one of the two V_L loci (of either the two V_κ or the two V_λ) is rearranged productively is $0.4 \times 0.4 = 0.16$. In other words, the vast majority of cells that begins to rearrange V-region genes – at least 84% – do not finish the rearrangements in such a way that an antibody molecule can be expressed. This low rate of success is compensated by the amazing frequency of rearrangement events in maturing B lymphocyte precursors, and the gigantic number of these precursors in bone marrow. It is thought that in the human bone marrow, about 100,000 antibody V-region gene rearrangements occur every second.

4.13 Of the three distinct types of segment that V_H loci contain (Section 4.6), which types rearrange first, or do all three rearrange simultaneously?

The **first event** in the sequence of events comprising rearrangement in the V_H locus (and, in fact, the very first event in the entire process of antibody V-region gene rearrangement) is the joining of one of the **D segments** with one of the J_H segments. This is followed by joining of one of the V_H segments to the formed "DJ_H unit," a continuum of segments that still cannot be transcribed. Such a "two-step" joining results in the formation of the $V_H DJ_H$ unit, which is the minimal transcriptional unit in the V_H locus. The movement of the chosen V_H segment towards the newly generated "DJ_H unit" proceeds even if the D to J_H joining resulted in the appearance of stop codons and/or in a shift of the reading frame.

4.14 What physical processes do actually happen during the rearrangement of discrete DNA segments?

It is thought that when a maturing B cell receives certain signals from its microenvironment, it activates otherwise dormant enzymes that catalyze the folding of the DNA of its antibody V-regions into loops. The process of such folding, as well as the excision of the intervening sequences, is greatly facilitated by the special **recognition sequences** located in the intervening DNA. These sequences serve as "docks" to which the enzymes attach. In the V_κ locus, where the mechanistic side of the antibody gene rearrangement is best understood, the recognition sequences are located 3' of each individual V_κ segment and 5' of each individual J_κ segment (Fig. 4-3). They include a highly conserved heptamer (a stretch of seven nucleotides), CACAGTG; two "spacers" of varying, nonconserved sequence, one of which is 12 and the other 23 nucleotides long; and a highly conserved nonamer (a stretch of nine nucleotides), ACAAAAACC. The heptamer flanks every individual V_κ segment on its 5' side (i.e., follows the V_κ segment immediately, without gap), and is followed by the 12-nucleotide spacer and then by the nonamer. The same CACAGTG heptamer flanks every individual J_κ segment on its 3' side (i.e., precedes it immediately, without gap); the heptamer itself is preceded by the 23-nucleotide spacer, which is, in turn, preceded by the ACAAAAACC nonamer (Fig. 4-3). When a rearrangement event occurs, the entire DNA between the recombining V_κ and J_κ segments folds in a loop, and the heptamer that follows the V_κ segment becomes juxtaposed to the heptamer that precedes the J_κ segment. The CACAGTG sequence hybridizes with the complementary GTGTCAC sequence of the opposite DNA strand within the boundaries of the other of the two recombining heptamers. This "locks" the loop of DNA on the side adjacent to the recombining V_κ and J_κ segments. The other side of the loop is being "locked" by the hybridizing nonamers (ACAAAAACC and its complement TGTTTTTGG). The 12- and 23-nucleotide spacers remain inside the loop. It is thought that the formed loop is actually the structure that is being recognized by special enzymes called **recombinases** (their structure and genetic determination remain unknown). A recombinase is thought to excise the entire loop between the recombining V_κ and J_κ segments and "stitch" (ligate) them together. It is thought that the described "looping" is only one of the two possible mechanisms of the V_κ–J_κ recombination; the other mechanism includes an inversion of DNA sequences (Fig. 4-3). Similar recognition sequences exist in the intervening DNA between the V_H, D, and J_H segments of the V_H locus, and similar recombination machinery is thought to operate there.

4.15 Would it be correct to say that only lymphocytes produce specific antigen receptors because only lymphocytes have the enzymes that catalyze the V-region gene rearrangement?

Not quite. Several different enzymes have been implicated into the V-region gene rearrangement, and some of them are not specific for lymphocytes. Several enzymes

CHAPTER 4 B Lymphocytes and Immuno Genes

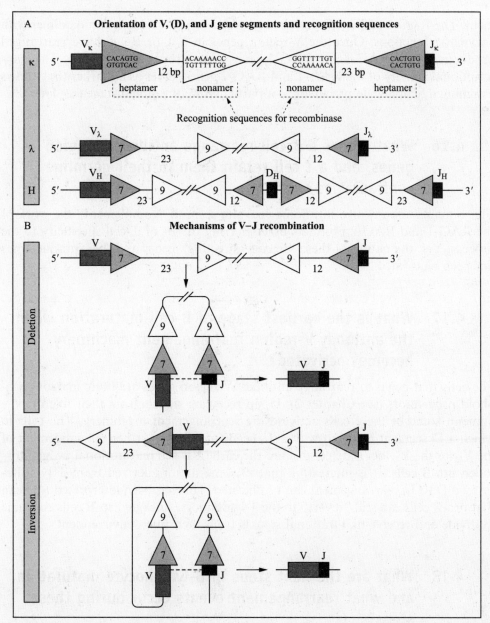

Fig. 4-3. Mechanism of antibody-variable region gene rearrangement. (A) DNA recombination recognition sequences for the V(D)J recombinase that mediates antibody gene recombination. Conserved heptamer (7 bp) and nonamer (9 bp) sequences, separated by 12- or 23-bp spacers, are located adjacent to V and J segments (for κ and λ loci) or to V, D, and J segments (for the heavy chain locus). The enzyme recognizes these regions and brings the segments together.

that may participate in the rearrangement belong to the ubiquitous family of enzymes that control DNA repair. Other enzymes are strictly lymphocyte-specific. For example, two enzymes that are coded by two different genes, called **RAG-1 and RAG-2** (from "rearrangement activation gene") are synthesized in immature T lymphocytes and in B lymphocyte progenitors that undergo V-region gene rearrangement. The RAG-1 and RAG-2 genes code for two proteins that form a dimer with enzymatic functions. Once the V-region genes in a T or B cell have rearranged productively, the RAG-1 and RAG-2 genes are irreversibly shut down. The transcriptional activity of the RAG-1 and RAG-2 genes serves as a marker that allows assigning a lymphocyte precursor to a particular maturational stage (see later).

4.16 What makes a B cell rearrange antibody V-region genes, and a T cell retain them in their germline configuration?

This remains unknown so far. Some enzymes seem to "complement" the work of the RAG-1 and RAG-2 products by conferring the B- or T-cell specificity to the process. Yet, the nature of these "lineage-directing" genes and their products has not been elucidated.

4.17 What is the earliest stage of B-cell maturation when the antibody V-region rearrangement machinery becomes activated?

The cells that begin to rearrange antibody V-region genes are bone marrow lymphoid progenitors (see Chapter 2). Upon receiving signals from their microenvironment, some of these cells activate the rearrangement machinery. The cells in which a D segment is joined with a J_H segment and no further rearrangement of the V_H or the V_L loci is detectable are the earliest B-cell maturational progenitors called **pro-B cells**. It is interesting that in some tumors derived from T lymphocytes, a D to J_H rearrangement can be detected. This has been interpreted to mean that pro-B cells can still "revert" to the T lymphocyte lineage. Pro-B cells continue to divide and receive maturational signals from their microenvironment.

4.18 What are the next steps in B-lymphocyte maturation, and what rearrangement events occur during these steps?

The next step in B-cell maturation includes joining of a V_H segment to the rearranged "DJ_H unit." If this event occurs and one of the two homologous V_H loci are rearranged productively, the lymphocyte is no longer called a pro-B cell, but "graduates" into the status of a **pre-B cell**. In pre-B cells, the rearranged $V_H DJ_H$ unit can be transcribed; after its message is joined with the message from the C_μ gene, the entire immunoglobulin heavy chain of the IgM isotype can be translated. The appearance of the μ chain in the cytoplasm marks a stage in the "natural history" of a lymphocyte when it becomes 100% committed to the B-cell lineage; a switch to the T-cell lineage is no longer possible. In some pre-B cells, the μ chain remains in the cytoplasm, but in other pre-B cells it can be expressed on the surface. To that end, the nascent μ chain needs to be joined with a special molecule

CHAPTER 4   B Lymphocytes and Immuno Genes   67

called a **surrogate light chain**. This molecule is coded by a gene that is outside of the antibody variable or constant region loci, and although it resembles antibody light chains, it has no variable region. Thus, even if the bona fide antibody heavy chain with its V-region is expressed, the cell at this point of its maturation is still unable to recognize antigens.

Pre-B cells continue to proliferate, and receive signals for further maturation. In those pre-B cells that respond to these signals and proceed to the next maturational step, the light chain V-regions rearrange. When the V_L (either V_κ or V_λ) rearrangement is complete and productive, the entire antibody molecule with its heavy and light chains can be synthesized and expressed. This marks the "graduation" of the lymphocyte into the status of an **immature B cell**. These cells express antibodies on their surface and can recognize antigens. However, the antibodies expressed by them belong exclusively to the IgM class. Also, an important feature of the immature B cell is that unlike the mature B lymphocyte, it **does not respond to the event of antigen recognition by activation**. Because of the immature state of its enzymes that control signal transduction (see later), the immature B cell either dies, or becomes functionally "silenced" (**anergic**) after the specific antigen is encountered.

4.19 How does an immature B lymphocyte become a mature B lymphocyte?

In an immature B cell, the V_H- and V_L-regions are already rearranged productively and no further rearrangements occur. Therefore, the "graduation" into the status of a mature B lymphocyte is marked not by an additional rearrangement event but by a shift in the way immunoglobulin gene messages are processed in the cell's mRNA. All antibody molecules in immature B cells belong to the IgM class. In mature B lymphocytes, however, antibody molecules of the IgM and IgD classes are synthesized and expressed simultaneously. This is possible because of the phenomenon well known to molecular biologists and termed alternative mRNA splicing.

4.20 How does the alternative mRNA splicing allow B lymphocytes to express IgM and IgD simultaneously?

C-region genes are located downstream of the V_H and the V_L loci. Unlike V-region genes, the C-region genes do not exist in the germline as segments interrupted by long stretches of irrelevant DNA. The C-region genes are uninterrupted exons, one exon per one heavy chain isotype. In other words, in the germline, there are C_μ, C_δ, C_γ, etc., exons that follow each other in a tandem array (Fig. 4.4). In the primary ("immature") mRNA molecule, the messages transcribed from these exons are located downstream of the message transcribed from the rearranged $V_H D J_H$ unit. The C_μ message is the most proximal to the $V_H D J_H$ message; the C_δ message is second proximal on the 3'-side. In immature B lymphocytes, all primary RNA transcripts are processed so that the excision of introns during the

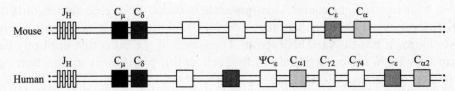

Fig. 4-4. Antibody-constant region genes. Genes encoding various heavy chain isotypes in humans and in mice are shown.

mRNA splicing leads to joining of the V_HDJ_H and the C_μ messages. The resulting transcripts are translated as antibody μ chains. However, with time, some immature B cells begin to process some of their primary mRNA transcripts differently. This alternative splicing includes excision of the introns between the V_HDJ_H and the C_μ messages as well as of the entire C_μ exon message, plus the message from the intron between the C_μ and C_δ. The message from the V_HDJ_H becomes thus joined to the message from C_δ, and the transcript can be translated as a delta heavy chain. As time elapses, the V_HDJ_H to C_δ joining takes place in a bigger portion of the mRNA transcripts. Eventually, the lymphocyte processes roughly 50% of its primary mRNA transcripts so that the μ chain can be translated, and the remaining 50% so that the delta chain can be translated. This point in the "natural history" of a B lymphocyte actually marks the beginning of its life as a **mature B lymphocyte**. From this point on, the B lymphocyte can not only recognize antigens, but may also respond to a recognition event by cellular activation.

4.21 Is IgD antibody crucially important for B-cell maturation? What are the unique functions of this antibody isotype?

In fact, no unique role of the IgD isotype in antigen recognition or signal transduction is known. The acquisition of this isotype expression due to the above-described shift in the mRNA splicing serves merely as a marker of the B-cell's maturity. It is believed that other molecules, in particular, signal transduction enzymes, are truly important factors that mediate the progression into the B-cell's maturity and ability to be activated by its specific antigen.

ANTIBODY GENE MUTATIONS AND THE GENERATION OF SECONDARY ANTIBODY REPERTOIRE

4.22 Why are antibodies generated during secondary responses "better" than those generated during primary responses?

In short, this is so because **memory cells** mediate secondary immune responses. The concept of immune memory and memory T and B lymphocytes has been introduced in previous chapters. As we discussed, memory cells are products of T- and

B-lymphocyte differentiation. Some B lymphocytes that have encountered their specific antigens are activated by these antigens and differentiate into memory cells. During B-lymphocyte activation induced by most protein (and maybe some polysaccharide) antigens, many of the activated B cells undergo the process of **somatic mutation of their antibody-variable region genes**. In a certain microenvironment created in the **germinal centers** (see Section 2.31), the B cells with mutated V-regions are tested for the **affinity** of their interaction with antigen (see Section 3.27). The mutants with the increased affinity are positively **selected**, i.e., are given survival stimuli, and allowed to differentiate into memory cells.

4.23 What are antibody V-region somatic mutations?

They are mostly point mutations (rarely inversions of deletions), strictly confined to the V-region genes. The two most important features of these mutations are their tight **regulation** and their astonishing **rate**. In the V-region genes of mature IgM^+IgD^+ B lymphocytes that have not encountered their specific antigen, the rate of mutation is not higher than the rate of spontaneous mutation in any genes of any somatic cells, i.e., the mutations are exceedingly rare. However, after the encounter with the antigen in secondary lymphoid organs and an interaction with other cells, notably activated T lymphocytes, the B cells are said to "turn on" their mutational mechanism. What exactly does this mechanism constitute, or what signals from the microenvironment induce it is not known so far. What is known is that after this mechanism is activated, the V-region genes begin to mutate at a rate of one out of 1,000 base pairs per one cell generation. This rate is thought to be 1,000 to 10,000 times higher than the rate of spontaneous mutation in all other known mammalian genes. The B lymphocytes that undergo such an intensive mutation proliferate very rapidly (approximately one division in 12 hours). Therefore, within several days after the antigen encounter and activation of this "hypermutational" mechanism, the antibody V-region genes accumulate a very large number of nucleotide substitutions.

4.24 What are the phenotypic manifestations of antibody V-region gene mutations?

Many of these mutations are **silent**, i.e., they do not lead to replacements of amino acids in the antibody V-region. Many other mutations are **replacement** mutations, i.e., they do manifest in amino acid changes. These changes may or may not affect the antibody affinity. In the case they do affect the affinity, it may increase or decrease in the mutated antibodies. It has been demonstrated that, sometimes, the decrease in affinity is so profound that the mutated antibody completely loses the ability to bind its antigen. It is important to remember that the process of somatic hypermutation in the antibody V-regions has no "brain;" it is not specifically aimed at increasing the antibody affinity. It is absolutely **random**, and stochastic as far as affinities are concerned. It is the selection of high-affinity mutants that actually makes the antibody affinity increase. Mutations are simply a "raw material" for this selection.

4.25 How does the selection of "good" mutants work?

It occurs exclusively in the microenvironment of a germinal center with its unique cellular architectonics. It is believed that activated B cells turn their hypermutational mechanism on before they enter the germinal center, i.e., in the parafollicular areas of the lymph node, PALS of the spleen, or analogous areas of other secondary lymphoid organs. When some of them enter the germinal center, the mutations continue to occur, and the rate of the cell proliferation increases to its maximum. Because of such a strong proliferation, one part or "pole" of the oval-shaped germinal center becomes filled with dividing cells very densely and is called the **dark zone**. The rest of the germinal center is called the **light zone**. The cells that fill the dark zone are called **centroblasts**; these are mostly activated and dividing B lymphocytes, with just a very few occasional T lymphocytes and dendritic cells found among them. The light zone is not nearly as dense, and the cells that populate it are of two types. The first of these types are **centrocytes**, which are B lymphocytes recruited from the pool of centroblasts. Unlike centroblasts, centrocytes do not divide (are out of cell cycle), and show signs of apoptosis. The other type of cells in the light zone are **follicular dendritic cells** (see Chapter 2). These are large, irregularly shaped cells that resemble dendritic cells by their morphology, although (as mentioned in Chapter 2) they differ from dendritic cells in that they are not bone marrow-derived.

The selection of high-affinity mutants actually takes place in the light zone, and follicular dendritic cells are major cellular mediators of this selection. They display antigens to the centrocytes using their surface Fc- and complement receptors. These receptors trap soluble antigen–antibody complexes by binding to the C-region of antibody molecules attached to antigens. It is thought that various antigens thus trapped by follicular dendritic cells are then "smeared" all over the large surfaces of these cells. Centrocytes randomly contact the follicular dendritic cells in the light zone and have a chance to "re-recognize" their specific antigens, i.e., establish a complementary bond between their V-region and the specific epitope. However, because antigens are "smeared," their local concentration is rather low. Therefore, only B lymphocytes that have increased the affinity of their antibodies due to somatic mutation are capable of establishing such a connection. These "lucky" lymphocytes receive a selective boost, which reverses the apoptotic processes that have begun to unwind in them. Eventually, they leave the germinal center and differentiate into memory or plasma cells. Other B lymphocytes, being unable to "re-recognize" their antigens, die by apoptosis. Thus, an average affinity of antibodies in the pool of B lymphocytes that leave the germinal center is much higher than in the B lymphocytes that enter the center.

4.26 Do all plasma and memory cells originate from germinal center cells that have undergone the above affinity selection?

No. Some of the plasma and memory cells originate from B lymphocytes that have not been selected in the germinal centers. Naturally, the affinity of surface-

CHAPTER 4 B Lymphocytes and Immuno Genes

expressed as well as secreted antibodies in such cells is lower than in the ones that passed the germinal center selection.

4.27 What other events, besides somatic hypermutation and selection of the "good" mutants, happen in the germinal centers?

In addition to these events, the phenomenon of **antibody class switch**, mentioned in Chapter 3, also happens there. After the encounter of its specific antigen and activation, the B lymphocyte soon loses the ability of a mature resting B cell to express IgD in addition to IgM, and for some time remains expressing IgM only. After a short while, many of the activated B lymphocytes that by this point have entered the germinal center, begin to express the so-called downstream antibody isotypes (mostly IgG, but also IgA and, occasionally, IgE). Initially, the mechanism of the expression of these isotypes involves the alternative mRNA splicing similar to the one described in Section 4.20. At this point, the message from the $V_H D J_H$ unit in one given B lymphocyte may be joining the messages from the C_μ gene in one part of the mRNA transcripts and the messages from C_γ, C_α, or C_ε genes in another part of transcripts. As expected, B lymphocytes that undergo such an alternative splicing may express IgM and one of the downstream isotypes (but not IgD). Later, however, an entirely different mechanism takes over. Due to this mechanism, called **switch recombination**, the B cells completely and irreversibly lose the ability to express both IgM and IgD and begin to express exclusively one of the downstream antibody isotypes.

4.28 How does switch recombination work?

It occurs not at the mRNA but at the DNA level and is, in a way, similar to antibody V-region gene rearrangement. Each of the C_H genes is preceded by a long (5–10 kb) intron, called a **switch (S-) region**. Activated B lymphocytes that enter the germinal centers receive external stimuli (especially signals from cytokines; see later) that initiate the transcription in the switch regions. This results in the appearance of the so-called "sterile" or "germline" transcripts of the C-region genes together with the messages from the S-regions themselves. Once the sterile transcripts appear, the DNA folds in such a way that the $V_H D J_H$ unit recombines with the particular C_μ gene that has formed the sterile transcript. All C-region genes that happen to be between this recombining C_μ and the $V_H D J_H$ are deleted (Fig. 4-5). Note that different signals, and, in particular, different *cytokines* (Chapter 10) activate sterile transcription of different C_μ genes, and thus *direct* the class switch.

4.29 What is the purpose of the class switch?

It has been shown that more mutations accumulate in $V_H D J_H$ units that are joined to the downstream C-region genes (especially IgG) than to the C_μ gene (or its

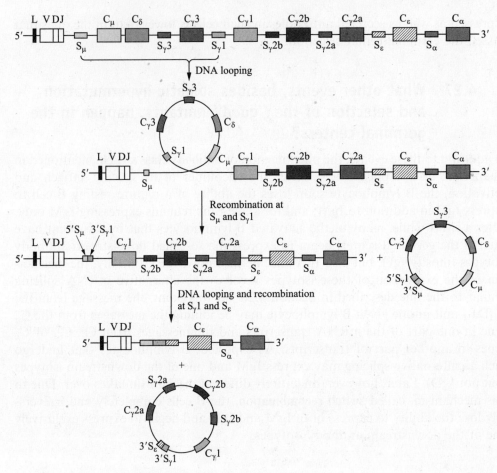

Fig. 4-5. Switch recombination. A proposed mechanism for class switching in rearranged antibody heavy chain genes is shown.

message). Because of that, more of mutational "raw material" is provided for the selection of high-affinity IgG mutants than IgM mutants. As a result, average affinity of the IgG antibodies is much higher than the average affinity of the IgM antibodies. To some extent, IgM antibodies compensate their low affinity by joining in pentamers and thus being of a greater avidity. Still, the affinity of most IgM antibodies elicited during primary immune responses rarely exceeds mediocre values (average $K_d \sim 10^{-8}$ M), while the K_d of many high-affinity IgG antibodies elicited during secondary immune responses is 10^{-11} M or less. Also, IgG and IgA antibodies perform some special effector functions (see Chapter 13).

4.30 Can antibodies switch to downstream isotypes, and yet not mutate?

It has been demonstrated that in some artificial tissue culture systems, B lymphocytes proliferate very extensively and switch to downstream isotypes, but do not turn on their hypermutational mechanism; thus, they acquire the so-called "partial germinal center phenotype." This indicates that signals necessary to turn on the hypermutational mechanism do not completely overlap with the signals necessary to induce the antibody class switch.

4.31 Are somatic mutations of antibody V-regions evenly distributed over these regions?

No. Many more mutations are usually found in CDR than in FR. One reason for that may be that CDR or, rather, parts of FR that are immediately flanking them, contain some mutational "hot spots," i.e., sequences that signal about the initiation of the mutation. Another reason for the preferential accumulation of mutations in CDR may be that perhaps many more B-cell clones with mutations in CDR are positively selected in germinal centers than clones with mutations in FR. Indeed, mutations in CDR directly affect the antibody interaction with its specific epitope. If one compares the nucleotide sequence of a specific antibody with the germline sequence of the corresponding V-region genes, one can see that often, many mutations concentrated in the CDR are replacement mutations, i.e., they lead to a substitution in the amino acid sequence. A high ratio of replacement to silent mutations (**R:S ratio**) in specific antibody CDR indicates that the selection in the germinal centers favored particular mutants over the initial unmutated B cells. Some antibodies specific to protein epitopes and elicited during secondary immune responses have an R:S ratio of more than 10:1. On the contrary, in some other antibodies the R:S ratio is extremely low, which indicates that the unmutated configuration was optimal for the epitope binding. This feature is especially characteristic to antibodies to polysaccharide epitopes.

Questions

REVIEW QUESTIONS

1. What is immunoglobulin allotype, and how different is it from immunoglobulin isotype?
2. Briefly summarize, what principally new information was obtained from Tonegawa's experiments.
3. A human V_H segment called V_H6 is located in the germline just 5′ of the first (or the most 5′) of the six J_H segments. You extracted DNA from an antibody-producing cell, treated it with a restriction enzyme, ran the digest on a gel, and prepared a Southern blot of this gel using a V_H6-specific probe. What conclusions will you make if you: (a) see the V_H6 band; or (b) do not see the V_H6 band in the antibody-producing cell but see it in a DNA prepared in the same manner from the same individual's granulocyte?

4. (a) Transgenic mice are mice that express a certain gene that has been injected into their zygote. Depending on a promoter that is injected together with the gene, it will be expressed in a certain cell type. Imagine that you have a mouse transgenic for RAG-1 gene, which is under such a promoter that it is expressed in all cells of this mouse. How different will this mouse be from its normal counterparts?

   (b) Knockout (KO) mice are mice in which a functional gene or a group of functional genes is "killed" by a special technique called homologous recombination. Imagine that you have a KO mouse that has her RAG-1 and RAG-2 genes "killed." How different will this mouse be from her normal counterparts?

5. If D segments did not exist, which mechanism of antibody repertoire diversification would it affect more – combinatorial diversity or junctional diversity? Provide some calculations that would back your answer.

6. Of the three CDR, which one is most likely to serve as a unique marker of a given lymphocyte clone? Why?

7. FACS analysis revealed that a lymphoid tumor cell does not express immunoglobulins on its surface. Yet, when a lysate was prepared and examined by Western blotting with an antibody specific to an immunoglobulin chain, a clear band was seen. What is the origin of this tumor? What kind of immunoglobulin chain was the antibody specific to?

8. Out of the ~80 human V_H segments, some are used for rearrangement more frequently than others. For example, a segment called V3-23 is used much more often than any other segment. Two hypotheses have been offered to explain this phenomenon: one suggests that this unusual frequency is due to some molecular mechanism that facilitates the use of this segment; the other states that lymphocytes that express a V3-23-coded antibody are somehow favored. How would you evaluate these hypotheses? Use the knowledge about stages of B-cell maturation presented in this chapter.

9. What is the most essential difference between light chains and surrogate light chains?

10. Read the following statement: "Unmutated antibodies are usually of low affinity. Because of the somatic mutation in antibody V-regions, antibody affinity substantially grows, which is called affinity maturation." What is wrong with this statement?

11. What would happen to the immune system of an animal with a genetic defect that would hamper the development of follicular dendritic cells?

12. In a nucleotide sequence of an antibody variable region, one finds 25 nucleotide substitutions in its CDRs compared to the germline sequence; of these one is a silent substitution and 24 are replacement mutations. In another antibody, ten substitutions are found in its CDRs, two of them are silent and eight are replacement mutations. What hypotheses can you offer regarding selection of the cells that made the two antibodies? Which of the antibodies is likely to be specific to a protein, and which to a polysaccharide?

MATCHING

Directions: Match each item in Column A with the one in Column B to which it is most closely associated. Each item in Column B can be used only once.

Column A

1. Allotype
2. J probe
3. D segment

Column B

A. Stop codons
B. Junctional diversity
C. Heavy chain

CHAPTER 4 B Lymphocytes and Immuno Genes

4. Independent pairing
5. Overlapping
6. Pro-B cell
7. CDR
8. Sterile transcript
9. Nonproductive rearrangement
10. Complement receptors

D. D to J rearrangement
E. Follicular dendritic cell
F. Mendelian fashion
G. Southern blotting
H. R:S ratio
I. Class switch
J. Heavy and light chains

Answers to Questions

REVIEW QUESTIONS

1. Allotype is a product of an *allele* that codes for an antibody chain. On the other hand, isotype is a product of a distinct antibody constant region *locus*.

2. These experiments showed that immunoglobulin genes are coded in the germline as "pieces," or segments, and that these segments physically move towards each other, juxtapose, during B-cell differentiation. This explains why only B cells can produce antibodies, and why different B-cell clones produce different antibodies.

3. If a V_H locus is rearranged and a V_H segment joins a DJ unit, everything that is in between is deleted. Therefore, if you see the V_H6 band (which indicates that the V_H segment has not been deleted), it means that at least one of the two V_H loci has not been rearranged. Since the cell you extracted the DNA from the cell that produces an antibody, the interpretation is that one of the two V_H loci has rearranged and the other has not. If you don't see the V_H6 band, it means that both V_H loci are rearranged. One of them must have been rearranged productively (because the cell produces an antibody), and the other nonproductively. The granulocyte control is important because it tells you that the reason you don't see the V_H6 band is not that your probe does not work.

4. (a) It might not be different from its normal counterparts. Although the RAG-1 gene will perhaps work in all cells of this mouse, by itself it is not sufficient to induce antibody gene rearrangement in the cells where it will be working.
   (b) It will differ from its normal counterparts in that antibody and TCR genes will not rearrange, and B and T lymphocytes will not be produced. This mouse is likely to suffer from severe infections and die early unless placed in sterile conditions.

5. Junctional diversity. The number of D segments is not all that great (not bigger than 30) to impede the combinatorial diversity. If D segments did not exist, the combinatorial diversity of heavy chains will be characterized by the number (~100 V_H segments × 6 J_H segments =) 600. This is clearly not enough to account for the diversity of antibodies in higher vertebrates. However, if all 30 D segments were present in the germline, the combinatorial diversity of heavy chains will be characterized by the number (~100 V_H segments × 30 J_H segments × 6 J_H segments =) 18,000, which is also clearly not enough to account for the diversity of antibodies in higher vertebrates.

Therefore, the real significance of D segments is to increase the junctional diversity and thus to increase the size of antibody repertoire.

6. CDR3. While the other two CDRs are coded entirely by V segments, the CDR3 is coded in part by V segments and in part by other segments used for antibody gene rearrangement. In particular, heavy chain V-region CDR3 is coded by a 3' part of the V_H segment, the entire D segment, and a 5' part of the J_H segment. Any two random B-cell clones can utilize the same V_H segments with a certain probability; the probability of these two random B-cell clones to utilize the same V_H, the same D, and the same J_H segments will be orders of magnitude lower; the probability of these two random B cells to not only to utilize the same segments but also to join them in the exactly same manner (with the same degree of overlap or gap) is so meager that it can be roughly counted as nonexistent. Therefore, it is very likely that no two B-cell clones will have identical CDR3s.

7. The tumor originates from a pre-B cell, and the kind of immunoglobulin chain is a heavy chain. It is produced in the cell but remains in the cytoplasm.

8. One should look at pre-B cells. If these have the same bias towards V3-23, hypothesis #1 is more likely to be correct.

9. Light chains have real V-regions while surrogate chains do not.

10. It presumes that any somatic mutation in the V-region will necessarily increase the affinity of the antibody. However, this is not the case. These mutations are random, they can increase the affinity or not have any effect on it or decrease it. It is not the mutation *per se* but the subsequent selection of useful mutants that accounts for affinity maturation.

11. There will be no selection of high-affinity mutants, and hence no antibody affinity maturation and no memory cell development as we know it.

12. The clone that made the first antibody was under a positive selection (mutants got a selective boost while clones that had no mutation were disfavored). The second clone was under a negative selection (unmutated variants were favored). The first antibody is likely to be specific to a protein, the second to a polysaccharide.

MATCHING

1, F; 2, G; 3, C; 4, J; 5, B; 6, D; 7, H; 8, I; 9, A; 10, E

CHAPTER 5

The Major Histocompatibility Complex

Introduction

The major histocompatibility complex (MHC) is a genetic region, and an array of its protein products whose primary role is to regulate immune responses, especially T-cell-mediated immune responses.

The name of this complex is historical, and reflects only a small "slice" of its function, namely, its involvement in the rejection or acceptance of tissue grafts (Greek, "histos" = tissue). Obviously, since no organ or tissue transplantations take place in nature, many investigators tried to understand the true role of the MHC since this complex system of genes and proteins was first discovered. However, the progress in this area was slow. Although the MHC genes and their protein products were implicated in the regulation of T-cell immune responses, no verifiable mechanism of their involvement in these responses was offered until the mid-1980s, i.e., more than 40 years after the MHC complex was discovered. Such a mechanism became testable only after the structure and the genetic determination of the TCR was deciphered (see Chapter 6), and some crucial details of the pathways of antigen processing and presentation became known. The discussion in this chapter will focus on the gradual transformation of the conceptual knowledge about the MHC. We will see how the perception of the MHC evolved from that of a polymorphic marker system useful for practical purposes (e.g., matching of the donor–recipient pairs during organ transplantation) to that of a system of polymorphic molecules that directly regulate antigen recognition by T lymphocytes, thus influencing on all of the subsequent events that occur during immune responses.

Discussion

5.1 How was the MHC discovered?

Initially, it was discovered through the methods of classical genetic analysis. Geneticists who were interested in the question, what gene or gene(s) control the rejection or acceptance of transplants, bred strains of mice in which all genetic loci were made homozygous because of the inbreeding (crossing littermates) for many generations. Within such inbred strains, all individuals are genetically identical, like identical (monozygotic) twins. Any two inbred strains, however, differ in their genetic make-up because they carry different allelic versions of their genes. By the end of the 1930s and early 1940s, many inbred murine strains were available, and it became possible to trace the fate of transplants (usually skin grafts) in different strains, or in individuals within a strain. After these experiments had been done, simple laws of transplantation were established. These laws stated that the skin grafted from one individual to another within any given inbred strain is always accepted, while the skin grafted from a mouse that belongs to one inbred strain to a mouse that belongs to another inbred strain is always rejected. In other words, a **syngeneic** transplantation (same allele in the donor and in the recipient) is always a success, while an **allogeneic** transplantation (one allele in the donor and a different allele in the recipient) is always a failure. Furthermore, an F1 hybrid from a cross between strain A and strain B always accepts skin grafts from either parent, but neither parent accepts skin grafts from the F1 hybrids. The conclusion made from these basic observations was as follows: (1) a gene or a linked group of genes controls the reaction against foreign tissues; and (2) these gene(s) exist as a large number of alleles in random-bred populations, and are inherited in a **codominant** fashion (that is why an F1 hybrid between the strains A and B has the AB phenotype; therefore, the expressed A allele is rejected by the B parent and the expressed B allele is rejected by the A parent). Further analysis showed that indeed, there exists a group of linked multiallellic (or polymorphic) genes that control the acceptance or rejection of transplants. The entire genetic region that is responsible for the acceptance or rejection of transplants in the mouse was, by the proposition of P. Gorer and G. Snell, called H-2 (where the "H" stands for "histocompatibility").

5.2 What individual genes (or loci) are included in the above-mentioned group of linked H-2 genes, and how were they identified?

A major tool used for the identification of these loci was the so-called **congenic mice**. Any two congenic mice are identical on all their genetic loci except those that belong to the H-2. The method of breeding the congenic mice (Fig. 5-1) is, in a way, similar to the Mendelian testcross, but repeated in many generations in a row. If an F1 hybrid between the strains A and B is back-crossed with one of the parents (i.e., the A parent), 50% of the F2 offspring resulting from this back-cross will be heterozygous on the H-2 locus (i.e., will have the A and the B allele), and

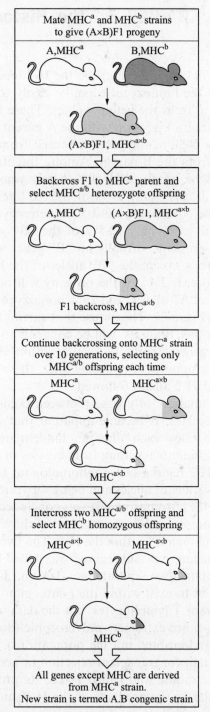

Fig. 5-1. Method of breeding congenic mice. Mice of two strains, strain A with genotype a and strain B with genotype b are crossed, and their F1 hybrid progeny, which are heterozygous at all loci including $MHC^{a\times b}$, are back-crossed to parental strain A. The F1 back-cross progeny, which are homozygous (AA) at 50% of their loci, are selected for expression of the other parental MHC (b); those that are $MHC^{a\times b}$ are back-crossed again to strain A. This continues for 10 back-cross generations, after which the mice are homozygous A at virtually all loci except MHC, where they are a/b. These mice are intercrossed and selected for homozygocity at the MHC for alleles of donor origin (MHC^b). Virtually all the rest of the genome is derived from strain A to which the MHC genotype MHC^b has been back-crossed. These mice are strain A congenic for MHC^b and are designated A.B. They can be used to determine if genetic traits that differ between strain A and strain B map to the MHC.

the other 50% will be homozygous on the H-2 locus (i.e., will have two A alleles). The hetero- and the homozygotic can be easily told apart because either parent will reject the skin from the heterozygotes. These heterozygotes are selected from the F2 and further back-crossed with the A parent. Note that in the F2, any given heterozygote has 50% of the genes inherited from the A parent and 50% of the genes inherited from the B parent. Among the offspring of this next back-cross, again, 50% will be homo- and 50% heterozygous on the H-2 locus, but the heterozygotes will carry only 25% of the "B" genes. By the same logic, the F4 generation will be 50% homo- and 50% heterozygous on the H-2 locus, but the heterozygotes will carry only 12.5% of the "B" genes. After more than 20 generations of back-crossing, virtually no "B" genes will remain in the mice resulting from the back-cross, except the "B" alleles of the H-2 genes. If such mice are now allowed to interbreed, 25% of the progeny will become homozygous on the H-2 loci, carrying two "A" alleles, and 25% homozygous on the H-2 loci and carrying two "B" alleles (Fig. 5-1). The mouse that carries two identical "B" alleles within its H-2, and no other B alleles, will be congenic to the mouse from the original A strain: all their genes, except the genes within H-2, will be identical. Once an investigator has a number of congenic mice, these can be crossed, and the inheritance of different H-2 alleles followed.

An extensive analysis of crosses between congenic mice revealed an internal complexity of the H-2. It became apparent that *H-2 is not a single locus, but a group of loci* that follow each other in a tandem array along the chromosome. The analysis of recombinants resulting from crosses of different congenic mice showed that within the H-2 region, two nonhomologous loci can be found, each of them showing a wide variety of alleles and being separated from the other by a relatively long segment of the chromosome. It was concluded that these loci are the *terminal* loci of the H-2 complex, while the segment of the chromosome that separates them is the *internal* "subregion" within the H-2. The two terminal loci were called **H-2K** and **H-D**, and the internal "subregion" was called **I**. Later, it was shown that the I "subregion" contains two separate loci, **I-A** and **I-E**. (An additional locus, called H-2S, was thought to exist within the I subregion and control the function of the so-called suppressor T lymphocytes, but the data were not confirmed.) All four of the mentioned loci are extremely polymorphic (dozens or even hundreds of their alleles exist in random-bred murine populations), and code for proteins that are extremely important for transplant rejection or acceptance. A fifth locus was later discovered in the mouse and called **H-2L**; its protein product turned out to be structurally similar to the products of the H-K and -D.

5.3 How were the human histocompatibility genes and proteins discovered?

Apparently, an analogue of the H-2 complex must exist in humans, but a different methodology should have been employed for the studies of its structure (for obvious reasons). A major tool in the early studies of the MHC in humans was sera collected from **multiparous women**, i.e., women who had many children of the same husband. When a woman becomes pregnant, the antigens that the fetus has

CHAPTER 5 The Histocompatibility Complex

inherited from the father stimulate her immune system, and the antibodies against these allogeneic antigens (alloreactive antibodies, or alloantibodies) are generated. Each successive pregnancy results in an increase of the alloantibody titer. The serum that contains alloantibodies is called **alloantiserum**. Alloantisera can be tested for their reactivity with cells taken from a large number of unrelated donors. This way, one can identify donors that share the antigenic alleles with the father of the woman's children. The alloantisera that are known to detect a certain allele can be further used for tracing of this allele in the donors' families. Again, through the analysis of recombinant phenotypes in these families it was possible to make conclusions about the internal complexity of the human MHC complex (which was called "HLA," from "human leukocyte antigens").

The analysis of the alloantisera reactivity patterns and the family studies showed that the **HLA complex** in humans is organized, principally, in the same way as the H-2 complex in mice. Similarly to the H-2, the HLA is a large region; it is mapped to the human chromosome 6. It contains three extremely polymorphic loci whose allelic products can be detected by antisera; these loci are called **HLA-A**, **HLA-B**, and **HLA-C**. Later studies showed that structurally, the products of the HLA-A, -B, and -C are very similar to the murine H-2K, H-2D, and H-2L. The HLA complex also includes loci that code for proteins whose allelic variations are detectable not by alloantisera, but rather by the mixed lymphocyte reaction (a proliferative reaction of cultured lymphocytes in the case of allelic mismatch; see later sections). These loci were collectively called the **D region**; a later mapping analysis showed that this region includes three separate loci called **HLA-DR** (from "D-related"), **HLA-DP**, and **HLA-DQ**. Subsequent structural studies showed that these three loci are very similar to the above-mentioned murine I-A and I-E loci. Because of the obvious similarity between the human and the murine systems, it was proposed to call the murine loci H-2K, H-2D, H-2L, and the human loci HLA-A, -B, and -C **MHC Class I genes**, and their products **MHC Class I antigens**. On the other hand, the genes I-A and I-E in the mouse and HLA-DR, -DP, and -DQ in the human were called **MHC Class II genes**, and their products **MHC Class II antigens**.

5.4 What is the extent of the MHC polymorphism, and is there a conventional way to denote MHC alleles?

Each of the MHC genes in the mouse, as well as in the human, exists in an extremely large number of allelic variations. The exact number of alleles is not known; but as the technology develops, new alleles are being discovered routinely. By the mid-1990s, international workshops recognized the existence of as many as 150 alleles of some MHC loci.

The conventional way to denote MHC alleles differs in the murine and human systems. Traditionally, small Latin letters on the right side of the capital Latin letter denoting the locus denotes the murine MHC alleles, in superscript. For example, there are H-2K^a (pronounced "H-2K of a"), H-K^b, H-K^d, H-2D^s, I-A^p, I-E^q alleles, etc. In the human, Arabic numbers following the name of the locus denote the alleles of the locus, for example, HLA-A2, HLA-A28, HLA-B7, HLA-DR1, HLA-DP14, etc. The constellation of alleles of all of the MHC loci on one chromosome is called a **haplotype**. Inbred mice, by definition, carry one allele, which is identical for all genes in all individuals within a strain (and differs from another allele that is carried by another strain). Therefore, it is acceptable to say that, for example, the BALB/cJ inbred strain has a haplotype H-2^d, which automatically means that every individual mouse of this strain has alleles H-$2K^d$, H-$2D^d$, H-$2L^d$, I-A^d, and I-E^d on both homologous chromosomes. Obviously, random-bred humans can have different alleles of different MHC loci, and can be either homo-, or heterozygous on any of these loci. For example, an individual can carry alleles HLA-A1, HLA-B2, HLA-C4, HLA-DR8, HLA-DP20, and HLA-DQ40 on one of the two homologous chromosomes, and alleles HLA-A7, HLA-B9, HLA-C1, HLA-DR8, HLA-DP3, and HLA-DQ30 on the other chromosome (note that this individual is homozygous on the HLA-DR locus and heterozygous on other MHC loci). Any two homologous alleles are **co-dominant**, which means that, for example, if a person inherits the HLA-A2 and the HLA-A28 alleles, both will be expressed. Thus, if a human individual is heterozygous on all six HLA loci, he or she can express at least 12 different allelic HLA antigens per cell. (In fact, more than 12 can be expressed because, as detailed later, the Class II antigens consist of two polypeptide chains, and more than six pairs of these chains can be expressed by the same cell.)

5.5 What is the structure of MHC antigens?

Class I and Class II MHC antigens have somewhat different structure. MHC Class I molecules in both human and mouse consist of two polypeptide chains that dramatically differ in size (Fig. 5-2). The larger (α) chain has a molecular weight of 44 kDa in humans and 47 kDa in the mouse, and is encoded by an MHC Class I gene. The smaller chain, called **β-2 microglobulin**, has a molecular weight of 12 kDa in both species, and is encoded by a nonpolymorphic gene that is mapped outside of the MHC complex. There are no known differences in the structure of the human MHC Class I antigen α chains encoded by the HLA-A locus compared to those encoded by the HLA-B or the HLA-C loci, or in the structure of the murine MHC Class I antigen α chains encoded by the H-2K locus compared to those encoded by the H-2D or H-2L loci. Regardless of which of these loci codes it, the α chain can be subdivided into the following regions, or domains (Fig. 5-2): (1) the peptide-binding domain; (2) the immunoglobulin-like domain; (3) the transmembrane domain; and (4) the cytoplasmic domain. The peptide-binding domain is the most N-terminal; it is the only region of the molecule where allelic differences in the amino acid sequence can be localized. As seen from its name, the peptide-binding domain of the molecule includes the site to which antigenic peptides bind. It makes much sense to have this site exactly where the allelic differences are, because different MHC alleles accommodate peptides better or worse, thus influencing on the magnitude of the T-cell response (see later sections). X-ray crystallography showed that the peptide-binding site in the MHC Class I molecules looks like a cleft that has a "floor" and two "walls" formed by spiral-shaped portions of the alpha chain, called alpha 1 and alpha 2. Since the "floor" of the peptide-accommodating cleft is closed, only relatively small

CHAPTER 5 The Histocompatibility Complex

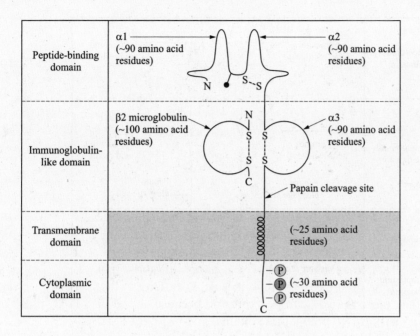

Fig. 5-2. Structure of MHC Class I antigens. See text for details.

peptides, consisting of 9 to 11 amino acid residues, can be "stuffed" there. The immunoglobulin-like domain is structurally conserved, and resembles a domain of an antibody C-region (see Chapter 3). It contains the binding site for the T-cell accessory molecule CD8 (Chapter 2; see also later sections of this chapter, and Chapter 6). The transmembrane and the cytoplasmic domains ensure that the alpha chain spans the membrane and is properly expressed by the cell. The β-2-microglobulin chain is also vitally important for the proper expression of the alpha chain. There are some mutant lymphoid cell lines (notably Daudi) that do not express MHC Class I molecules because of the defect in the β-2-microglobulin gene.

Class II MHC molecules (Fig. 5-3) in both human and mouse consist of **two polypeptide chains** that have a **similar**, albeit not identical size. One of them is called **alpha (α)** and the other **beta (β)**. The molecular weight of the α chain is 32–34 kDa, and of the β chain 29–32 kDa. A separate gene controls each of the chains. Thus, the murine I-A locus actually consists of the Iα and Iβ genes, the human HLA-DR locus of the HLA-DRα and HLA-DRβ, etc. Both the α and the β genes are polymorphic. The β genes of some of the MHC Class II loci can be tandemly duplicated, so, instead of one gene per homologous chromosome, a cell can have two or three. Because of that, one cell can simultaneously express more than two allelic products of each of the MHC Class II loci. For example, a cell can express allelic products of its HLA-DR molecule that can be identified as HLA-DRα1 – HLA-DRβ1; HLA-DRα2 – HLA-DRβ2; HLA-DRα1 – HLA-DRβ2; HLA-DRα2 – HLA-DRβ1; etc. Overall, one cell can simultaneously express as many as 20 different MHC Class II gene products because of this tandem duplication phenomenon.

CHAPTER 5 The Histocompatibility Complex

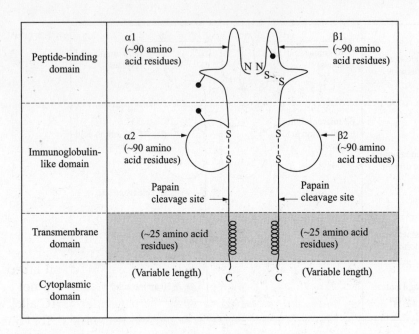

Fig. 5-3. Structure of MHC Class II antigens. See text for details.

The structure of the α and the β chains of the MHC Class II molecules resembles that of the alpha chain of the MHC Class I molecules in that the former can be also divided into the peptide-binding, the immunoglobulin-like, the transmembrane and the cytoplasmic domains. One important difference, however, is that the peptide-binding cleft in Class II molecules is formed by both alpha and beta chains. Although positioned close to each other in space, the spirals of the alpha and the beta chains that form the cleft are not physically bound to each other. Because of that, the "floor" of the peptide-accommodating cleft in Class II MHC molecules is "open," or "has a hole" in it (Fig. 5-3). That allows MHC Class II molecules to accommodate peptides that are larger than those that fit MHC Class I molecules. The immunoglobulin-like domain of the MHC Class II molecules contains the binding site for a T-cell accessory molecule, CD4 (see later chapters). This site cannot bind the above-mentioned CD8 molecule. We will discuss the functional implications of the specificity of CD4 to Class II molecules and CD8 to Class I molecules in later chapters.

5.6 How do peptides interact with the MHC molecules?

The interaction of peptides with MHC molecules is in a way similar to the interaction of antibody or TCR V-regions with their respective ligands, but it has some properties that distinguish it from the above. First, the *affinity* of peptide–MHC interaction is *very low*; the K_d of this interaction does not fall below 10^{-6} M. Second, *the "on rate" of the peptide–MHC interaction is much faster than the "off rate."* Studies performed with the help of the equilibrium dialysis technique have shown that it takes approximately 15–30 minutes for a peptide to saturate MHC Class II molecules; once bound, however, the peptide will remain in the

CHAPTER 5 The Histocompatibility Complex

peptide-binding cleft for many hours or even many days before it dissociates. This is so because the peptide-binding cleft has many "pockets" that change their conformation after the peptide binds, making it more difficult for the peptide to fall off. Finally, the allelic variations in the sequence of amino acids in the peptide-binding domain determine, which peptides will be bound by an MHC molecule well, which will be bound poorly, and which will not be bound at all.

5.7 Is differential peptide binding the explanation of the regulation of T-cell immunity by MHC?

Yes. T-cell responses can be stronger or weaker depending on the allele of the MHC molecule that presents the antigen to the T cell. In fact, this idea appeared much earlier than the data about peptides and MHC being the ligand for the TCR. As early as the early 1970s, B. Benacerraf and his co-workers showed that inbred guinea pig T cells respond to synthetic polypeptides stronger or weaker (or not at all) depending on the particular inbred strain carrying the particular MHC allele. Subsequent experiments showed that the genetic entity responsible for this phenomenon mapped to the central (I-) region of the guinea pig MHC. Benacerraf and others obtained similar results in mice and rats. Benacerraf proposed to call the genes that regulate T-cell-mediated immunity "**Ir-genes**" (from "immune response" or "immune regulation"). Later, it was shown that the exact mechanism of this regulation includes a direct interaction of peptides with the corresponding domain of the MHC products. The allelic specificity of this interaction determines the magnitude of T-cell response to a wide variety of protein antigens.

5.8 Does the allelic specificity of the binding mean that only one polypeptide can bind to any given MHC allele?

No. Although the total number of MHC alleles is very big, it is still not nearly as big as the total number of antibody or TCR specificities. Therefore, the available MHC alleles are not numerous enough to accommodate all the tremendous variety of existing peptides. What does happen is that groups of peptides can be preferentially accommodated by a particular MHC allele. These groups can be quite large, but all peptides within such a group possess certain similarities in their amino acid sequence and structure.

5.9 How many representatives of peptide "groups" can bind one MHC molecule at a time?

Only one. It does not mean, however, that an MHC molecule is either occupied by the "right" peptide or remains "bare." The MHC molecules need to be "stuffed" with peptides all the time, because without peptides filling their peptide-binding clefts, they are structurally unstable. After one peptide has dissociated from the peptide-binding cleft, another one occupies it almost immediately.

5.10 Are only foreign peptides able to associate with the MHC molecules?

Not at all. In fact, the majority of MHC molecules expressed by all cells of the body are at any given moment of time occupied by peptides generated from self proteins. Only a tiny minority of the MHC molecules shows foreign peptides to T cells. The T lymphocytes do not react with self peptides not because these are not presented to them by the MHC molecules, but because the cells with self peptide-reactive TCR have been eliminated from their pool by negative selection in the thymus (see later chapters).

5.11 What methods are used for detection of MHC alleles?

Traditionally, the detection of allelic differences between the MHC molecules expressed by different individuals relied on serological techniques. Soon after alloantisera were collected and patterns of their reactivity described, a technique called **microlymphocytotoxic test** was developed. The essence of this test, as proposed by the laboratory of P. Terasaki in the early 1970s, was to minimize the volume of reactants, thus saving the precious alloantiserum and allowing other laboratories or clinics to repeat the test many times in order to share the results obtained on different individuals. The Terasaki protocol, in brief, is as follows. Wells of a miniature polystyrene tray (volume 20–30 µl) are filled with mineral oil. A tiny drop of the donor's lymphocyte suspension (1 µl) is dispensed underneath the oil, and mixed with 1 µl of the alloantiserum. After a short incubation, a 5-µl drop of rabbit serum (a source of complement) is added. After another short incubation, the mixture inside the oil drop is supplemented by a dye that stains only those cells whose plasma membrane is damaged. If the majority of lymphocytes in the well appear to be stained during a microscopic examination, the alloantiserum is considered to be reactive with the donor's lymphocytes. Thus, if it is known that the used alloantiserum detects, for example, the product of the HLA-A2 allele, the donor of the lymphocytes used in the test is considered to bear the HLA-A2 allele. The Terasaki test played a very important role in matching the MHC alleles during the screening of donor–recipient pairs in organ transplantation. However, its obvious limitations are the scarcity of well-characterized alloantisera and their polyreactivity.

As we have already mentioned, the products of the I subregion of the MHC complex in the mouse and the D subregion in man were initially characterized through the use of the so-called **mixed lymphocyte reaction** (**MLR**). In this reaction (Fig. 5-4), lymphocytes of two donors are cultured together for several days, and the proliferative response of lymphocytes that respond to a foreign MHC allele is monitored by incorporation of a radioactive isotope (^{3}H-thymidine) into the newly synthesized DNA. Usually, the MLR is made "one-way," because the lymphocytes of one of the donors are irradiated so that they can survive but not divide. Such irradiated lymphocytes are called **stimulators**, and the nonirradiated lymphocytes **responders**. If the alleles in the I or D region of the stimulators' and the

CHAPTER 5 The Histocompatibility Complex

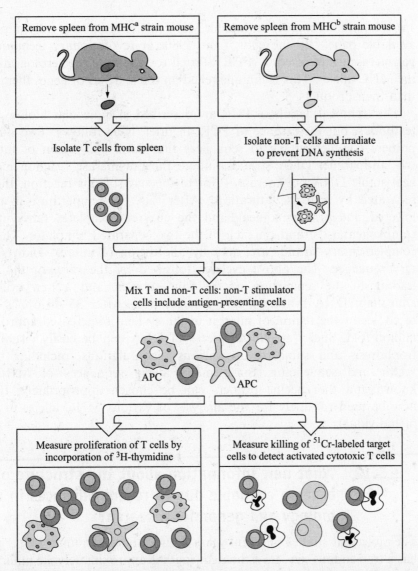

Fig. 5-4. Mixed lymphocyte reaction. The figure shows how a mismatch in MHC between two populations of mouse cells can trigger a proliferative reaction (lower left) or a cytotoxic reaction (lower right; see also Chapter 12). Similar reactions can be triggered in humans and used for detecting histocompatibility or histoincompatibility.

responders' MHC are identical, the incorporation of the isotope will be close to background values. If these alleles differ, the incorporation of the isotope will be substantial, indicating a strong proliferative reaction of the responders. By testing different combinations of stimulators and responders, it is possible to identify a set of alleles of the HLA-DR, -DP, and -DQ loci in large populations of donors. Some xenogeneic (rabbit) antisera and mouse monoclonal antibodies detecting these alleles are now also available.

In the 1980s, hybridoma-derived monoclonal antibodies began to replace alloantisera as tools for the detection of MHC alleles. However, the monoclonals specific to many important human MHC alleles remained sadly missing from the available panel. By the end of the 1980s, a revolutionary technique called the **polymerase chain reaction** (**PCR**) started to take over the serological detection of the MHC alleles. This technique relied on the analysis of genes themselves, rather than their products.

The technical details of PCR are described in molecular biology and genetics textbooks, but the essence of PCR, in brief, is as follows. Two oligonucleotide primers complementary to sequences that flank the region of interest on the two antiparallel DNA strands initiate the polymerase reaction mediated by a heat-stable DNA polymerase. Prior to the start of this reaction, the strands are separated by heat (denaturation). After this, the temperature in the system is lowered, the primers anneal, and the enzyme completes the synthesis of the complementary strand on each of the two separated templates. The newly built complementary strands are, actually, the region of interest and its complementary sequence. The above cycle is followed by the raise of the temperature, leading to the separation of the four strands, and a new cycle of primer annealing, DNA synthesis, and denaturation. After 35–40 cycles, the number of copies of the region of interest in the system is increased approximately one million-fold. Such "**amplified**" pieces of DNA can be easily visualized by electrophoresis and subjected to various manipulations, including cloning in a vector and sequencing. Since many allelic sequences of MHC genes are known and the specific primers can be chosen appropriately, the PCR can now be used routinely for the analysis of various MHC alleles in large groups of individuals.

5.12 What new information about the structure of MHC became available due to recent advances in molecular biology and genomics research?

The molecular studies of recent years showed that the human MHC gene complex has a tremendous size. Its estimated length is approximately 3.5 million base pairs, which is close to the length of the entire *Escherichia coli* genome. In humans, this region maps to the short arm of the Chromosome 6 and contains approximately 40 discrete loci (Fig. 5-5). In addition to the above-described MHC Class I and Class II genes whose products regulate T-cell responses, these include: (a) the so-called *Class III* MHC genes (whose products are components of *complement* – see later sections); (b) the genes that code for cytokines (e.g., lymphotoxin, tumor necrosis factor); (c) the genes that code for molecules involved in the MHC-peptide joining (TAP, DM, and the so-called proteasome genes; see later chapters); and (d) heat shock protein genes. Interestingly, the large gap between the HLA-C and the HLA-A loci contains several genes that are classified as "MHC-Class I-like." Many more of these genes are mapped outside the true MHC region, telomeric of the HLA-A locus. Many of these genes are pseudogenes (i.e., they contain stop codons and cannot be expressed), but some are functional genes that code for nonpolymorphic HLA-A-like molecules collectively called "**IB**." The function of the IB molecules is one of the "hot spots" of today's

CHAPTER 5 The Histocompatibility Complex

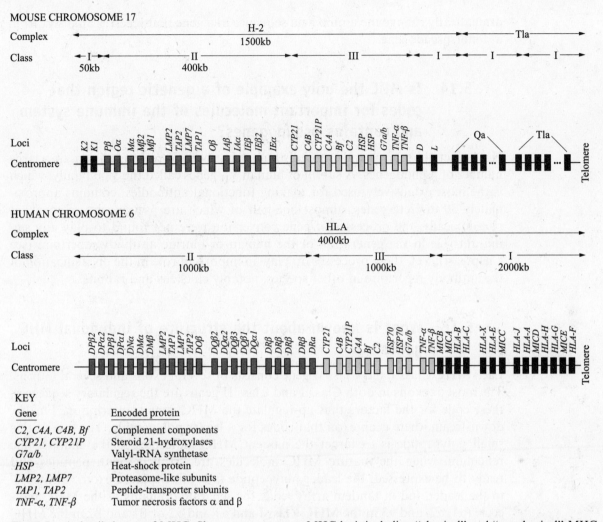

Fig. 5-5. A detailed map of MHC. Shown are numerous MHC loci, including "classical" and "nonclassical" MHC Class I and Class II loci as well as Class III loci (encoding complement components, see Chapter 13).

immunology and immunogenetics; some of these molecules were recently implicated into a number of physiological mechanisms as well as disease conditions.

5.13 Is there any sense in keeping pseudogenes within MHC?

It is thought that the main purpose – and the main advantage – of pseudogenes is that they serve as donors of nucleotide sequences in the process of **gene conversion**. This process includes modifications of functional genes through insertion of short stretches of DNA "borrowed" from pseudogenes. The gene conversion is very strongly implicated in the mechanism of generation of allelic MHC polymorphism. It is currently thought to play a more important role in the generation of the MHC polymorphism than point mutations, because it can much more dramatically change the amino acid sequence in a gene that receives an insert from a donor pseudogene.

5.14 Is MHC the only example of a genetic region that codes for important molecules of the immune system and contains pseudogenes?

No. In fact, the murine and especially human antibody V-region loci contain a large number of pseudogenes. A subset of human V_H genes called the $V_H 3$ family, which is the most extensively used for making functional antibodies, contains approximately 30 discrete genes, almost one-half of which are pseudogenes. Somewhat paradoxically, the processes of gene conversion were not found to play any significant role in the generation of the human or murine antibody repertoire (see Chapter 4). Yet, these processes do play an important role in the diversification of the antibody repertoire in other species, notably chickens and rabbits.

5.15 What is known about the structure of individual MHC genes?

Each MHC Class I and Class II gene consists of several exons and several introns. The most 5′ exons in both Class I and Class II genes are the **regulatory sequences**; these code for the factors that up-regulate the MHC gene transcription. Further downstream, there is an exon that codes for a **leader** polypeptide. The role of this small polypeptide is to target the nascent MHC molecules to the endoplasmic reticulum; when the mature MHC molecules are "stuffed" with peptides and ready to be expressed, the leader polypeptide falls off. Separate exons, located 3′ to the leader and in tandem array, code for separate domains of the MHC molecule ($\alpha 1$, $\alpha 2$, and $\alpha 3$ in the MHC Class I and $\alpha 1$ and $\alpha 2$ or $\beta 1$ and $\beta 2$ in the MHC Class II molecule). Finally, the exons that code for the MHC molecule's transmembrane and cytoplasmic domains are also separate and located the most 3′ in an individual gene.

5.16 How is the expression of MHC molecules regulated?

The answer is dramatically different for the MHC Class I and for the MHC Class II molecules. The former are expressed **constitutively** and can be found on the membranes of all nucleated cells. The latter are expressed only by the so-called "**professional antigen-presenting cells**" – macrophages, dendritic cells, and B lymphocytes, and show wide variations in the level of their expression. This makes much sense in view of the existence of two different subsets of T lymphocytes that are involved in the recognition of peptides in the context of two different classes of MHC molecules, which we shall describe in detail in subsequent chapters. The most essential features of the MHC regulation are that the expression of these molecules is controlled, first and foremost, by the level of their **transcription**, and is responsive to a number of **cytokines**. For example,

CHAPTER 5 The Histocompatibility Complex

interferon alpha enhances the transcription (and, hence, the expression) of the Class I molecules and decreases the expression of the Class II molecules; interferon gamma enhances the expression of both Class I and Class II molecules, etc. The expression of Class II molecules can be **induced**. This means that incubation with certain cytokines enables the cells that are not the "professional" antigen-presenting cells to express these molecules. The induction of the Class II molecules is especially characteristic for monocytes and macrophages. Some cells for example, neurons, are resistant to the cytokines and other factors that induce the Class II molecules in other cells, and remain negative for the expression of the Class II molecules. Functional implications of the expression of Class II MHC and its regulation will be discussed later.

Questions

REVIEW QUESTIONS

1. Why is it inadequate to say that the role of MHC is to control the fate of transplants?
2. What observation implied that the inheritance of MHC alleles is co-dominant?
3. In two to three sentences, explain the principle of breeding congenic mice.
4. Which allele on the H-2K locus must a mouse bear in order to accept the transplant from a mouse that expresses an I-E^b allele?
5. Which isotope will you employ to monitor the mixed lymphocyte reaction, and why?
6. You have a panel of alloantisera and a suspension of lymphocytes from a donor. What other reagents and equipment do you need to set up a microlymphocytotoxic test?
7. What is common between the following molecules: HLA-DR7, I-A^k, I-E^d, HLA-DP36?
8. Explain the concept of Ir genes.
9. In mechanistic terms, how exactly do different MHC alleles regulate the T-cell immunity?
10. Why do bigger peptides fit into the peptide-binding cleft of the MHC Class II molecules than of the MHC Class I molecules?
11. What is the role of β-2 microglobulin, and what cell line is the proof of such a role?
12. Summarize what is known and what is not known about the content of the MHC gene complex.
13. What is the role of the MHC leader peptide?
14. At what stage of the information flow in the cell are the MHC genes predominantly regulated?
15. Compare the role of pseudogenes and the role of point mutation in the generation of MHC polymorphism.

MATCHING

Directions: Match each item in Column A with the one in Column B to which it is most closely associated. Each item in Column B can be used only once.

Column A

1. Multiparous
2. Congenic
3. H-2K
4. HLA-DP6
5. Proliferation
6. Tandem duplication
7. Gene conversion
8. Haplotype
9. Peptide binding
10. Cytokine

Column B

A. Transcription
B. Allele
C. Locus
D. Pseudogene
E. Chromosome
F. Allele specificity
G. Back-cross
H. Alloantiserum
I. Mixed lymphocyte reaction
J. β-chain

Answers to Questions

REVIEW QUESTIONS

1. Because there are no transplantations in nature.
2. Rejection of the transplants taken from F1 hybrids of two inbred murine strains by their both parents.
3. Back-cross F1 hybrid with a parent (P), and test the F2 offspring for acceptance of a skin graft from F1 and P. Select the subset of F2 that rejects both F1 and P grafts and back-cross with F1. Repeat the procedure for 20 generations, and then subject the offspring to inbreeding; the resulting strain will be congenic to the P.
4. H-2K^b.
5. ^{3}H-thymidine – because it incorporates into the newly synthesized DNA.
6. A source of complement; a dye that selectively stains cells with damaged membranes; miniature polystyrene trays; mineral oil; a microscope.
7. They all refer to Class II genes or gene products.
8. They were described as genes whose alleles determine the magnitude of T-cell-dependent immune responses of guinea pigs to synthetic polypeptides; currently thought to be identical with the MHC genes, particularly, with those of the Class II.
9. Peptides can bind better, worse, or not at all to an MHC molecule depending on the allelic variation of the amino acid sequence in its peptide-binding domain. If they bind poorly, the T-cell-mediated immune responses will be meager; if they bind well, these responses will be strong.
10. Because the former is built with the help of two separate polypeptide chains with a gap between them, while the latter is built by one polypeptide chain with no gap.

CHAPTER 5 The Histocompatibility Complex

11. Final assembly and surface expression of the Class I molecule; Daudi (has a defect in the β-2 microglobulin gene – hence, no surface Class I MHC expression).
12. The size and the function of many of its genes are known; the exact number of loci and the function of some genes are not known.
13. Targets the newly synthesized MHC molecules to the endoplasmic reticulum.
14. Transcription.
15. Pseudogenes donate several nucleotides while point mutations substitute, insert, or delete only one nucleotide. Hence, the variation of the resulting amino acid sequence will be more dramatic in the case of gene conversion mediated by pseudogenes than in the case of random mutation.

MATCHING

1, H; 2, G; 3, C; 4, B; 5, I; 6, J; 7, D; 8, E; 9, F; 10, A

Antigen Processing and Presentation

Introduction

In previous chapters, we introduced the concept that antigen recognition by T cells necessarily includes **processing** of antigens in such a way that peptides are generated and joined with the MHC molecules. The resulting peptide–MHC complexes are being **presented** to T cells. This is so because the TCR can recognize the peptide–MHC complex and (with few exceptions) nothing else. In other words, the peptide presented by an antigen-presenting cell "in the context" of an MHC molecule is the ligand that the TCR specifically recognizes. How are the peptides generated? What enables cells to generate these peptides, to traffic them towards the MHC molecules, and to display the MHC–peptide complexes on their surface? These and similar questions will be discussed in this chapter. It should be mentioned, however, that the concept of antigen presentation and the term "antigen-presenting cell" became known long before the notion about the TCR recognizing peptides and MHC was formed. We will discuss the early studies that demonstrated the existence of the phenomenon of antigen presentation, and see how these studies helped to develop the current concepts of antigen presentation and TCR antigen recognition.

Another important experimental finding that we have mentioned previously is that two different classes of MHC molecules interact with two different T lymphocyte subsets. This is so because the $CD4^+$ T lymphocytes can associate exclusively with the MHC Class II, and the $CD8^+$ T lymphocytes with the Class I molecules. Apparently, this has a profound biological significance because the $CD4^+$ T lymphocytes are predominantly T helpers, while the $CD8^+$ lymphocytes are predominantly CTL (see Chapter 2). In this chapter, we will follow the route of peptides from the site of their generation to the MHC molecules with which they

CHAPTER 6 Antigen Processing

join, and will try to understand exactly how this trafficking of peptides determines the involvement of the two distinct functional populations of T cells in the immune response.

Discussion

6.1 What early observations and experiments led to the idea of antigen presentation?

Soon after the cellular components of the immune system were discovered in the 1960s, it became apparent that nonlymphoid "accessory" cells are needed for the induction of the T cell-mediated proliferative responses to antigens. T lymphocytes develop a proliferative response to their specific antigen when cultured with it *in vitro*. If macrophages, which are adherent (attach to the bottom of the flask or culture tray), are removed from these cultures, the T-cell response to antigens *in vitro* will be weakened or even abolished. It has been hypothesized that macrophages, and perhaps some other adherent nonlymphoid cells, somehow "present" antigens to T lymphocytes. Later, experiments showed that the above T-cell response could also be diminished or blocked if the cultures are supplemented with antibodies to what we now call MHC Class II molecules. This observation implied that the MHC Class II molecules are somehow involved in the antigen presentation. These experiments did not, however, point to the idea that antigens are processed in accessory cells.

6.2 How was the antigen processing initially demonstrated?

It has been observed that antigens injected into an animal accumulate in the animal's macrophages. Such observation can be made, for example, if an antigen is radioactively labeled. In the 1960s and 1970s, the following experiments, based on this observation, were performed. An antigen was injected into a mouse and allowed to accumulate in the macrophages. The macrophages with the accumulated antigen were then taken from the mouse and injected into intact mice. In parallel, the same antigen was injected into another group of intact mice. The T-cell responses to the antigen were monitored in both groups. It transpired that the antigen which accumulated in the macrophages was an extremely powerful immunogen (see Chapter 3). On a molar basis, up to 1,000 times less of the antigen in macrophages than of just antigen was required to induce the same T-cell response. The conclusion from these experiments was that in both groups of mice, the T cells respond to the antigen that is processed by macrophages. The reason why intact antigen needs to be injected in a much bigger dose than the "pre-processed" antigen is that only a small portion of the free antigen actually ends up in macrophages and undergoes processing.

6.3 What made immunologists consider the MHC molecule to be the entity that actually presents the processed antigen to the TCR?

A crucial step here was to show that the MHC molecule is itself recognized by the TCR, or, in other words, serves as a ligand for the TCR. In this regard, the experiment designed by P. Doherty and R. Zinkernagel in 1974 was of major importance. These investigators sought to determine the role of MHC alleles in the response of cytolytic T cells to a virus called the lymphocyte choriomeningitis virus (LCMV). It was very convenient to study this response because the cells that are infected by the LCMV, the murine fibroblasts, can be easily cultured and infected by the virus *in vivo*. Zinkernagel and Doherty injected a mouse with the LCMV, and in parallel prepared cultures of fibroblasts from another mouse and infected those with the LCMV. After 7–8 days, the T lymphocytes were isolated from the spleen of the injected mouse and placed in the culture of the infected fibroblasts. A strong cytotoxic reaction of the T lymphocytes occurred only in the case when the immunized mouse and the mouse that had donated the fibroblasts belonged to the same inbred strain. In other words, **the T-cell recognition was restricted to the same MHC allele as the one expressed by the donor of the T cells**. Later, a similar phenomenon was demonstrated for the T helper cells. Notably, while the cytolytic T-cell responses were restricted to the alleles of the MHC Class I molecules, the T_h responses were restricted to the MHC Class II alleles. Regardless of which of the two MHC classes was involved, however, the phenomenon of the MHC restriction prompted immunologists to think that T cells recognize the MHC molecule itself together with the specific antigen. T cells are maturing in such an environment that they do not "see" MHC molecules that bear nonself alleles; because of that, they "learn" to see only the self MHC molecule as target, or ligand, for their antigen receptors. The role of the antigen was not clear. Some immunologists tended to interpret the Doherty–Zinkernagel experiment as pointing to the "**dual recognition**." According to the "dual recognition" hypothesis, a T cell has two kinds of TCR, one for antigen and one for the self MHC molecule. This hypothesis was not consistent with newer experimental data, however, and was completely abandoned after the structure of the TCR was elucidated (see later chapters). Another way to interpret the MHC restriction was to hypothesize that the TCR actually recognizes "**altered self**," i.e., the MHC molecule that changes its structure or conformation after an interaction with processed antigen.

6.4 Did the "altered self" hypothesis turn out to be true?

It was proven partially correct. The part of it that is correct is the notion that there is just one kind of the TCR, not one kind for antigen and the other for MHC. The wrong part of it, though, is that the TCR does not in fact recognize merely the altered MHC molecule, but the complex of a foreign peptide and self MHC molecule.

CHAPTER 6 Antigen Processing

6.5 How did immunologists develop the notion of peptide?

During the late 1970s and early 1980s, several groups of immunologists began to work with standard preparations of proteins whose amino acid sequence had been by that time completely known. Hen egg lysozyme (HEL) was one of the most popular among these "model proteins" used for the T-cell epitope mapping studies. It was shown that the T-cell epitope, or the part of HEL that is physically recognized by the TCR of the T-cell clone responding to it, is just 10 amino acids long. Further, it has been demonstrated that the T-cell epitope is linear and that it is hidden within the globular structure of the HEL before the latter is processed. Finally, the 10 amino acid-long HEL-derived peptide that serves as the T-cell epitope physically associates with a Class II MHC molecule of the antigen-presenting cell, and can be eluted from it.

6.6 Where do peptides associate with the MHC molecules?

The peptides join the MHC molecules before the latter are expressed on cellular surfaces. This can be demonstrated by isolating MHC molecules from intracellular compartments and eluting peptides from their N-terminal domains. In particular, the above-mentioned HEL-derived peptide, and other peptides generated from "model" foreign antigens were eluted from murine I-A molecules that had been isolated from intracellular vesicular compartments. By using the so-called "pulse–chase" technique, i.e., tracing the movement of MHC molecules from the place where they are synthesized to the place where they are expressed on the cell surface with the help of a radioactive label, it can be also shown that *the MHC molecules are first joined with foreign peptides and then expressed on the surface.*

6.7 How does intracellular antigen processing and intracellular association of peptides with MHC molecules benefit immunity?

The key to the answer to this question is the fact that there exist two different pathways of antigen processing, each of which results in joining of the generated peptides with one of the two existing classes of MHC molecules. The peptides that are generated from **extracellular** proteins are joined with **MHC Class II molecules**, and the peptides that are generated from **intracellular** proteins with **Class I MHC molecules**. This makes profound biological sense. It is beneficial to involve different classes of T lymphocytes in responses to these two different classes of antigens. As was discussed in Chapter 2, T_h are more efficient against extracellular microorganisms because they stimulate antibody production, microbicidal functions of macrophages, and inflammation; T_c are more efficient against intracellular microorganisms because they directly lyse the infected cells. Yet, the TCR cannot distinguish between antigens of extracellular microorganisms (e.g., bacteria that live in extracellular spaces or body fluids) and antigens expressed by cells infected by intracellular microorganisms (e.g., intracellular bacteria or viruses). There is nothing intrinsically different between TCR expressed by T_h and those expressed by T_c.

Therefore, the differential involvement of the two functional T-cell subsets is dictated by the way intracellular and extracellular protein antigens are processed and "trafficked" inside the antigen-presenting cell.

6.8 How exactly do the two pathways of antigen processing and presentation differ?

The principal difference between the two pathways can be understood better if one takes a closer look at the intracellular site where peptides are generated. This site differs for peptides generated from extracellular proteins and those generated from intracellular proteins. The peptides derived from extracellular proteins are generated and "trafficked" inside the cell in the so-called **acidic vesicular compartment** (also called the **endosomal-lysosomal compartment**). The peptides derived from intracellular proteins are generated and "trafficked" in the **cytosol**.

6.9 What is the acidic vesicular compartment?

It is, essentially, a collection of vesicles that enclose the internalized antigens. When antigen-presenting cells (dendritic cells, macrophages, and B lymphocytes) internalize extracellular proteins (often in a form of molecular aggregates), they form an invagination of their plasma membrane, and enclose the protein aggregate into the forming deep "cleft." (It should be mentioned that the above concerns healthy cells with intact plasma membrane.) The edges of the cleft then join each other, and the aggregate becomes completely enclosed in the formed vesicle. In the cell, the vesicle fuses and joins with other vesicles called **endosomes**. The latter are not true organelles because they vary in size, shape, and internal content. The endosomes are thought to be formed by budding of the plasma membrane and other membranous organelles (mostly endoplasmic reticulum). On the most part, they remain attached to the plasma membrane. Most, or at least some of the endosomes have an acidic internal reaction and contain proteolytic enzymes that begin the process of proteolytic cleavage of the enclosed antigens. As the endosomes continue to fuse with each other, the partially hydrolyzed antigens are "relayed" from endosome to endosome. At the end of this "relay race," the antigens are dispatched to **lysosomes**. These are more true organelles with a defined morphology and enzymatic content (see biology or cytology textbooks for details). The proteolysis of the antigens and the generation of peptides is thought to be completed in these organelles.

6.10 Can intracellular proteins, or peptides generated from these proteins, penetrate into the endosomes or lysosomes?

Normally, the proteins that are synthesized by the protein biosynthetic machinery of the antigen-presenting cell do not penetrate into the endosomes or lysosomes. However, some bacteria (notably, *Mycobacterium tuberculosis*) are

CHAPTER 6 Antigen Processing

adapted to survive within the acidic vesicular compartment. The proteins synthesized by these bacteria are, formally speaking, intracellular, but they are processed in the endosomal-lysosomal compartment, just like the extracellular proteins.

6.11 How do the peptides generated in the above vesicular compartments reach the MHC Class II molecules?

Immediately after their biosynthesis in polyribosomes, the nascent MHC Class II molecules are transported through the channels of the granular endoplasmic reticulum to the Golgi apparatus. On their way, their separate subunits are assembled into heterodimers, and joined with two nonpolymorphic polypeptides. One of these polypeptides, the so-called **invariant (I-) chain**, associates with them, "locking" their peptide-binding domains (Fig. 6-1). In addition to serving this "locking" function, molecules of the I-chain target vesicles that contain nascent Class II molecules to endosomes. A special class of "trans-Golgi" vesicles called Class II **storage vesicles** brings nascent Class II molecules to endosomes. These vesicles bind to endosomes and fuse with them. The I-chain is thought to take an active part in the attachment of the storage vesicles to endosomes and in the fusion between them and the endosomes. Once Class II molecules have penetrated into the endosomes, the I-chain falls off and degrades. MHC Class II molecules spend approximately 2–3 hours in the endosomes, which allows the peptides that have been generated there to associate with their peptide-binding domains. After that, the MHC Class II–peptide complexes are transferred to the outer plasma membrane, span it, and become expressed on the surface of the antigen-presenting cell.

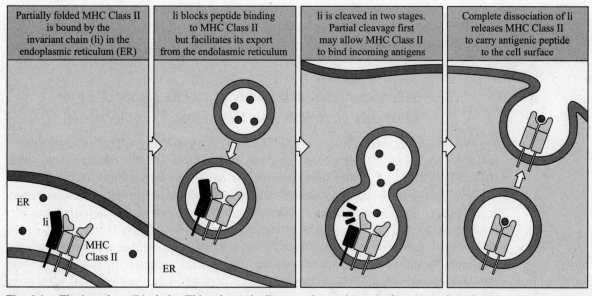

Fig. 6-1. The invariant (I-) chain. This schematic diagram shows the main functions of the invariant chain.

6.12 Can peptides derived from intracellular proteins be generated in the granular endoplasmic reticulum and in the Golgi apparatus? If so, why do they not join the MHC Class II molecules there?

Most peptides that derive from intracellular, endogenous proteins are generated in the cytosol (see Sections 6.16 and 6.17). However, these peptides are being constantly "pumped" into the channels and cisterns of the granular endoplasmic reticulum and Golgi apparatus. This is needed for the sake of their association with the MHC Class I molecules. The reason why these peptides cannot associate with the MHC Class II molecules is that the latter are "locked" by the I-chain. Parts of the I-chain occupy the peptide-binding domain of the Class II MHC molecules and prevent peptides from reaching these molecules until the I-chain falls off, which, as already mentioned, happens only in the endosomes. On their way, their separate subunits are assembled into heterodimers, and joined with two nonpolymorphic polypeptides: The I-chain, which associates with these peptides, targets their transfer from the cisterns of the Golgi apparatus to the endosomes, and a protein called **calnexin** facilitates their transport and assembly. A special class of "trans-Golgi" vesicles that mediates the joining of the Class II molecules with the endosomes in macrophages is called "MHC Class II compartment" or MIIC, and in B lymphocytes "Class II vesicles" (CIIV). These bind to endosomes or lysosomes and fuse with them. The I-chain takes an active part in the attachment of these vesicles to some (not all) endosomes or lysosomes. It is thought that the amino-terminal portion of the I-chain binds only to those endosomes or lysosomes that are suitable for antigen processing because of their enzymatic content and other properties.

Once the Class II molecules have penetrated into the endosomes, the I-chain falls off and degrades. The MHC Class II molecules spend approximately 2–3 hours in the endosomes, which allows the peptides that have been generated there to associate with their peptide-binding domains. After that, the MHC Class II–peptide complexes are transferred to the outer plasma membrane, span it, and become expressed on the surface of the antigen-presenting cell.

6.13 What enzymes are involved in the generation of peptides that join the MHC Class II molecules?

As one would expect, these enzymes belong to the family of proteases. Several different proteases are involved (or have been implicated) in the gradual, controlled hydrolysis of protein antigens in the endosomal-lysosomal compartment. The involvement of such proteases as **cathepsin** and **leupeptin** is best documented. Interestingly, leupeptin is also involved in the hydrolysis of the I-chain necessary for "vacating" the peptide-binding domain of the MHC Class II molecules in the endosomes (see next section). Naturally, the work of proteases requires physiological temperatures and an intact cellular respiration. Because of that, incubating the antigen-presenting cells at low temperatures or in the presence of metabolic inhibitors like sodium azide stops antigen processing. Importantly, the enzymatic

CHAPTER 6 Antigen Processing

activity of proteases requires a low pH. This is the reason why the pH in those endosomes and lysosomes that are responsible for antigen processing is maintained at a low level, and they are collectively called an "acidic" vesicular compartment. Drugs that raise the pH in vesicles, e.g., chloroquine and ammonium chloride, can abrogate antigen processing.

6.14 How does leupeptin help "vacate" the peptide-binding domain of the MHC Class II molecule?

Once the freshly assembled MHC Class II molecules penetrate into the endosome, leupeptin cleaves the I-chain, "cutting off" a large piece of it, but leaving intact a smaller piece, called the "Class II-associated invariant chain peptide" (**CLIP**). The latter is approximately 24 amino acids long, and sits inside the peptide-binding cleft of the MHC Class II molecule, imitating the peptides derived from antigens that it should associate with. Leupeptin is unable to remove the CLIP. For that, another protein factor is needed, and, paradoxically, this factor is coded within the MHC region itself. It is called **HLA-DM** in humans, and H-2M in the mouse. The HLA-DM (or H-2M) molecules are nonpolymorphic and exclusively cytoplasmic. Interestingly, they not only remove the CLIP, but also serve as catalysts for the association of the MHC Class II peptide-binding domains with antigenic peptides. It has been shown that while in the absence of HLA-DM the saturation of an MHC Class II molecule with a peptide takes up to 48 hours, but in the presence of HLA-DM it occurs within as little as 20 minutes.

6.15 How are the MHC Class II molecules that have been saturated with peptides further moved to the plasma membrane to be expressed on the cell surface?

The MHC Class II–peptide complexes are transported from the acidic vesicular compartment to the plasma membrane by the process of sequential fusion of the acidic vesicles with each other and, eventually, with the plasma membrane itself.

6.16 What is known about the generation of peptides from intracellular (cytosolic) proteins, and how are these peptides trafficked to MHC Class I molecules?

These peptides are generated from proteins that have been synthesized in the cell's own polyribosomes (either naturally, or because of the viral infection, or because of such artificial laboratory phenomena as gene transfection), or from proteins that have been introduced into the cytosol bypassing the vesicular compartment of the cell (e.g., by microinjection or diffusion through damaged external membrane). The major cellular entity that is responsible for the generation of cytosolic peptides is called the **proteasome** (Fig. 6-2). This is a cylinder-like multiprotein complex with a molecular weight of 1500 kDa. It consists of the proteasome *per se*,

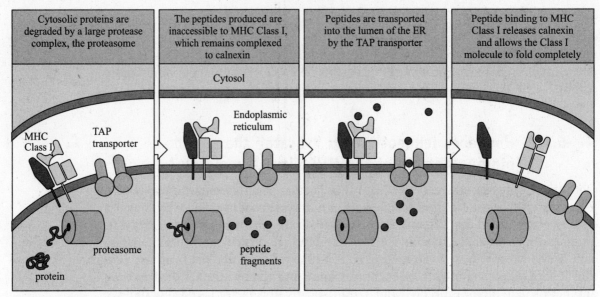

Fig. 6-2. Proteasome and TAP. This diagram shows the functions of the proteasome and the TAP in the Class I MHC pathway of antigen processing

which has a molecular weight of 700 kDa, and several "catalytic subunits," of which the best known are the so-called LMP2 and LMP7. These subunits are coded by the genes that map within the MHC complex, and are regulated by the same cytokines as the MHC Class I molecules (notably interferon gamma). The proteasome performs the proteolytic cleavage of those intracellular proteins that are complexed with a small protein called **ubiquitin**. Such "ubiquitinated" proteins are targeted to the proteasomal subunits and cleaved to peptides with the help of its enzymes and catalytic subunits.

6.17 What facilitates the delivery of these peptides to Class I MHC molecules?

Indeed, there must exist a special mechanism that is responsible for such a delivery, because nascent cytoplasmic MHC Class I molecules cannot bind peptides. To do that, they need first to associate with the β-2 microglobulin (see Chapter 5). Such an association takes place in the granular endoplasmic reticulum and in the Golgi apparatus. Therefore, something needs to transport the cytosolic peptides to the cisterns and sacks of the reticulum and Golgi. This transport is active (requires the energy of ATP) and is catalyzed by two molecules, called "transporters associated with antigen presentation" or **TAP**. There are two distinct kinds of TAP, called TAP-1 and TAP-2, both coded by the genes mapped within the MHC complex (see Chapter 5). These two molecules form a heterodimer that serves as an active transporter of peptides across the phospholipid layer of the reticular and Golgi membranes. Interestingly, the TAP heterodimer cannot accommodate proteins or even large peptides; the "upper limit" on the size of peptides it can transport is 12 amino acid residues. This is important in view of the structure of

CHAPTER 6 Antigen Processing

the MHC Class I binding cleft which, as discussed in Chapter 5, can associate only with relatively small peptides.

6.18 Does the TAP heterodimer dissociate and release the transported peptides before they associate with the MHC Class I molecules?

No. In fact, the TAP-peptide complex stays intact for a while because of the other important function of TAP. The latter is needed to stabilize the association between the alpha chain of the MHC Class I molecule and the β-2 microglobulin. Only after this is done can the peptides be transferred into the peptide-binding cleft if the fully assembled MHC Class I molecule, and the TAP heterodimer is "recycled" to transport other peptides.

6.19 How many of the expressed MHC Class II-peptide complexes actually display foreign peptides?

Rough estimates show that this portion is very small. In the case of the MHC Class II molecules expressed on the surface of the "professional" antigen-presenting cells (macrophages, dendritic cells, and B lymphocytes), this portion perhaps does not exceed 0.1% of all MHC–peptide complexes expressed by an antigen-presenting cell at any given moment of time. This may seem amazing because seemingly, it lessens the probability that a T lymphocyte will encounter a foreign peptide–self MHC complex and not a self peptide–self MHC complex on the cellular surface. It has been demonstrated, however, that **a single T lymphocyte can sequentially engage many peptide-MHC complexes** and disengage from them. This gives the T cell enough opportunity to "screen" multiple cell-surface entities until it finds one of the 100–200 peptide–MHC complexes that contain foreign peptides. On the other hand, there is only a small risk that a T lymphocyte will recognize, and be activated by, self peptide–self MHC complexes, because more than 99.9% of the T cells with receptors specific to such entities are eliminated during the selection in the thymus (see next chapters).

6.20 Can endogenously synthesized proteins enter the acidic vesicular compartment and generate peptides there, to be joined with the MHC Class II molecules?

Yes. Some proteins synthesized on the cellular polyribosomes can be taken into the endosomes and lysosomes by virtue of a mechanism called **autophagy**. This is mediated by special vesicles called autophagosomes. These can surround and enclose many cytoplasmic proteins and then fuse with endosomes and lysosomes, thus ensuring the processing of the enclosed antigens in the acidic vesicular compartment and the joining of the derived peptides with the MHC Class II molecules. Also, endogenously synthesized and membrane-expressed proteins can be internalized by the cell through a process of the so-called **endocytosis**. In this case, again, the proteins will be processed in the acidic vesicular compartment. Because of the above two mechanisms, some important foreign antigens, e.g., viral proteins, could be made subject to not only $CD8^+$, but also $CD4^+$ T cell recognition.

6.21 Are all peptides that can be generated from a complex protein molecule equally immunogenic for T cells?

No. Experiments performed in the 1970s and early 1980s firmly established that in complex multiepitope protein molecules, some parts of the molecule are more immunogenic to T cells than others. These epitopes (peptides) were called **dominant**, and the less immunogenic ones cryptic or **subdominant**. The current understanding of dominant epitopes is as follows: they are the peptides that can associate with the particular alleles of MHC molecules better than other peptides originating from the same protein molecule. It is important to remember that the ability of peptides to associate with MHC molecules depends on the nature of the particular peptide as well as on the allelic variant of the MHC molecule itself. Thus, the same peptide may be dominant in individuals who bear certain MHC alleles, and subdominant in individuals that bear other alleles. For example, the HEL peptide mentioned in Section 6.5 is dominant compared to other fragments of the HEL molecule when presented in the context of the $I\text{-}A^k$ molecule. In those inbred mice strains that express other I-A alleles, however, this peptide is subdominant. In humans, one can observe some clinically important antigens display T-cell epitopes that are either dominant or subdominant depending on the person's HLA haplotype. For example, a peptide generated from the hepatitis B virus is subdominant to individuals homozygous on the HLA-B, HLA-DR, and HLA-DQ loci and carrying the alleles HLA-B8, DR3, and DQw2a. If an individual is heterozygous on the above loci, however, the same peptide behaves as dominant. The understanding of molecular mechanisms of antigen processing and presentation is thus important for the understanding of the immunogenicity of various proteins derived from infectious agents, and is used in the generation of new vaccines.

Questions

REVIEW QUESTIONS

1. A mouse is immunized with a protein antigen. After several days, macrophages from this mouse are put in culture together with the T cells from the same mouse. The antigen used for the immunization is added in the culture medium.
   (a) What subpopulation of the murine T cells – $CD4^+$ or $CD8^+$ – will proliferate in this culture?

CHAPTER 6 Antigen Processing

(b) Will there be a change in the proliferating T-cell subset if the culture medium is temporarily replaced by distilled water?
(c) What happens if the antigen is microinjected in the macrophages instead of being fed into the culture medium?
(d) If the antigen is in the medium, what agents will you use to prevent its processing?
(e) If the antigen is microinjected, what agents will you use to prevent its processing?

2. In two to three sentences, describe how the use of radioactively labeled antigens helped to establish the current notion of antigen processing.

3. If a mouse is immunized with xenogeneic (human, rabbit, etc.) MHC molecules, it generates antibodies against many different epitopes present on the latter, including the epitopes specific to the MHC allele of the donor of these molecules ("polymorphic epitopes"), and the species-specific or "monomorphic" epitopes. Both categories of antibodies can be produced by hybridomas obtained after the immunization (see Chapter 3). Which category of hybridoma-derived antibodies – anti-polymorphic, anti-monomorphic, both, or neither – would you use to block the lysis of LCMV-infected fibroblasts in the Doherty–Zinkernagel experiment?

4. In three to four sentences, summarize the advantage of having two distinct pathways of antigen processing.

5. Knockout mice are mice in which certain genes are rendered irreversibly nonfunctional through the technique of homologous recombination (see genetics textbooks for details). Which knockout mice would you use to study the two pathways of antigen processing and presentation?

6. What is the function of the I-chain?

7. What is the function of TAP proteins?

8. How would you imagine processing and presentation of tumor antigens?

9. What is the approximate number, per one antigen-presenting cell, of the MHC Class II molecules displaying foreign peptides, and what is the relation of this number to the total number of the MHC Class II molecules per one antigen-presenting cell?

10. Why are the studies on immunodominant peptides thought to be important for making new vaccines?

MATCHING

Directions: Match each item in Column A with the one in Column B to which it is most closely associated. Each item in Column B can be used only once.

Column A

1. Leupeptin
2. LMP-2
3. Invariant chain
4. Acidic vesicular compartment
5. Cytosolic peptides
6. Peptide carrier
7. Removes CLIP
8. $CD4^+$ T cells
9. Bind to the MHC Class II molecules
10. Subdominant

Column B

A. Respond to peptide + Class II MHC
B. Cannot bind the MHC allele
C. HLA-DM
D. Trafficked to Class I MHC
E. Works at low pH
F. Peptides in endosomes
G. Part of the proteasome
H. TAP
I. Locks MHC Class II molecules
J. Includes lysosomes

Answers to Questions

REVIEW QUESTIONS

1. (a) $CD4^+$, because they will recognize the peptides from the actively internalized protein. These peptides will be generated in the acidic vesicular compartment and trafficked to the MHC Class II molecules. (b) Yes. $CD8^+$ T cells will respond, because the osmotic shock makes cellular membranes penetrable for proteins. The diffused antigen will be in the cytosol, where the peptides will be generated and trafficked to the MHC Class I molecules. (c) Same as described under (b). (d) Anything that raises the pH in the acidic vesicular compartment (e.g., chloroquine, ammonium chloride). (e) Inhibitors of proteasome function (e.g., peptide aldehydes).

2. The labeling allowed calculating how much of the antigen accumulates in macrophages. If the macrophages with the accumulated antigen are put back in the animal, it stimulates the animal's T cells as strongly as the 1,000 excess of the antigen injected without macrophages. This leads to the conclusion that only a small portion of antigen is subject to processing and presentation to T cells.

3. Neither. The T-cell recognition in the Doherty–Zinkernagel system depends on the matching of the *murine* alleles. Hence, to block the T-cell reaction, one needs to have antibodies to the murine (H-2) allospecific determinants. Such antibodies can be generated after immunizing one inbred murine strain with cells derived from another inbred murine strain (*allogeneic* immunization).

4. TCR is unable to discern between extracellular and intracellular microbes, and yet somehow the immune system needs to involve $CD4^+$ T cells in the reactions against the former and the $CD8^+$ T cells in the reactions against the latter. To solve this problem, the peptides generated from extracellular proteins and the peptides generated from intracellular proteins are segregated in the cell and trafficked to different classes of MHC molecules. These different classes, in turn, involve different T-cell categories: $CD4^+$ T cells respond to Class II, and $CD8^+$ to Class I.

5. To study the processing and presentation of peptides that associate with the MHC Class II molecules, one needs knockouts that lack I-chain, H-2M, and various Class II loci. To study the processing and presentation of peptides that associate with the MHC Class I molecules, one needs knockouts that lack LMP-2, LMP-7, genes coding for other proteasome subunits, ubiquitin, TAP-1, TAP-2, various MHC Class I loci and β-2 microglobulin.

6. It has two important functions: (1) to target nascent MHC Class II molecules to the acidic vesicular compartment, and (2) to lock the peptide-binding cleft of the MHC Class II molecules until they reach the acidic vesicular compartment, thus preventing them from association with the peptides that are pumped into the endoplasmic reticulum from cytosol.

7. They form a heterodimer that serves as: (1) an active transporter of peptides from cytosol to the endoplasmic reticulum, and (2) a promoter of the association between the MHC Class I molecule alpha chain and the β-2 microglobulin.

8. Since they are synthesized inside the cell (based on the information from viral oncogenes or deregulated cellular protooncogenes), they must be processed in the cytosol, and the peptides generated from these antigens must be trafficked to the MHC Class I

CHAPTER 6 Antigen Processing

molecules. Yet, some of these antigens perhaps can be processed in the acidic vesicular compartment because of endocytosis or autophagy.
9. One to two hundred, which is ~0.1% of all peptide–MHC Class II complexes.
10. Because synthetic peptides that show immunodominance can turn out to be better vaccines than "natural" viral peptides that happen to be subdominant.

MATCHING
1, E; 2, G; 3, I; 4, J; 5, D; 6, H; 7, C; 8, A; 9, F; 10, B

CHAPTER 7

T-Lymphocyte Antigen Recognition and Activation

Introduction

In previous chapters, we introduced the concept that T lymphocytes recognize antigens through the use of their antigen receptors (T-cell receptors for antigen, or TCRs). Like antibodies, TCR are specific to antigens and clonally distributed; moreover, they share many structural similarities with antibody molecules and belong to the so-called immunoglobulin superfamily of molecules mentioned in Chapter 3. However, TCRs are not just another class of antibodies. They are a separate entity, differing from antibodies in a number of aspects. Most importantly, the character of their antigen recognition differs from that of antibodies. *The only form of antigen that the TCR can recognize is a peptide associated with a self MHC molecule.* Such pattern poses a special demand on the structure and conformation of the TCR V-region. In this chapter, we will discuss how the TCR was discovered; how it is coded genetically and what its biochemical features are; how T lymphocytes are activated after the TCR is triggered; what types or subsets of TCR exist; and finally what molecules expressed by the T lymphocyte and the antigen-presenting cell are important for the proper activation of T lymphocytes.

CHAPTER 7 T-Lymphocyte Antigen Recognition

Discussion

7.1 What was the reason for the fact that the structure and genetic determination of the TCR was elucidated much later than the structure and genetic determination of antibodies? What played a decisive role in solving the TCR puzzle?

The reason for difficulties in the elucidation of the TCR structure lies in the fact that unlike antibodies, TCRs are never secreted. The necessity to deal with structures embedded in the cellular membrane was a major obstacle, further aggravated by the lack of monoclonal T-lymphocyte populations of known specificity. However, after the discovery of immunoglobulin rearrangement in the late 1970s and the extensive follow-up of this discovery made in the first half of the 1980s, it became possible to use the acquired knowledge of the antibody V-region genes. In addition, the methodology of propagating T-lymphocyte-derived monoclonal cell populations *in vitro* greatly improved, and served for the characterization of TCR expressed by these populations.

7.2 What kinds of monoclonal T-cell populations were used, and how can they be obtained?

Two kinds of monoclonal T-cell populations were instrumental in the discovery of the TCR structure: **T-cell hybridomas** and normal factor-dependent **T-lymphocyte clones**. Hybridoma technology was described in Chapter 3. As applied to T-cell hybridomas, it includes fusion of normal T lymphocytes of immunized rodents with HGPRT-deficient tumors of T lymphocyte origin (e.g., thymomas), and selecting the hybrids in the HAT medium. The grown and cloned hybrids are then screened for the ability to recognize peptides derived from the antigen used for the immunization in complex with MHC molecules of the T-lymphocyte donor. Normal factor-dependent T-lymphocyte clones can be grown from murine splenic or human peripheral blood T lymphocytes derived from immunized donors. Such "primed" T lymphocytes must be periodically "boosted" by antigen-presenting cells derived from the same donor and pulsed with the antigen used for the donor's immunization. In addition, these T lymphocytes must be continuously supplied by cytokines that serve as their survival and growth factors. The cytokines can be added individually or, better, as a "cocktail" in a form of the supernatant of T-lymphocyte cultures stimulated by polyclonal T-cell activators (e.g., phytohemagglutinin, PHA, and other potent inducers of cytokine secretion).

7.3 How were these cell populations used for the TCR identification?

The approach was to immunize laboratory animals with these growing clones, obtain antibodies specific to the epitopes unique to one particular T-lymphocyte clone, and use these antibodies as tools to isolate and biochemically characterize the molecules that they were binding. It was assumed that the T-cell molecules containing such unique clonospecific determinants (idiotypes) would prove to be TCR. In the early 1980s, the laboratory of J. Allison generated a large panel of murine hybridomas, some of which produced T-cell idiotype-specific monoclonal antibodies. Molecules containing these idiotypic determinants were isolated from the membranes of the T-cell clones that expressed them. An extensive biochemical characterization of these molecules showed that they were homologous to antibodies and, like antibodies, had constant and variable regions. The latter showed amino acid sequence that substantially differed from one T-lymphocyte clone to another. Thus, the molecules that contained the idiotypic determinants appeared to be good candidates for TCR. Yet, a more complete characterization of the TCR was still in order.

7.4 What completed the characterization of the TCR initiated by the above-described approach?

A decisive experiment was performed in the laboratory of S. Hedrick and M. Davis in the mid-1980s, later to be reproduced and extended in many other laboratories. The principal idea of the experiment can be summarized by the term "subtractive hybridization." The logic of the subtractive hybridization is as follows (Fig. 7-1). The entire mRNA of a T-cell clone is used for the preparation of cDNA, which is radioactively labeled and hybridized to the entire mRNA of a B-cell clone. The portions of T-lymphocyte cDNA that do hybridize with the B-lymphocyte mRNA are discarded, and those that do not are analyzed further. This way, an investigator gains access to the cDNA clones that contain messages from genes expressed in T cells and not in B cells. Hedrick and Davis determined the complete nucleotide sequence of several of these clones, and showed that some of them have V- and C-regions homologous to those of immunoglobulins. The Southern blot analysis of these clones (see Chapter 4) revealed that, just as in the case of immunoglobulins, V and J segments that code for these clone's variable regions are separated by intervening DNA in the germline. They undergo a rearrangement in maturing lymphocytes; the latter, however, takes place exclusively in T lymphocytes. Further, the deduced amino acid sequence agreed very well with Allison's data on the protein molecules identified by anti-T-cell idiotypic antibodies (see above). It was concluded that the cDNA clones thus identified coded for the TCR. The notion was finally established that TCR is an entity that acquires its antigen specificity due to stochastic processes of V-region gene rearrangement in maturing T lymphocytes, the same way as antibodies acquire their antigen specificity due to stochastic processes of V-region gene rearrangement in maturing B lymphocytes.

CHAPTER 7 T-Lymphocyte Antigen Recognition 111

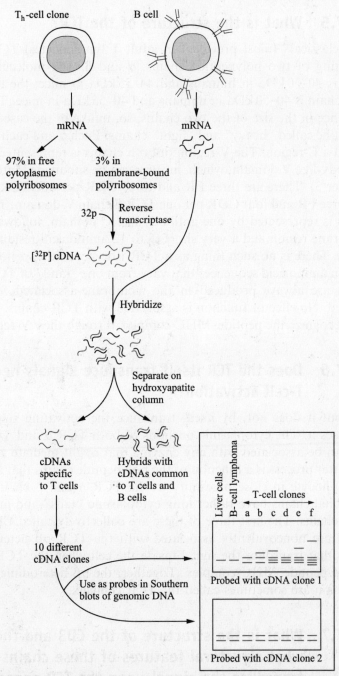

Fig. 7-1. The experiment of Hedrick and Davis with the use of "subtractive hybridization" technique. cDNA clones from a T_h clone were hybridized to mRNA extracted from a B-cell line. The cDNAs that did not hybridize (a total of ten, shown on the left) were used as probes in Southern blotting of genomic DNA isolated from various T and non-T cells. The cDNA clone 1 corresponds to a gene that is rearranged in T cells. This gene was later sequenced and shown to encode the b chain of the TCR.

7.5 What is the structure of the TCR?

The "classical" (most prevalent in adult T lymphocytes) TCR is a **heterodimer**, consisting of two polypeptide chains, **α** and **β**. The molecular weight of the α chain is 40–60 kDa in humans and 44–55 kDa in mice; the molecular weight of the β chain is 40–50 kDa in humans and 40–55 kDa in mice. There is no dramatic difference in the size of the two chains, so, unlike in the case of antibodies, these cannot be called "heavy" and "light" chains. Each α and each β chain consist of a V- and a C-region. The V-region of both chains is represented by one N-terminal antibody-like V domain, which, in turn, can be subdivided into FR and CDR (see Chapter 3). There are three FR and three CDR per one TCR α chain V domain, and three FR and four CDR per one TCR β chain V-domain. The C-region of both chains is represented by one antibody-like C domain, followed by a short transmembrane region and a very short (5 to 12 amino acid residue long) cytoplasmic region. There is no such thing as a TCR isotype: neither α nor β chain C-regions contain amino acid sequences that vary from one "kind" of TCR to another. TCR chains are always produced in the membrane-associated form and are never secreted. No effector function is associated with TCR chains. Their only function is to recognize the peptide–MHC complex through their V-regions.

7.6 Does the TCR itself transduce signals necessary for T-cell activation?

Yes, but it does not, by itself, transduce the activating signal in the cell that expresses it. The cytoplasmic portions of both TCR α and TCR β chain are too short to be associated with any enzyme that might mediate such a transduction. The latter process is a function of the polypeptide chains that are expressed by the T lymphocyte in close proximity to the TCR α and β chains (Fig. 7-2). These additional chains have rather long cytoplasmic "tails" and are called the γ, δ, ε, and ζ chains. The first three of these are collectively called **CD3**. The CD3 and ζ chains are noncovalently associated with the TCR αβ heterodimer. It is these chains that transduce the signal inside the cell when the TCR is triggered by its specific peptide–MHC complex. Together, the αβ heterodimer, the CD3 and the zeta chain are sometimes called the **TCR complex**.

7.7 What is the structure of the CD3 and the ζ chain, and what structural features of these chains allow them to transduce the signal upon the TCR engagement?

The chains of the CD3 and the ζ chain have a rather similar structure (Fig. 7-2). In the human, the molecular weight of the CD3 γ chain is 25–28 kDa, the CD3 δ chain is 20 kDa, the CD3 ε chain is 20 kDa, and the ζ chain is 16 kDa. In the mouse, the molecular weights of these chains are very similar, although they differ slightly from the human chains. In both species, the CD3 complex is expressed on the membrane as two heterodimers, one of which consists of one gamma and one epsilon chain, and the other of one delta and one epsilon chain. Each of the three

CHAPTER 7 T-Lymphocyte Antigen Recognition

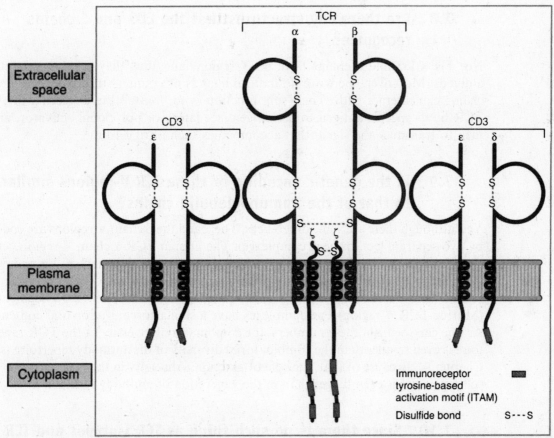

Fig. 7-2. The T-cell receptor (TCR) complex. A schematic diagram. From Abbas, K., A.H. Lichtman, and J.S. Pober, *Cellular and Molecular Immunology*, fourth edition, W.B. Saunders, 2000, p. 108 (their Fig. 6-5).

CD3 chains contains one N-terminal immunoglobulin-like domain, which makes these chains belong to the immunoglobulin superfamily. A remarkable feature of the CD3 chains is their long cytoplasmic "tails." These contain from 44 to 81 amino acid residues. Within these cytoplasmic residues, there are conserved sequence motifs called **immunoreceptor tyrosine-based activation motifs (ITAMs)**. An ITAM consists of two copies of the following sequence: tyrosine-X-X-leucine, where X is an unspecified amino acid; these copies are separated by six to eight varying amino acid residues. There is one copy of ITAM per each of the CD3 chains. The ζ chain has a molecular weight of 16 kDa both in humans and in mice. It does not have an immunoglobulin-like domain; its N-terminal extracellular portion is linear and very short – only nine amino acids. On the other hand, a ζ chain has even longer cytoplasmic tail than the CD3 chains; in the ζ chain, it consists of 113 amino acid residues. The cytoplasmic tail of the ζ chain contains not one but three copies of ITAM. When the TCR is engaged, the noncovalently associated CD3 and ζ chains induce the enzymatic activity of tyrosine kinases that "dock" to the ITAMs. This initiates a complex cascade of enzymatic events, which, as we will discuss in later sections, leads to activation of the T lymphocyte.

7.8 Are there any structures that the CD3 and ζ chains recognize?

No. The CD3 and ζ chains have no V-regions and thus they cannot recognize antigens. Moreover, there are no natural ligands or counter-structures that these chains can interact with. Their sole function is to "sense" the triggering of the TCR by its specific antigen, superantigen (see later), or polyclonal activator, and then to transduce the signal that accompanies such a triggering.

7.9 Is the genetic encoding of these TCR V-regions similar to that of the immunoglobulin chains?

Yes, although there are some differences. The α and the β chain V-regions are coded by two separate loci (or gene complexes). The human TCR α chain V-region gene complex is located on chromosome 14 and contains approximately 50 Vα and 50–70 Jα segments. The human TCR β chain V-region gene complex is on the chromosome 7 and contains 75 Vβ segments, two Dβ segments and 12 Jβ segments. (Murine TCR V-region gene complexes have a similar organization and content.) As one can see from these numbers, the combinatorial diversity of the TCR repertoire is even smaller than the combinatorial diversity of the antibody repertoire (see Chapter 4). Because of that, the role of junctional diversity in the generation of T-cell repertoire is even bigger than in the generation of antibody repertoire.

7.10 Since there is no such thing as TCR isotypes and TCR class switch, does it mean that there is just one TCR C-region gene?

Actually, there is only one C-region gene for the TCRα chain (Cα) and two C-region genes for the TCR β chain (Cβ1 and Cβ2). The use of either of them seems to be completely random, and there are no known differences in the "quality" of the T-cell clones that utilize one of them versus the other.

7.11 Is the way α and β TCR chains are expressed dramatically different from the way antibodies are expressed?

No. The mechanisms of the TCR V-region gene rearrangement are thought to be quite similar to the mechanisms of the antibody V-region gene rearrangement. Some minor differences should be mentioned, however. The β chain V-region rearranges prior to the α chain V-region, and begins with the joining of one of the two Dβ segments with one of the 12 Jβ segments. Interestingly, the Dβ segments do not follow each other in a tandem array, but are separated from each other by six of the 12 Jβ segments (grouped into a cluster called Jβ1) and one of the two Cβ genes. Because of this peculiar organization of segments, Cβ1 can be deleted during the Dβ-Jβ rearrangement. Also, if a rearrangement between the Dβ1 and one of the

CHAPTER 7 T-Lymphocyte Antigen Recognition

Jβ segments from the Jβ1 cluster appears to be nonproductive, a second rearrangement, involving the other Dβ and one of the remaining six Jβ segments (collectively called Jβ 2) is possible. This gives a T cell one more chance to produce a functional TCR. If the β chain V-region cluster rearranges productively, the beta chain is made through the use of RNA splicing, as detailed in Chapter 4. This event sends the signal for the homologous TCR β chain V-region to stop its rearrangement (allelic exclusion), and to the TCR a chain V-region to rearrange. Interestingly, the nascent TCR β chain may pair with the molecule called **pre-T α chain (pTα)**, which resembles the event of pairing between the antibody μ chain and the surrogate light chain in pre-B lymphocytes. Also, the allelic exclusion during the rearrangement of the two homologous TCR α chain V-region loci is rather "lenient." Because of this "leniency," up to 30% of T lymphocytes in the human as well as in the mouse contain two different functional TCR α chains, but only one TCR β chain. The functional significance of such a "dualism" is unknown.

7.12 What parts of the TCR V-region are coded by what segments, and what allows the TCR to recognize the peptide and the MHC simultaneously?

Like in the case of antibodies, the V segment codes most of the TCR V-region. That includes the N-terminal FR1, the CDR1, the FR2, the CDR2, and the FR3. The CDR3, which in a linear sequence is the most C-terminal part of the V-region and in the folded TCR V domain is on the outside, is coded by the most C-terminal part of the V segment, the D segment (for the TCR beta chain) and an N-terminal part of the J segment. The antigen-binding site of the TCR is a flat surface formed by the juxtaposed CDR of the alpha and beta chains. As shown by X-ray crystallography, in some TCR the CDR3 physically contacts the peptide while the CDR1 and the CDR2 interact with the MHC molecule; in other TCR, all three CDRs make contact with the MHC molecule but, again, it is the CDR3 that also interacts with the peptide. An important feature of the TCR–peptide interaction is that only one or two amino acid residues of the MHC-bound peptide make contact with the TCR. This points to an amazing ability of the TCR to distinguish between antigens based on very slight differences in their amino acid sequence. While the discriminating ability of the TCR is so high, the affinity of its interaction with the peptide and MHC is very low (K_d 10^{-5} to 10^{-7} M). Because of that, the TCR must stay in close proximity to the peptide- and MHC-expressing cell for a rather long time. A number of **accessory molecules** expressed on the T cell interact with their counterparts on the antigen-presenting cell, providing the adhesion that is necessary for such a prolonged association (see later).

7.13 What events follow the binding of the peptide and MHC by the TCR?

That crucially depends on whether or not the above-mentioned accessory molecules are bound by their APC-expressed counterparts called co-stimulators. If these are available and the accessory molecules are bound, a sequence of signal transduction events eventually results in the **T-cell activation**. If the TCR binds the peptide and MHC and the accessory molecules are not simultaneously bound by co-stimulators (because the latter are absent or blocked), a different sequence of signal transduction events follows and eventually results in the specific unresponsiveness of the given T lymphocyte, or the T-cell **tolerance**.

7.14 What is the sequence of the "activator" events?

It includes clustering of the TCR and accessory molecules together on the cell surface; phosphorylation of cytoplasmic portions of the CD3 and ζ chains; activation of adaptor proteins and biochemical intermediates; activation of certain intracellular enzymes and, eventually, transcriptional activation of genes that are silent in resting T lymphocytes (Fig. 7-3). All of these events, except the activation of gene transcription, occur very rapidly – within seconds to minutes

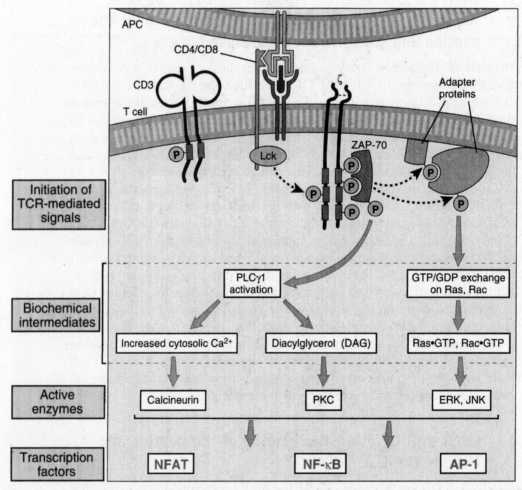

Fig. 7-3. T-cell receptor-mediated cell signaling. An overview of events that follow the triggering of the TCR and CD4/CD8 is presented. See text for details. From Abbas, K., A.H. Lichtman, and J.S. Pober, *Cellular and Molecular Immunology*, fourth edition, W.B. Saunders, 2000, p. 171 (their Fig. 8-8).

CHAPTER 7 T-Lymphocyte Antigen Recognition

after triggering of the TCR. The phosphorylation of cytoplasmic portions of the CD3 and ζ chains is an especially rapid and also complex phenomenon that involves a family of enzymes called **protein tyrosine kinases** (**PTK**).

7.15 What comprises the family of PTK?

PTK catalyze the transfer of the terminal phosphate group from ATP to the hydroxyl group of a tyrosine residue in a substrate protein. In contrast to serine or threonine phosphorylation, induced tyrosine phosphorylation does not happen often and is reserved for crucial regulatory steps of intracellular events. The PTK involved in signal transduction in immune cells are classified into four families: the so-called **src, Csk, Syk/Zap70**, and **Janus** kinases. Of these three families, the first family, **src** kinases, is the most important for T-cell activation. Eight individual members of the src family were identified so far: Src (named after the Rous sarcoma virus where the gene for this enzyme was first described), Yes, Fgr, Fyn, Lck, Lyn, Hck, and Blk. The amino-terminal region of its molecule determines the function of each of these PTK; this region has an affinity towards the cytoplasmic portion of a certain specific proteins. For example, the N-terminal region of lck is able to bind the cytoplasmic tails of CD4 or CD8. As will be detailed later, both CD4 and CD8 are typical accessory molecules; this illustrates that not only CD3 and ζ chain, but also accessory molecules are absolutely essential for the activator signal transduction. Each PTK from the src family contains two internal domains, called SH2 and SH3 (where "SH" stands for "src homology"), which have a unique spatial structure and enable the PTK to bind noncovalently ("dock") to other proteins. As we will see, this "docking" is crucial for activation of a number of molecular intermediates and further transduction of activator signals.

The second PTK family, **Csk** (so far represented by the only individual member, Csk-1) phosphorylate PTK from the Src family, thus making them inactive. The third group, Syk/Zap70, includes two individual members, Syk and Zap70. Syk operates predominantly in B lymphocytes and will be discussed in later chapters. Zap-70 (from "70-kDa zeta-associated protein") is, as its name tells, associated with the cytoplasmic portion of the TCR complex ζ chain, also has SH2 and SH3 domains, is activated when docks to the zeta chain, and phosphorylates a number of adapter proteins (see later). The fourth of the above-mentioned groups, Janus kinases, work in pathways of cytokine-mediated signaling and will be discussed in the appropriate chapters.

7.16 When phosphorylated, PTK become inactive. Does this mean that when they are dephosphorylated, they become active?

This is true at least for the PTK from the src family. A special group of enzymes called **protein tyrosine phosphatases** (**PTP**) are responsible for the removal of the Sck-1-induced inhibitory phosphate from the src tyrosine residues. One of the best-known PTPs is a molecule called **CD45**. It is a membrane-expressed T-cell protein that has two isoforms, CD45RA and CD45RO (the second one is slightly smaller in size; both have identical enzymatic activity). Like CD4 and CD8, CD45 also belongs to the family of accessory molecules and enhances the T-cell activation when involved.

7.17 How exactly do the activator src PTK work?

This is best studied for one of the PTK of the src family, **lck**. When the TCR is triggered by its specific peptide–MHC complex, either CD4 or CD8 bind to their appropriate binding sites. As detailed in Chapters 2 and 5, these sites are in the immunoglobulin-like domains of the MHC Class II and the MHC Class I molecules, respectively. Once that happens, lck, which is physically associated with CD4 or CD8, is brought into close proximity of the ITAMs of the CD3 chains and of the ζ chain. Simultaneously, it is dephosphorylated by CD45 and thus activated. In addition, a similar enzyme, called Fyn, is brought into close proximity with the ITAMs of the CD3 chains. Within seconds, lck tyrosine-phosphorylates the ζ chain ITAMs and thus makes them able to "dock" on the Zap-70. This strongly activates Zap-70 and, in turn, enables the latter enzyme to act on a number of cytoplasmic signaling molecules.

7.18 What are cytoplasmic signaling molecules, and what do they do?

It has been shown recently that Zap-70 acts directly on a special class of molecules called **adapter proteins**. Two of these are considered especially important for T-lymphocyte activation: a membrane protein called **LAT** (from "linker of activation of T cells"), and a cytoplasmic protein called **SLP-76** (from "SH domain-containing leukocyte protein with the molecular weight of 76 kDa"). The adapter proteins become activated when they are tyrosine-phosphorylated by Zap-70. Their activation leads to their binding to a number of molecules, including both membrane-associated molecules (which serve as other adapters, thus continuing the "relay race" of signal transduction), and cytosolic proteins with enzymatic activities. In short, the involvement of adapter proteins leads to recruitment of important enzymes and to triggering of two principal pathways of T-lymphocyte activation. The first of them is the **Ras pathway of T-cell activation**, and the second is the **inositol phospholipid pathway of T-cell activation**. Both pathways end with the activation of gene transcription, leading, eventually, to cytokine production, cytokine receptor expression, and the entrance of the T lymphocyte into cell cycle.

7.19 How does the Ras pathway of T-cell activation work?

Ras is, essentially, a collective name for a system of ubiquitous cellular signal transducers whose main feature is the ability to bind phosphorylated guanine nucleotide. In T lymphocytes, the main representative of the Ras family is a

CHAPTER 7 T-Lymphocyte Antigen Recognition

21-kDa Ras protein that is loosely bound to the cytoplasmic side of the membrane through covalently attached lipids. In its "resting" state, Ras contains an attached guanosine diphosphate (GDP); when activated by adapter proteins, it exchanges its GDP into a guanosine triphosphate (GTP). The GTP-associated Ras (**Ras·GTP**, for short) becomes an allosteric activator of various cellular enzymes, of which the so-called mitogen-activated protein kinases (**MAP kinases**) are most important for the subsequent signal transduction in T cells.

7.20 What are MAP kinases, and what do they do?

MAP kinases are a large family of serine-threonine protein kinases that are present in a wide variety of cells and serve as intermediates between cytoplasmic and nuclear events of cellular activation. In T lymphocytes, at least three MAP kinases are involved in signal transduction. The first of them used to carry a "Byzantine" name "MAP-kinase kinase kinase" (MAPKKK), but is now known under a more palatable name **Raf**. Ras·GTP directly activates this enzyme. The activated Raf, in turn, phosphorylates another representative of the MAP kinase family, called MEK-1, and that enzyme passes the relay of phosphorylation to yet another MAP kinase called **ERK** (from "extracellular receptor-activated kinase"). The phosphorylated ERK itself phosphorylates a protein called **Elk**. This protein possesses the ability to penetrate inside the nucleus of the T lymphocyte and to serve there as a transcription factor for a gene called **Fos**. The latter, when its transcription is activated, controls the production of Fos protein, which is an important inducer of the transcription of the interleukin-2 (IL-2) gene (see later). Thus, Ras-activated MAP kinases pass the activator signal from the cytoplasm of the T lymphocyte to its nucleus.

In parallel with the Raf-MEK-ERK enzymatic cascade, the adapter protein activity leads to another "relay race," also resulting in the activation of the IL-2 gene transcription. This other cascade of phosphorylation begins with adapter-mediated activation of a GTP exchange protein called vav. This protein, when activated, triggers the exchange of GDP into GTP in another 21-kDa guanine nucleotide-binding, Ras-like protein called **Rac**. The latter stimulates a MAP kinase called JNK (from "c-Jun N-terminal kinase," where c-Jun is a cellular proto-oncogene that codes for a protein that this kinase phosphorylates). The phosphorylated c-Jun, like Elk, is able to penetrate into the nucleus and serves as another activator of the IL-2 gene transcription.

7.21 How does the other of the two above-mentioned pathways, the inositol phospholipid pathway of T-cell activation, work?

The key enzyme of this pathway is a cytosolic enzyme called **phospholipase C γ1** (**PLCγ1**). It is inactive in its cytosolic status but, when recruited by a Zap-70-activated adapter protein, translocates towards the cellular membrane and, through phosphorylation by Zap-70, becomes active. In this new status, it becomes able to catalyze the hydrolysis of a membrane-bound phospholipid called phosphatidylinositol 4,5-biphosphate (PIP2). The two products of this hydrolysis are called **inositol 1,4,5-triphosphate (IP3)** and **diacylglycerol (DAG)**. The function of IP3 is to rapidly increase the concentration of free ionized cytosolic calcium $[Ca^{2+}]_i$. In resting T lymphocytes, this concentration does not exceed 100 nM; within minutes after TCR triggering and Zap-70 activation, it reaches levels of 600 to 1,000 nM (Fig. 7-4). Initially, this raise is explained by a rapid release of calcium from its cellular depots (mostly endoplasmic reticulum). It is then sustained for approximately 1 hour because of the opening of a membrane calcium channel and an influx of Ca^{2+} from outside. The net result of such a dramatic and sustained increase in $[Ca^{2+}]_i$ is the formation of complexes between calcium ions and a ubiquitous cellular protein called **calmodulin**. Calcium–calmodulin complexes activate a number of enzymes, of which for T-cell activation the most important one is the so-called **calcineurin**. This latter molecule plays an important role in the activation of IL-2 gene transcription, acting in concert with the above-described Fos, Jun, and other molecules (see later).

Another effect of the increased $[Ca^{2+}]_i$ is the translocation of a cytosolic enzyme called protein kinase C (PKC) to the cellular membrane. The other product of PIP2 hydrolysis, DAG – a hydrophobic, membrane-bound molecule – strongly activates the translocated PKC. The activation of this enzyme, which belongs to the family of serine-threonine protein kinases, is important for a further progression of the T-cell activation. The exact mechanisms of PKC-mediated signal transduction in the T lymphocyte activation are not clearly elucidated so far. However, the importance of this enzyme is underscored by the fact that pharmacological agents that are direct PKC activators (e.g., phorbol esters) stimulate T lymphocytes and, in combination with calcium ionophores, induce powerful proliferation of T cells. Indirect evidence suggests that PKC also participates in the up-regulation of the IL-2 gene transcription.

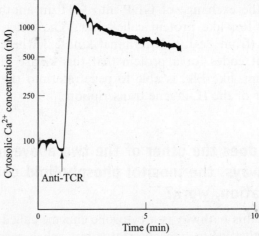

Fig. 7-4. Growth of $[Ca^{2+}]_i$ in T cells due to T-cell receptor engagement. This oscillogram shows an abrupt and sharp rise of $[Ca^{2+}]_i$ within a short time after T-cell stimulation through the TCR.

7.22 How exactly do the above-described pathways of signal transduction culminate in the activation of IL-2 gene transcription?

In fact, the IL-2 gene is just one (albeit important) gene whose transcription is activated by various signal transduction pathways in T cells. As we will detail in later chapters, the IL-2 gene is silent in resting T cells, and its transcriptional activation is a cornerstone of T-lymphocyte response to antigens or polyclonal stimulators. The IL-2 gene is under the control of a large (300-bp) promoter-enhancer region that can bind several different transcription factors. Three of these factors appear to be activated in T cells by antigen recognition and critical for most T-lymphocyte responses. These factors are nuclear factor for activated T cells (**NFAT**), **AP-1**, and nuclear factor κB (**NFκB**). The above-described signal transduction pathways in T lymphocytes lead to the activation of these factors and, through this activation, to the synthesis of IL-2. The latter, as we will discuss later, further drives T lymphocytes to proliferation and differentiation into effector or memory cells.

7.23 What are the transcription factors participating in antigen-dependent T-cell activation, and how do they work?

NFAT is actually a family of molecules. Two representatives of this family, called NFATp and NFATc, are found in the cytoplasm of T lymphocytes. There, they are serine-phosphorylated and inactive; however, when calcineurin is activated as a result of antigen-triggered $[Ca^{2+}]_i$ increase (Section 7.12), it dephosphorylates NFAT and thus activates it. The activated NFAT is translocated into the nucleus and binds to regulatory sequences of the IL-2 (and also IL-4 and some other cytokine) genes. This binding is necessary but not sufficient to activate the IL-2 gene; the AP-1 and NFκB binding must supplement it. The importance of this transcription factor was recently underscored by the discovery that immunosuppressive drugs cyclosporine A and FK-506 form complexes with some cytosolic proteins and directly inhibit calcineurin. Such an inhibition results in a profound suppression of the T-lymphocyte activation and is extensively used in clinics to stop allograft rejection.

AP-1 is also a family of molecules that are protein dimers held together by the so-called "leucine zipper" (see molecular biology textbooks for details). Unlike NFAT, AP-1 proteins are not stored in the cytoplasm in their inactive form. For example, the representative of the AP-1 family that is featuring most prominently in the T-lymphocyte activation consists of two proteins, Fos and Jun. As mentioned in Section 7.20, Fos is a nuclear protein that is activated when phosphorylated by Elk protein, a product of the MAP kinase pathway of T signal transduction in T lymphocytes. On the other hand, Jun is a cytoplasmic protein but, after being phosphorylated by JNK (a component of the Rac·GTP, MAP kinase-related pathway of signal transduction), it translocates into the nucleus and binds Fos, forming the AP-1. After being thus formed, the AP-1 binds to regulatory sequences in IL-2 and other cytokine genes and acts in concert with NFAT and NFκB to activate them.

NFκB is a dimeric protein molecule that is homologous to the product of an oncogene called c-*rel*, and present in the cytoplasm. In resting T lymphocytes, it is covalently attached to a larger molecule called "inhibitor of κB" (**iκB**). When T cells are activated, iκB becomes phosphorylated; the precise mechanism of this phosphorylation is unknown but it seems to depend on the inositol phospholipid pathway of T-cell activation described in Section 7.21. The phosphorylated iκB is complexed with ubiquitin and targeted to proteasome (see Chapter 6), where the NFκB is cleaved from the iκB. When the NFκB is is thus released, it translocates into the nucleus and binds the regulatory sequences of the IL-2 gene, supplementing the action of NFAT and AP-1.

7.24 What other accessory molecules affect T-lymphocyte activation?

Indeed, many additional molecules expressed on the T-cell surface play the role of accessory molecules. Interaction of these molecules with the **co-stimulatory molecules** expressed on the antigen-presenting cells is necessary to achieve the T-lymphocyte activation.

The best described of the above pairs is the **CD28-B7** pair (Fig. 7-5). CD28 is a member of immunoglobulin superfamily that consists of two subunits and is expressed on >90% of $CD4^+$ and on approximately 50% of $CD8^+$ T

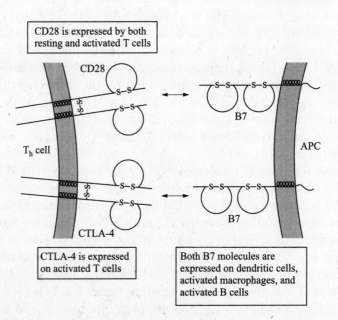

Fig. 7-5. Co-stimulation of T cells through CD28 and negative signaling through CTLA-4. See text for details.

CHAPTER 7 T-Lymphocyte Antigen Recognition

lymphocytes. B7 co-stimulators are expressed on activated professional antigen-presenting cells (dendritic cells, macrophages, and B lymphocytes) and exist in the form of two distinct molecules, B7-1 (CD80) and B7-2 (CD86). As we will discuss in more detail in the appropriate chapter, the TCR triggering in the absence of interaction between CD28 and B7-1 or B7-2 leads to profound and irreversible T-cell tolerance. The exact way in which CD28 triggering supplements or modifies the TCR-CD4/CD8-mediated signal transduction is subject of extensive research. Evidence indicates that when bound by B7-1 or B7-2, CD28 is clustered and activates the inositol phospholipid pathway of T-cell activation as well as the Ras and Rac pathways of T-cell activation. CD28 triggering activates a form of NFκB that binds to a site in the IL-2 gene promoter called "CD28 response element." Also, CD28 triggering seems to be necessary for the phosphorylation and function of ERK and JNK kinases, because the T cells that are triggered through their TCR complex in the absence of CD28 not only do not activate these enzymes, but lose the ability to activate them in response to subsequent "dual" TCR-CD28 triggering. A second receptor for B7-1/2, called **CTLA-4** (CD152), is structurally similar to CD28, but sends a clear inhibitory signal to the T cell upon its binding. This molecule is expressed predominantly on recently activated T lymphocytes, and is thought to prevent or suppress their further activation through blocking the phosphorylation of ζ chains. Another example of an accessory molecule that aids to T-cell activation is CD45, a phosphatase mentioned in Section 7.16. Several other molecules are important for T-cell activation, but their role in the TCR-associated signaling is less clear; it is believed that their primary function is to increase the adhesion between the T-cell and the co-stimulator-expressing, antigen-presenting cell.

7.25 If the "classical" (most prevalent in adult T lymphocytes) TCR is a heterodimer, consisting of the two polypeptide chains, α and β, what is a "nonclassical," "minority" TCR, and why is it needed?

The alternative TCR is called "γδ T cell receptor." Although this has an overall structure of a TCR, it consists of two distinct polypeptide chains, called γ and δ. Like the α and the β chains of the αβ TCR, the γ and the δ chains have C- and V-regions, but they are coded by genes distinct from those that code the α and the β chains of the "classical" TCR. Like the αβ TCR, the γδ TCR is also a heterodimer associated with the CD3 and ζ chains. (The γ and the δ chains of the γδ TCR should not be confused with the gamma and delta chains of the CD3.) The pathways of cellular signaling activated by the γδ TCR triggering are very similar to those activated by the "classical" αβ TCR triggering. Furthermore, the cornerstone of the TCR-mediated signal transduction in the γδ TCR-expressing T lymphocytes is the transcriptional activation of the IL-2 gene, and its mechanisms are similar to the ones described in Section 7.23.

There are some important features of the γδ TCR and the cells that expressed it, however, that make them differ from the conventional αβ TCR-expressing T lymphocytes. The expression of γδ TCR is abundant in fetal T lymphocytes, and diminishes very dramatically soon after birth. In adult humans and mice, less than 5% of all T lymphocytes express this receptor. In the adult, γδ TCR-expressing T lymphocytes are usually CD4- and CD8-negative, and tend to accumulate in particular anatomical locations (especially the epidermis and mucosal epithelium of the gut, tongue, and vagina). It is important to stress that the γδ TCR-expressing T lymphocytes are a *lineage* that is separate from the "conventional" αβ TCR-expressing lineage. A γδ TCR-expressing T lymphocyte cannot become an αβ TCR-expressing T lymphocyte, and vice versa. This can be deduced from the way the V-regions of the γδ TCR-expressing T lymphocytes are encoded genetically (see Fig. 7-4). The entire TCR γ chain V-region locus is positioned between the cluster of the TCR alpha chain V segments and the cluster of the TCR alpha chain J segments. Therefore, if the TCR alpha chain V-region gene segments rearrange, the entire TCR gamma chain V-region locus becomes deleted. Isolated TCR delta chains can be expressed together with the assembled alpha and beta chains of the αβ TCR, but they have no known function in this case.

The biological role of the αβ TCR and the γδ TCR-expressing T lymphocytes remains quite mysterious. Apparently, since they are so abundant in the fetus, they must play some role in the ontogeny of immunity, perhaps because they recognize some ubiquitously expressed fetal ligands. The diversity of the γδ TCR repertoire is very limited compared to the conventional T-cell repertoire. Finally, it has been shown recently that the γδ *TCR-expressing T lymphocytes do not recognize complexes of peptides and MHC*. Rather, they recognize some unprocessed nonpeptide molecules, like bacterial lipoglycans, presented by the nonpolymorphic MHC Class I-like molecule called CD1. The significance of this pattern of antigen recognition awaits further elucidation.

Questions

REVIEW QUESTIONS

1. In three to four sentences, describe how J. Allison and his co-workers used T-cell clones for the TCR identification.
2. In their experiments, S. Hedrick and M. Davis hybridized the cDNA prepared from a T-cell tumor with the mRNA extracted from a B-cell line. In your opinion, why did they bother to go through the step of cDNA preparation – would it not have been easier simply to hybridize the B-cell mRNA with the genomic DNA extracted from T cells?

CHAPTER 7 T-Lymphocyte Antigen Recognition

3. Outline the main differences between antibodies and TCR.
4. Using the data discussed in this chapter as well as in Chapters 5 and 6, briefly outline the structural relationships between MHC, peptide, and TCR.
5. Monoclonal antibodies to CD3 trigger a powerful proliferative response of lymphocytes *in vitro*, but if the cultures are rid of adherent cells, the response weakens. Explain this occurrence.
6. Murine T lymphocytes proliferate when phorbol myristate acetate (PMA) is added in their culture together with a calcium ionophore. However, when added separately, neither PMA nor the calcium ionophore induce a significant proliferative response of the cells. Explain this phenomenon in molecular terms.
7. Offer an experiment that would utilize gene transfection methodology and prove that there exists a "division of labor" within the TCR complex.
8. What exactly is the link between lck, Zap-70, adapter proteins and the MAP kinase pathway of T-cell activation?
9. Cadmium ions are known to be selective blockers of membrane calcium channels. How would you imagine the TCR-triggered signal transduction in T lymphocytes cultured in the presence of cadmium chloride?
10. What two functions of PKC are most crucial for T-lymphocyte activation?
11. Describe the role of Ras·GTP and Rac·GTP in the formation of active AP-1.
12. What is the role of ubiquitin in T-cell activation?
13. If active NFAT, active AP-1, and active NFκB are translocated in the nucleus of a T cell by some mechanism, will that allow the T cell to respond to TCR triggering in the absence of B7-1 and B7-2? Why, or why not?
14. Explain why T cells that express an αβ TCR can also express the delta chain, but not the gamma chain, of the γδ TCR.
15. Propose a hypothesis explaining why the γδ TCR-expressing T lymphocytes tend to accumulate in epithelial tissues.

MATCHING

Directions: Match each item in Column A with the one in Column B to which it is most closely associated. Each item in Column B can be used only once.

Column A

1. Idiotype
2. Vβ locus
3. Vγ locus
4. ε chain
5. ζ chain
6. Lck
7. LAT
8. Calcineurin
9. Elk
10. iκB

Column B

A. Adapter protein
B. Phosphorylates Fos
C. Present only on a given clone
D. Processed in the proteasome
E. Contains D segments
F. Part of CD3
G. Activates NFAT
H. Used by most fetal T cells
I. Associated with CD4
J. Has three ITAMs

CHAPTER 7 T-Lymphocyte Antigen Recognition

Answers to Questions

REVIEW QUESTIONS

1. They immunized mice with these clones and obtained hybridomas, which they then selected for the ability to bind the clonospecific epitopes (or idiotypic determinants). When such hybridomas were obtained, the antibodies that they produced were used to isolate the T-cell surface molecules that carried the idiotypic determinants. These molecules were characterized biochemically and shown to contain V- and C-regions homologous to, but distinct from, those contained in antibodies.

2. It would be easier, but it would not lead to the desired result. Genomic DNA of any eukaryotic cell contains all genes; therefore the genes that code for unique T-cell-specific molecules would not be identified. On the other hand, cDNA contains only those gene sequences that are expressed in a given cell.

3. Unlike antibodies, TCR are never secreted; show no isotypes or class-switch; possess a significantly lower affinity of their binding; their molecules are heterodimers that cannot be viewed as "heavy" and "light" chains because they are of a similar molecular weight; they do not mutate their V-region genes and show no affinity maturation; allelic exclusion in TCR V-region loci operates with a greater leniency.

4. Peptides bind to peptide-binding domains of MHC molecules, using their side chains to fit into peptide-binding pockets. Only very small portions of peptides (one to two amino acids) are contacted by V-regions of the TCR α and β chains, mostly by the CDR3. Simultaneously, the TCR V-regions contact the MHC molecule. In some TCRs, this contact is mediated by CDR1 and CDR2, and in others by all three CDRs.

5. The T cells need co-stimulatory molecules like B7-1 and B7-2, which are presented to them by dendritic cells and macrophages. The latter are adherent; therefore, if they are removed, the response weakens.

6. A plausible mechanism of this proliferation is activation of the IL-2 gene. This activation is achieved when PKC is strongly activated; perhaps PMA alone, in the absence of calcium release from the intracellular depots, is unable to fully activate PKC. Similarly, an increase in $[Ca^{2+}]_i$ alone, without DAG or its mimic, cannot fully activate this enzyme.

7. Isolate TCR genes from a clone with a known peptide-MHC specificity. Co-transfect a cell line either with these TCR chain genes only, or with CD3 and ζ chains only, or with both TCR and CD3 + ζ. Neither the first nor the second transfectant will respond to the original MHC–peptide complex, but the third transfectant will.

8. Lck phosphorylates ζ chains, thus activating Zap-70; the latter activates adapter proteins; these activate Ras and Rac, which initiates the MAP kinase cascade.

9. There will be a brief increase of $[Ca^{2+}]_i$ due to the IP3-induced release of Ca^{2+} from intracellular depots, but it will not be sustained if Ca^{2+} channels are blocked.

10. Its ability to act synergistically with Ca^{2+} ions in the induction of cellular responses, and its indirect influence of the iκB phosphorylation.

11. Ras·GTP activates ERK kinase, leading to phosphorylation and nuclear translocation of Elk protein; the latter, when translocated, activates c-*fos* gene and thus helps the production of Fos protein. Rac·GTP activates JNK kinase, leading to phosphoryla-

CHAPTER 7 T-Lymphocyte Antigen Recognition

tion and nuclear translocation of Jun protein. When Fos and Jun proteins meet in the nucleus, they form a dimer, which is AP-1.

12. Without ubiquitin, iκB would not be cleaved in the proteasome, and active NFκB would not be formed. This, in turn, would preclude the transcriptional activation of the IL-2 gene.
13. No, because for the IL-2 gene to be fully activated, the CD28 response element in its regulatory region must be occupied, and this cannot happen in the absence of CD28-B7 interaction.
14. Because the Vγ locus is positioned between the Vα and the Jα gene clusters and is deleted when the Vα locus rearranges. The Vδ is located elsewhere and remains intact.
15. They may recognize some ubiquitous antigen that is present at the border between the organism's tissues and its environment.

MATCHING

1, C; 2, E; 3, H; 4, F; 5, G; 6, I; 7, A; 8, G; 9, B; 10, D

CHAPTER 8

B-Lymphocyte Activation and Antibody Production

Introduction

T and B lymphocytes are the two indispensable "arms" of the immune system. Together with accessory cells and other nonlymphoid cells, they mount an efficient immune response to a wide variety of antigens, including dangerous microbes. In previous chapters, we showed how the diversity of the B- and T-lymphocyte repertoire is generated. We also discussed how T lymphocytes are activated after they recognize antigens. In this chapter, we will see how B lymphocytes are activated after they recognize antigens and, importantly, what signals from helper T lymphocytes are necessary to drive the B-cell activation after the recognition of protein antigens. In fact, the concept of T helper cells and the concept of "**T cell help**" that has been introduced in Chapters 1 and 2 will be further analyzed in precise cellular and molecular terms. We will show why the T cell help is needed, and how is it being generated, delivered, received, and responded to.

Like T lymphocytes, B lymphocytes can be activated by the combination of antigen and co-stimulators (signal 1 plus signal 2). However, there is a profound difference between B and T lymphocytes in the exact way of antigen signal delivery. Unlike TCR, individual antibody molecules on the surface of a B lymphocyte must be **cross-linked** by the antigen. Many antigens, however, cannot cross-link surface antibody molecules. As we will see, the meaning of the T cell help is,

CHAPTER 8 B-Lymphocyte Activation

essentially, to bypass this requirement of surface antibody cross-linking. In this regard, we will see why certain antigens need this help and are therefore called **T-cell-dependent**, while other antigens do not need it and are called **T-cell-independent**. We will also discuss in depth, how the signal transduction in T and B lymphocytes during their activation by T cell-dependent antigens leads to B-cell proliferation and differentiation.

Discussion

8.1 How exactly did immunologists develop the idea that B lymphocytes may need T lymphocyte help?

This idea is quite old, and its beginning in fact preceded the classification of lymphocytes into T and B cells. As early as the late 1960s, immunologists began to analyze antibody production with the help of an **adoptive transfer** technique (see also Chapter 1). In these experiments, sublethally irradiated mice were injected with cells derived from various sources. The recipients who had "adopted" these transferred cells were then immunized with sheep red blood cells, and the antibody response of the "adopted" cells to the antigen was monitored. A crucial experiment performed by J.F.A.P. Miller and G. Mitchell, who showed that adoptively transferred rat thoracic duct or bone marrow lymphocytes did not respond to sheep red blood cells, but a mixture of the two cell types did. Miller and Mitchell offered a simple and very far-sighted explanation for their data. Thoracic duct lymphocytes are mostly of the thymic origin (T cells), and bone marrow-derived lymphocytes contain little or no cells of the thymic origin but express antibodies on their surface (B cells). Based on this fact, Miller and Mitchell proposed that B lymphocytes are activated and produce antibodies to sheep red blood cell antigens when they receive a "help" from T lymphocytes.

8.2 Do B lymphocytes always need T cell help?

No. As we mentioned in the introduction to this chapter, some antigens are T-cell-independent. Again, the notion about T-cell-dependent and -independent antigens came to being even earlier than the classification of lymphocytes into T and B became a convention. Immunologists observed that in mutant animals or humans who had a congenital aplasia (dramatic underdevelopment) of the thymus, immune responses to many antigens are missing, but responses to some antigens develop normally. For example, in the mice that carry the so-called "**nude**" mutation (nu/nu mice), the immune responses to sheep red blood cells and a variety of protein antigens cannot be triggered; yet, these mice produce high titers of antibodies when immunized with polysaccharides, glycolipids, or DNA. Apparently, T cell help is not needed for the response to these antigens to develop. These antigens were called T-cell-independent antigens, and the antigens that needed T cell help (mostly protein antigens) were called T-cell-dependent.

8.3 What structural features make antigens T-cell-independent or T-cell-dependent?

As we have already mentioned, they are mostly polysaccharides, glycolipids, and DNA. The feature that unites them and makes them T-cell-independent is the following: **they all have a "cartridge" of repetitive epitopes that follow each other on the molecule in close proximity.** For example, a carbohydrate structure Gal alpha 1-3Gal beta1-4GlcNAc-R (where Gal stands for galactose, Glc for glucose, Ac for acetyl group and R for a radical), termed the alpha-galactosyl epitope, is repeated many times in capsular polysaccharide antigens of various encapsulated bacteria. It is thought that because of these structural features, T-cell-independent antigens present closely positioned identical epitopes and allow them to be bound by closely positioned identical antibody molecules on the B-lymphocyte surface. As we will discuss later, this pattern of epitope binding may send a signal that, under certain conditions, will activate B lymphocytes in the absence of T cell help.

Most T-cell-dependent antigens are proteins that acquire a certain tertiary structure by folding their polypeptide chains in spirals, sheets, or globules. Because of this natural folding, an individual protein molecule usually presents only one epitope for antibody binding. This makes surface antibody cross-linking difficult or impossible. In the absence of the latter, B lymphocytes cannot be fully activated unless they receive a T cell help.

8.4 Why is surface antibody cross-linking so important for B-cell activation?

The answer lies in the way the signal from surface antibody is transduced inside the B lymphocyte. Membrane IgM and IgD, which serve as B-cell antigen receptors, do not transduce the signal by themselves. Like TCR chains (see Chapter 7), the μ and δ chains have very short cytoplasmic tails and cannot associate with cytoplasmic signal-transducing enzymes. Surface antibody-mediated signals are actually transduced by other molecules, called **Igα** and **Igβ**. These have long cytoplasmic tails that contain the same ITAMs that are present in the cytoplasmic portions of TCR-associated CD3 and ζ chains (see Chapter 7). A distinctive feature of further signaling in B lymphocytes is that for the surface antibody-mediated signal to be transduced, several ITAMs must be brought into proximity. This makes simultaneous triggering of more than one surface antibody molecule a must.

8.5 What signal transduction events follow the cross-linking of surface antibodies?

Generally, these events are not very different from those associated with T-lymphocyte activation and described in Chapter 7. The earliest event is the phosphorylation of juxtaposed Igα and Igβ chains on their ITAMs, which is thought to be mediated by Src family PTK like Fyn, Lyn, and Blk (see Chapter 7). Once these chains are phosphorylated, they begin to serve as a "docking site" for a B-lym-

CHAPTER 8 B-Lymphocyte Activation

phocyte-specific PTK called **syk**. The latter is a B-lymphocyte analogue of the T-cell-specific Zap-70 (Chapter 7). When syk binds to the phosphotyrosine residues of the phosphorylated Igα and Igβ, it either is phosphorylated by Fyn, Lyn, and Blk, or phosphorylates itself, and becomes active. When activated, syk, together with other B-cell antigen receptor-associated tyrosine kinases activate several downstream signaling molecules. One of these is PLCγ1, which is the same enzyme as the one operating in T lymphocytes (Chapter 7). PLCγ1 catalyzes the breakdown of PIP2, generating IP3 and DAG (Chapter 7). IP3 mobilizes Ca^{2+} from intracellular depots, and Ca^{2+} facilitates the membrane translocation of cytoplasmic PKC, while DAG activates the translocated PKC. In parallel, syk, through some intermediates that are not yet fully identified, activates Ras and the MAP kinase transduction pathway. The net result of all these events is the transcriptional activation of a group of genes, and the entrance of the B lymphocyte into the cell cycle.

8.6 If the principle of the necessity of two signals is true for B lymphocytes, what constitutes the second, or co-stimulatory, signal for their activation?

The current notion is that there exists a **B-cell co-receptor complex**, which in a way is similar to an "aggregate" of accessory molecules that deliver the necessary "second signal" for B-cell activation. This co-receptor complex consists of four parts. The first of these is a molecule called **CR2** (or **CD21**), which serves as a receptor for one of the principal breakdown products of complement. (See Chapter 1 for a definition of the complement and later chapters for details of complement functions.) CR2 is a receptor of a protein called **C3d**, which is generated by the proteolysis of the key component of the complement system, called **C3**. Such proteolysis, as detailed later, can occur as a consequence of the C3 binding to some bacterial cell wall proteins, or as a consequence of its binding to pre-existing antibodies bound to their specific antigen (antigen–antibody complexes). The resulting C3d must be bound to the antigen to which the B cell is specific, or to the antigen–antibody complex in order to be recognized, and bound, by CR2.

The second component of the B-cell co-receptor complex is a B-cell-specific surface molecule called **CD19**. Immediately after the C3d is attached to CR2, their complex forms an association with CD19, and the latter becomes phosphorylated on its ITAM, which is within the CD19's cytoplasmic tail. This phosphorylation is mediated by PTKs associated with the B-cell antigen receptor complex, especially by syk. It greatly enhances the activator signal transduced after the B-cell antigen receptor triggering.

The third component of the B-cell co-receptor complex is a small, ubiquitously expressed molecule called **TAPA-1**, or **CD81**. TAPA-1 belongs to the so-called **tetraspanin** family of molecules, the distinctive feature of which is that these proteins are "curled" so that they span the plasma membrane four times. The fourth component of the co-receptor complex is a molecule that is also small and not limited to B cells in its expression. This is called **Leu-13**. The principal role of

TAPA-1 and Leu-13 in the B-cell co-receptor complex is thought to be to facilitate the adhesion between B lymphocytes (homotypic adhesion) or between B lymphocytes and nonlymphoid stromal cells (heterotypic adhesion). These proteins also play a certain role in cognate T–B-cell interaction (see later sections).

8.7 What are the functional consequences of the signaling through the B-cell antigen receptor and the B-cell co-receptor complexes?

After the signal transduction pathways are fully activated and their activity results in transcriptional activation of a group of genes by NFAT, NFκB, Ap-1, and other factors, the B lymphocyte enters the G_1 stage of the cell cycle. This is accompanied by an increase in cell size, synthesis of new RNA, and an increase in the number of cellular organelles (notably ribosomes). Very importantly, at this stage **the B lymphocyte up-regulates the expression of co-stimulatory molecules**. Of these, the earliest up-regulated molecule is B7-2 (CD86). Its enhanced expression is followed by an increase in the expression of B7-1 (CD80) after several hours. The expression of cytokine receptors, is also increased at this point. Finally, the B lymphocyte activated through the B-cell antigen receptor and the B-cell co-receptor complexes **strongly up-regulates the expression of MHC Class II molecules**. This is a very important factor for the B-cell-mediated antigen presentation, which provides a link with T-lymphocyte activation and T–B-cell interaction, and will be discussed next.

8.8 How do B cells present antigen?

Indeed, B lymphocytes are very powerful antigen-presenting cells. Essentially, they present peptides generated in their endosomal-lysosomal (or acidic vesicular) compartment and displayed in the context of MHC Class II molecules (see Chapter 6). However, unlike macrophages or dendritic cells, **B lymphocytes process and present antigens that are first recognized by variable regions of their surface antibodies and then internalized by receptor- (antibody-) mediated endocytosis.**

8.9 What are the main features of this antibody-mediated processing and presentation?

This phenomenon was best studied on haptens conjugated to protein carriers and exposed to hapten-specific B lymphocytes. In this case, the presentation of the antigen to the antigen-specific T lymphocytes by hapten-specific B lymphocytes is, first of all, extremely sensitive: 10^4 to 10^6 smaller concentrations of the hapten–carrier conjugate are required to activate the T cells compared to the situation when the antigen-presenting cell is a macrophage. Second, although B lymphocytes recognize the hapten, they process the carrier and present peptides generated from the protein carrier. Nevertheless, the T cells presented with such peptides exert help to the hapten-specific B cells.

CHAPTER 8 B-Lymphocyte Activation

8.10 How exactly do the T lymphocytes exert their "help?"

In the 1970s, it was shown that the presence of T lymphocytes augments antibody responses of B lymphocytes that share the antigenic specificity (cognate interaction) as well as the MHC Class II allele (MHC Class II restriction). As for the exact molecular mechanism of such an augmentation, it remained unknown for a rather long time. Evidence indicated, however, that the cognate T cell help consists of two parts. One part of it is a signal or signals delivered by **cytokine**(s), while the other part is a signal transmitted from T to B lymphocytes during their **contact**. Two lines of evidence supported this notion. First, T and B cells specific to the same protein antigen do not show signs of cognate interaction when separated by a membrane that allows cytokines to diffuse but precludes a membrane-to-membrane contact of the T and B cells. Second, the antibody response of B lymphocytes can be strongly augmented if, in addition to cytokines, they are exposed to a membrane fraction of activated T lymphocytes.

8.11 What component(s) of the T-cell membrane is important for T cell help delivery to B cells?

The protein that plays the most important role in the T-cell-mediated enhancement of B cell responses is called **CD40 ligand**. This molecule was discovered after several years of search for the ligand of **CD40**. The latter is a transmembrane glycoprotein with a molecular weight of 45 kDa that is constitutively expressed on B lymphocytes (as well as on dendritic cells, macrophages, and some neurons). Antibodies to CD40 were shown to cause a weak proliferative response of B cells, and to strongly enhance their proliferative response to antigen-mimicking anti-immunoglobulin antibodies. On the other hand, these antibodies very strongly stimulated proliferation and differentiation of B lymphocytes in combination with a cytokine produced by activated T lymphocytes, IL-4 (see later chapters for the characterization of this cytokine). Based on these data, it was suggested that anti-CD40 antibodies mimic a natural ligand of the CD40 molecule, and that this ligand is likely to be expressed on activated T lymphocytes. It has been shown that the CD40 ligand (also called CD40L or CD154) transiently appears on the membrane of the activated T lymphocyte and, upon binding to the CD40 expressed on the nearby B lymphocyte, delivers a powerful activator signal to the B cell.

8.12 How is the CD40-mediated signal transduced, and how does the CD40-expressing cell respond to it?

CD40 belongs to a large family of molecules that includes, among other members, the tumor necrosis factor (TNF) receptor and a protein called Fas (see later chapters for their description). CD40L belongs to a related family of molecules that includes TNF and Fas ligand. All members of both families are responsible for a balance between cellular survival and programmed death. The signaling pathway used by both CD40 and TNF receptor begins after binding of these molecules by their respective ligands, and their oligomerization. This is followed by an association of cytoplasmic proteins, called TNF receptor-associated factors (**TRAF**). To date, six different TRAF proteins have been identified, of which the so-called TRAF2 and TRAF3 interact with CD40. After TRAF2 or TRAF3 are recruited to the cytoplasmic portion of the CD40, they bind and activate a cytoplasmic serine-threonine kinase called "receptor interacting protein" (**RIP**). This is followed by activation of the transcription factors mentioned in Chapter 7 and in earlier sections of this chapter: TRAF2 activates JNK kinase and (through the latter) AP-1, while RIP activates NFκB. The activated transcription factors are thought to trigger the expression of a variety of genes, but the details of their action in CD40-activated B lymphocytes are still unknown.

8.13 What are the ultimate consequences of CD40 signaling?

As already mentioned, triggering of the CD40 alone can cause a weak (but detectable) proliferation of B lymphocytes. However, this response is short-termed, and unless signaling through the B-cell antigen receptor complex or IL-4 receptor supports the CD40 triggering, leads to apoptotic death of the CD40-triggered B lymphocytes. When supported by signaling through the B-cell antigen receptor complex or IL-4 receptor, however, the CD40-mediated activation leads to a strong **increase in the expression of co-stimulatory molecules** like B7-1 and B7-2 and of the MHC Class II molecules, in arrest of the apoptotic programs and, ultimately, to B-cell **clonal expansion** and increased **antibody secretion**. Also, the combined signaling through CD40 and cytokine receptors leads to antibody class-switch.

8.14 If the engagement of surface antibody or IL-4 receptor rescues the B lymphocyte from CD40-mediated cell death, what is the mechanism of such a rescue?

Indeed, the signaling through surface IgM or IL-4 receptor does rescue the B cell. The mechanism of this rescue is not completely elucidated, but some details were recently revealed. It has been shown that the engagement of CD40 by CD40L strongly increases the expression of **Fas** (also known as CD95). The latter is structurally related to CD40, but triggers a different signaling pathway when bound by its ligand, called **Fas ligand**. This ligand is expressed by activated T lymphocytes, especially by the cells that belong to the subset of T_H1 lymphocytes (see Chapter 10). Binding of the Fas ligand to Fas triggers the assembling of a group of cytoplasmic proteins collectively called **FADD** (Fas-associated death domain). The assembled FADD bind a number of adaptor proteins and ultimately activate an enzyme called **caspase-8**. The latter activates a number of **effector caspases** that ultimately cause apoptotic cell death by destroying the nucleus and its DNA. The parallel signaling through B-cell antigen receptor complex

CHAPTER 8 B-Lymphocyte Activation

and IL-4 receptor, however, seems to create a state of *resistance to the Fas-mediated killing*. The details of this mechanism are not known, but it does not seem to be connected to a reduction in the Fas expression. Rather, it includes intracellular processes. It has been shown, for example, that the engagement of surface IgM increases the resistance to Fas-mediated killing through phosphorylation of some isoforms of PKC and other protein kinases.

8.15 What other molecules, besides CD40L, are important for contact-mediated T cell help?

A number of other molecules that belong to the TNF-TNF receptor family of proteins also play certain (though as yet not well-defined) roles in **the T–B-cell interaction** (or "**cross-talk**"). Among them are CD30 and OX40, which are T-cell molecules, and their B-cell ligands, CD30 ligand and OX40 ligand; the so-called nerve growth factor receptor (which is expressed on neurons and T lymphocytes), lymphotoxin receptor, etc. The feature that unites all these molecules is their ability to recruit TRAF proteins through their cytoplasmic tails and thus to initiate cascades of intracellular signaling, ultimately leading to further activation of lymphocytes. In addition, a number of the so-called **adhesion molecules** plays an important role in the T–B-cell cross-talk.

8.16 What are adhesion molecules, and what do they do?

The main function of these molecules, as can be seen from their name, is to ensure that a T lymphocyte and a B lymphocyte (or other antigen-presenting cell, or a nonlymphoid cell) remain in close contact, adhering or "sticking" to each other. There are three major classes of adhesion molecules (Fig. 8-1): integrins, selectins, and CD44. Of these three, the **integrins** are most essential for T–B-cell cross-talk, while selectins and CD44 are more important for lymphocyte homing and migration, and will be discussed in the appropriate chapters. The integrin superfamily includes approximately 30 heterodimeric proteins that consist of noncovalently linked α and β chains. The extracellular regions of these chains end with globular "heads" that bind various ligands found on other cells, or in the extracellular matrix. The cytoplasmic regions of these chains are always associated with components of cytoskeleton (including actin, α-actinin, vinculin, tropomyosin, and other components). The name of this large family of homologous molecules comes from the hypothesis that because of their association with cytoskeleton, they coordinate or "integrate" the binding of extracellular ligands and the subsequent changes in the shape and motility of the cell that expresses them. The most important feature of the adhesion mediated by integrins is that *it is weak in the absence of the TCR occupancy, but dramatically increases when the TCR is occupied* by the peptide–MHC complex presented by the B cell (or other APC). Since the TCR-mediated signal is sent inside the T cell, but affects the properties of the surface (i.e. "outside") adhesion molecules, this mode of regulation is called "the **inside-out signaling**."

CHAPTER 8 B-Lymphocyte Activation

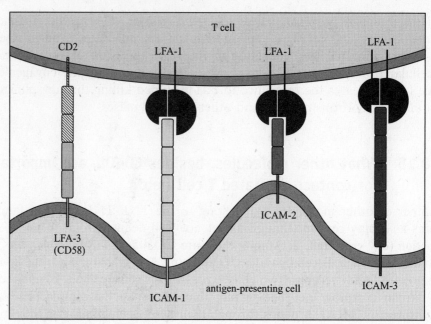

Fig. 8-1. Adhesion molecules. This figure shows interactions between some important adhesion molecules expressed on antigen-presenting cells (including B lymphocytes) and their counter-receptors on T lymphocytes.

The best-studied integrin involved in the T–B-cell cross-talk is called **LFA-1** (or CD11aCD18). It is expressed on both T and B lymphocytes, although primarily, it functions as a T-cell-associated adhesion molecule. Its main ligand on the B lymphocyte surface is a molecule called **ICAM-1** (CD54); this molecule is also expressed on T cells and on a variety of nonlymphoid cells. The LFA-1–ICAM-1 interaction seems to be of a major importance for the establishment of a firm adhesion between the interacting T and B cells, and also between T lymphocytes and other antigen-presenting cells. It is thought that before the TCR is occupied by the peptide presented by the B cell's MHC molecule, T and B cells interact through LFA-1–ICAM-1, which leads initially to weak adhesion. As soon as the TCR is occupied, this adhesion grows dramatically, allowing the T and B cells to stay close together for a long time. That greatly facilitates the interaction of CD40L with CD40, and other molecular interactions described in previous sections. Another subgroup of the integrin family, **VLA** proteins, are more important for the binding of T lymphocytes to endothelial cells (see Chapters 11 and 12).

8.17 Do adhesion molecules transduce their own signals?

Currently, it is believed that they do transduce certain activator signals, but the role of these signals for the B-cell activation is minor compared to the role of B-cell receptor complex- and CD40-mediated signaling. Perhaps it would be correct to say that the major function of the adhesion molecules during B-cell activation is simply to help win time that is necessary for other molecules to deliver and transduce their signals.

CHAPTER 8 B-Lymphocyte Activation

8.18 Are there molecules and signals that inhibit B-cell activation?

Yes. The principal mediator of signals that inhibit B-cell activation is the so-called type IIB receptor for the Fc portion of IgG antibody (**FcγRIIB**, or **CD32**). When secreted IgG antibodies bind antigen, and the forming antigen–antibody complexes are simultaneously bound to the FcγRIIB, the B cell activation stops. This phenomenon has been known for a rather long time and is called **antibody feedback**. The mechanism of the antibody feedback (Fig. 8-2) is related to the ability of a cytoplasmic region of the FcγRIIB molecule to "dock" two major inhibitory enzymes. One of these enzymes is called **SHP-1**; it serves as a protein tyrosine phosphatase, and removes phosphates from activator signaling molecules, thus impairing their function. The other enzyme is called **SHIP**; it is a direct inhibitor of PIP3 (see Chapter 7). When a secreted IgG forms a complex with the antigen, which simultaneously binds the B-cell antigen receptor (through the antigen) and the FcγRIIB (through the IgG's Fc fragment), it brings SHP-1 and SHIP close to the B-cell antigen receptor complex, thus facilitating the inhibition of its positive signaling.

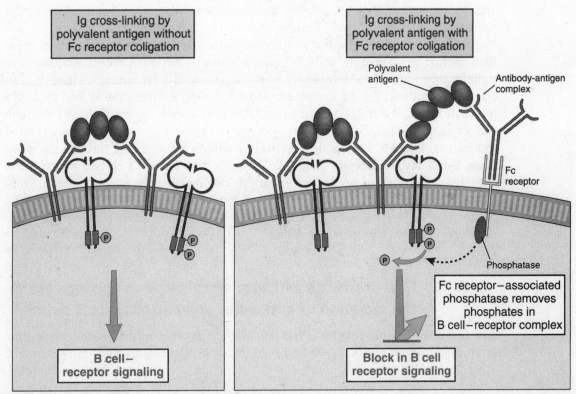

Fig. 8-2. Antibody feedback. Co-ligation of the B-cell antigen receptor complex and the antibody bound to the FcγRIIB inhibits B-cell activation because of the inhibitory nature of the FcγRIIB-associated phosphatase. From Abbas, K., A.H. Lichtman, and J.S. Pober, *Cellular and Molecular Immunology*, fourth edition, W.B. Saunders, 2000, p. 206 (their Fig. 9-17).

8.19 How are B lymphocytes activated by T-cell-independent antigens?

When T-cell-independent antigens cross-link surface antibodies, the B cells are activated by the signaling through the B-cell antigen receptor complex and the CR2-CD19-TAPA-1-Leu13 complex, as described in Section 8.6. This second signal can indeed be delivered, because many T-cell-independent antigens (especially polysaccharides expressed on bacterial cells) are able to activate complement through the alternative pathway and thus to generate C3d (see Section 8.6). Also, some circumstantial evidence indicates that during responses to T-cell-independent antigens, B lymphocytes still do interact with some T lymphocytes and are affected by some cytokines. The precise nature of these T lymphocytes and cytokines is not known, however.

8.20 Is there any difference in the way the two major subsets, B-1 and B-2 are activated?

Yes. B lymphocytes that belong to the "minority" B-1 subset produce antibodies that bind to a wide variety of antigens, usually with low affinity. Although their exact function *in vivo* is not completely elucidated, they are thought to react with, and to be naturally activated by, some ubiquitously expressed environmental agents, for example, polysaccharides of bacteria that colonize the gastrointestinal tract early in life. The signaling requirements for their activation differ from those that operate in "conventional" B-2 lymphocytes. It has been recently shown that signaling through surface antibody in adult splenic B-1 lymphocytes leads not to cellular activation, but to apoptosis. However, in the mice that lacked the CD5 molecule (see Chapter 2) because of the CD5 gene knockout, the signaling through surface antibody led to the activation of the B-1 cells. These data were interpreted to mean that **the CD5 molecule negatively regulates the signaling through the membrane B-cell antigen receptor complex**. It may be that *in vivo*, a certain ligand binds the CD5 molecule and prevents it from exerting its negative effect of the antigen-mediated signaling. The exact role of T lymphocytes, CD40L-CD40 signaling pathway, cytokine signaling and Fas-Fas ligand interactions in the B-1 cell responses has not yet been elucidated.

8.21 How exactly do activator signals lead to the increase in the secretion of antibodies and antibody class switch?

The fact that B lymphocytes activated by their specific antigen begin to secrete large amounts of antibodies was firmly established decades ago. However, the exact molecular mechanisms of this phenomenon are still not fully understood. Obviously, since anti-CD40 antibodies and CD40 ligand, especially in combination with IL-4, enhance the secretion of antibodies, the CD40L-CD40 pathway of signaling is involved in this phenomenon. Recently, the differential role of various TRAF proteins and cytokines involved in the promotion of antibody secretion

CHAPTER 8 B-Lymphocyte Activation

began to be dissected. It was shown that TRAF proteins called TRAF3 and TRAF6 are especially important, and that TRAF6 enhances antibody secretion by stimulating the production of the cytokine called IL-6 by CD40-activated B lymphocytes (see Chapter 10). The precise molecular mechanism that underlies the effect of these factors is not known, but it is thought to be related to their influence on the process of alternative mRNA splicing similar to that described in Chapter 4 (see sections about simultaneous IgM and IgD expression). This process allows a simultaneous expression of the membrane-bound and secreted forms of the µ chain. The signaling through TRAF3, TRAF6, and IL-6 perhaps results in an increase of the proportion of mRNA transcripts where the exon responsible for the µ chain membrane embedding is deleted.

The CD40L-CD40 signaling pathway also seems to be responsible for the switch to downstream antibody isotypes, and various TRAF proteins as well as cytokines have been implicated in this phenomenon. Again, the details of the mechanism are not understood, but it is generally agreed that while the CD40L-CD40 signaling pathways stimulate the class switch in general, **the cytokines give it its particular direction**. As will be detailed in the chapter about cytokines, IL-4 is known to direct the class switch toward the production of IgE, while interferon γ (IFN-γ) tends to stimulate the switch to IgG2a (at least in mice). Again, the above concerns mostly conventional (B-2) lymphocytes that respond to T-cell-dependent antigens. B-1 lymphocytes and B lymphocytes that respond to T-cell-independent antigens switch to downstream antibody isotypes rarely, and the molecular mechanisms responsible for that are largely unknown.

Questions

REVIEW QUESTIONS

1. Explain in two to three sentences, why *H. influenzae* polysaccharide can activate B lymphocytes directly, while tetanus toxoid cannot.
2. What structures in T lymphocytes are homologous to Igα and Igβ, and which feature of signaling through those homologues is not characteristic to Igα and Igβ?
3. A protein called decay-accelerating factor (DAF, see Chapter 13 for details) is a powerful inhibitor of complement activation. What can you predict about immune responses in individuals who overproduce DAF?
4. You want to grow an antigen-specific T-lymphocyte clone (see Chapter 7). Propose a protocol of maintaining such a clone in such a way that only B lymphocytes serve as the cells that present the specific antigen to this clone.
5. Recall from Chapter 6 that chloroquine and peptide aldehydes can be used to block antigen processing. Which of these two agents will impair the cognate T–B-cell interaction?

CHAPTER 8 B-Lymphocyte Activation

6. A fibroblast cell line transfected with an unknown gene has the following properties: it induces the growth and antibody secretion of B lymphocytes in culture and, in the presence of IL-4, stimulates the production of IgE. What is the gene that was used for transfection? What direct test would you use to prove the nature of the gene and its product?

7. Briefly explain how the signaling through CD40 can simultaneously induce B-cell activation and programmed B-cell death. What genes would you propose to knock out in mice to study the relationship between these two processes?

8. What will happen to immune responses of mice transgenic for caspase-8?

9. Human B lymphocytes are incubated with monoclonal mouse anti-human immunoglobulin antibody and polyclonal rabbit anti-mouse immunoglobulin antibody. The measurement of $[Ca^{2+}]_i$ shows that it is significantly elevated. If, however, the above two reagents are added together with an anti-LFA-1 antibody, no increase in $[Ca^{2+}]_i$ is registered. Explain this situation.

10. Design an experiment aimed at testing whether OX40 is functionally similar to CD40L.

11. If a mouse is injected with sheep red blood cells (SRBC), it will develop a strong specific antibody response, which can be quantified in a hemolytic plaque assay. The number of plaques (sites of local hemolysis) in dishes with the immune mouse's lymphocytes and SRBC suspended in agar will be equal to the number of specific SRBC antibody-producing cells. If a mouse is injected with SRBC together with anti-SRBC IgG, the number of plaques will increase compared to the response of the mouse immunized with SRBC only. If, however, the mouse is injected with SRBC and anti-SRBC IgM, the plaque response will increase. Explain these findings.

12. Splenectomy (surgical removal of the spleen) often results in an increase of acute and chronic infections caused by encapsulated bacteria. Can you explain why?

13. A subpopulation of B-1 cells, called "B-1b" or "sister cells," contains mature CD5 mRNA transcripts, but does not translate the CD5 protein. How would you use these cells to study the outcomes of CD5 signaling in B lymphocytes?

14. J. Banchereau's laboratory showed in the mid-1990s that if B lymphocytes are cultured in the presence of CD40L and IL-10 (a T-cell-derived cytokine), they become large, egg-shaped, and secrete very high numbers of antibodies. If, however, IL-10 is replaced by IL-4, the cells become smaller, hand mirror-shaped, and secrete moderate numbers of antibodies. What pathways of B-cell differentiation does this remind you of?

15. A protein called BclX is a strong inhibitor of apoptosis. Propose experiments evaluating its role in B-lymphocyte activation.

MATCHING

Directions: Match each item in Column A with the one in Column B to which it is most closely associated. Each item in Column B can be used only once.

Column A

1. Glycolipid
2. Polypeptide
3. Igβ
4. C3
5. Leu-13

Column B

A. Gives class switch a direction
B. Ensures T–B-cell adhesion
C. Creates a "dock" for syk
D. A phosphatase
E. T-cell-dependent antigen

CHAPTER 8 B-Lymphocyte Activation

6. TRAF3
7. Integrin
8. FcγRIIB
9. SHP-1
10. Cytokine

F. Binds IgG
G. The precursor of C3d
H. Binds CD40
I. Part of B-cell co-receptor complex
J. T-cell-independent antigen

Answers to Questions

REVIEW QUESTIONS

1. The polysaccharide presents a "cartridge" of identical epitopes and thus can cross-link surface antibody molecules, bringing the ITAMs of the associated Igα and Igβ into proximity. This is sufficient to initiate the cascade of activator signaling. The tetanus toxoid is a protein and, as such, folds so that only one epitope per molecule is presented; this is insufficient to cross-link surface antibodies and to initiate B-cell signaling.

2. CD3 and ζ chains. The necessity to bring their ITAMs into proximity.

3. There will be few or no responses to T-cell-independent antigens, because there will be not enough C3d to activate the B cell through the B-cell co-receptor complex (signal 2). The responses to T-cell-dependent antigens will likely remain intact, because signaling through the B-cell co-receptor complex is not needed for their initiation.

4. Immunize a mouse with an antigen of choice. After 7–10 days, isolate T lymphocytes and accessory cells from the spleen. From the latter, purify surface antibody-positive B cells by cell sorting, and use them as the antigen-presenting cells.

5. Chloroquine, because it affects the acidic vesicular compartment (the pathway of antigen processing that leads to the joining of peptides with MHC Class II molecules).

6. CD40L (or CD154). Antibody specific to CD40L should abrogate the B-cell growth and antibody production.

7. The default pathway of CD40 signaling (in the absence of B-cell antigen receptor signaling) leads to the up-regulation of Fas and its binding by Fas ligand; the parallel activator signaling through CD40 and TRAF is not strong enough to overcome this effect. The CD40-mediated activation of B lymphocytes can take place only if the cell is rendered resistant to Fas-mediated killing, which is exactly what the signaling through the B-cell antigen receptor and cytokine receptors does. The genes to be knocked out for the analysis of this phenomenon are CD40, CD40L, Fas, FasL, a number of genes that code for PTK, adaptor proteins (including precursors of FADD), and transcription factors.

8. Perhaps they will be indistinguishable from normal, for two reasons: (1) The gene codes an inactive form of caspase-8, which becomes active only after it "docks" to the assembled FADD. (2) Even if an active form of the enzyme is made, its negative effect on lymphocyte survival will be counteracted by the signaling through the antigen–receptor complex.

CHAPTER 8 B-Lymphocyte Activation

9. If the antibody prevents LFA-1 from binding to ICAM-1, the lymphocytes cannot form clumps (adhere to each other). It may be that in the absence of homotypic adhesion, PLCγ1 is not activated strongly enough, and the phospholipid breakdown does not occur even if surface antibodies are cross-linked.
10. An interaction between OX40-expressing T lymphocytes should lead to B-cell proliferation and differentiation. An antibody to OX40 should prevent the class-switch to IgE in the presence of OX40 protein and IL-4.
11. Because the IgM binds complement (thus activating the B-cell co-receptor complex), but does not bind FcγRIIB. The IgG may also bind complement, but they simultaneously bind FcγRIIB and thus inhibit the signaling (antibody feedback).
12. Because the marginal zone of the spleen is a unique anatomic site where polysaccharide antigens are stored and recognized by B cells. Most surface antigens in encapsulated bacteria are polysaccharides.
13. First, one can study B-cell antigen receptor signaling in these cells. If it leads to B-cell activation, that would mean that the CD5 protein and not RNA (or the transcriptional activity of the gene) serves as a negative signaling protein in B-1 lymphocytes. If, however, the above signaling in "sister" B cells leads to apoptosis, one can suggest that it is the transcriptional activity of the CD5 gene that plays the negative regulatory role. To test this hypothesis, one might attempt to transfer the CD5 mRNA to B-2 lymphocytes and study the above signaling in these cells.
14. The first mimics plasma cell differentiation; the second, the memory cell differentiation pathway.
15. Test its expression in B lymphocytes stimulated through the CD40 ligand in the presence or absence of B-cell antigen receptor and cytokine signaling.

MATCHING

1, J; 2, E; 3, C; 4, G; 5, I; 6, H; 7, B; 8, F; 9, D; 10, A

CHAPTER 9

Immunologic Tolerance

Introduction

Immunologic tolerance can be defined as **a state of unresponsiveness to an antigen that is induced by prior exposure to that antigen**. Obviously, such a sharp "weapon" as the immune system must be kept under tight control in order to prevent it from developing injurious responses to self antigens. As was discussed in earlier chapters, the intrinsic molecular mechanisms that create antigen receptors and their diversity do not depend on external factors like antigens. Because of that, antibodies and TCR specific to self are being constantly produced and something must either eliminate them, or render them functionally silent. It is exactly this purpose that tolerance to self antigens serves. In addition, the immune system may develop tolerance to foreign antigens. As we will detail in the Discussion, there are several different ways in which lymphocytes can become tolerant to self as well as foreign antigens. In all cases, **tolerance is strictly antigen-specific and develops only after the recognition of the antigen**. Many more details about the development of immunologic tolerance have become known recently, after the introduction of such powerful tools as transgenic and knockout animal models. Nonetheless, many details that are essential for the understanding of the molecular mechanism of tolerance induction are still not known.

We will begin our discussion with a description of early experiments performed on animal models, especially nonidentical twin cattle and newborn mice. These experiments showed convincingly that tolerance to antigens is specific to these antigens, and that it can be induced by an experimenter, thus putting the studies of immunologic tolerance into the framework of experimental dissection of the functions of the specific immune system. We will continue with more recent studies on the development of immunologic tolerance to self which, as we will see, takes place in the central (or generative) as well as in peripheral lymphoid organs. We will then proceed to a characterization of tolerance to foreign antigens. Throughout the chapter, we will try to pay special attention to molecular mechanisms that underlie the phenomena comprising the immunologic tolerance.

Discussion

9.1 Who discovered immunologic tolerance, and how was it discovered?

At the very beginning of the 20th century, Paul Ehrlich postulated that a mechanism should exist that would not allow organisms to produce antibodies to their own antigens. Ehrlich called this putative mechanism "horror autotoxicus" ("a fear to poison (or damage) oneself"), but he did not elaborate on its nature or address it experimentally. Later, Traub showed that mice that had been inoculated with a virus *in utero*, after their birth developed a lifelong infection and never produced anti-virus antibodies. These data suggested that the immune system of the mice could develop a state of unresponsiveness to specific antigens if these were introduced in the organism during its embryogenesis. In 1945, Owen reported an observation that he made on nonidentical cattle twins. If the calves that grew in the same uterus developed vascular anastomoses that resulted in an exchange of their blood, they later did not produce antibodies to each other's red blood cells in spite of the fact that their ABO blood groups were different. Like Traub's experiment, Owen's observation suggested that a fetal contact with an antigen renders subsequent specific lack of responsiveness, or tolerance, to that antigen in the developing adult.

In 1953, the laboratories of P. Medawar (Great Britain) and M. Hašek (Czechoslovakia) reported that specific immunologic tolerance could be induced in animals that received injections of allogeneic cells during the recipient's very early postnatal development. In Medawar's experiments, inbred mice that belonged to a strain A received injections of cells of mice that belonged to a strain B. When the A recipients grew to adulthood, they could accept skin grafts from the strain B mice, but not from mice that belonged to a third strain. These experiments firmly established the notion that specific immunologic tolerance can be induced and that the time of the antigen encounter is crucial for its development. Further, these experimental findings were accommodated in F.M. Burnet's clonal selection theory (see Chapter 1). Burnet interpreted the above data to mean that lymphocyte clones, which encounter antigens (usually self antigens, but also experimenter-induced foreign antigens) early in the animal's life, become "forbidden clones," i.e., a mechanism exists that results in the deletion of these clones. Later, Lederberg modified this view, suggesting that the factor that determines whether a clone would be "forbidden" is not the age of the animal but the maturational status of the clone itself. According to Lederberg, immature lymphocytes would always be deleted when they encounter their antigens. This hypothesis later received a solid experimental support, but was also modified. Currently, it is

CHAPTER 9 Immunologic Tolerance

thought that immature T and B lymphocytes are indeed rendered unresponsive to their specific antigens if they encounter them, but their fate is not necessarily a physical deletion (see later sections). Although these studies and hypotheses pertained mostly to self-tolerance and were not always correct in detail, they laid a solid foundation for a further development of the concept of immunological tolerance in general.

9.2 How can immunologic tolerance be classified?

As already mentioned, the tolerance can be subdivided into **tolerance to self** and **tolerance to foreign** antigens. Depending on where does it develop, it can be also subdivided into **central** tolerance and **peripheral** tolerance. The former develops in central lymphoid organs (the thymus and the bone marrow), and the latter in the peripheral lymphoid organs. Tolerance to self can be central or peripheral: most of the lymphocyte clones that are specific to self antigens are made tolerant in central lymphoid organs, but some escape developing tolerance there and are made tolerant in the periphery. Tolerance to foreign antigens is almost exclusively peripheral, because central lymphoid organs contain almost no foreign antigens.

9.3 What is the mechanism of central T-cell tolerance?

To understand this, one needs to take a closer look at the events of **selection** that accompany the maturation of T-cell precursors in the thymus (Fig. 9-1). In this organ, T-lymphocyte precursors undergo a vigorous selection that has a dual purpose: to ensure that the resulting mature T cells will be able to react with self-MHC and not able to react with self peptides, at least with high affinity. The first purpose is achieved through the process of **positive selection**. This selection is, essentially, a screening of T-cell precursors that enter the thymus for their ability to associate with self MHC. The precursors that emigrate from bone marrow and enter the thymus begin to rearrange their TCR V-region genes, express TCR, and use the expressed TCR for binding to peptides presented to them in the context of self MHC molecules expressed by thymic nonlymphoid cells. (These nonlymphoid cells are largely a special class of epithelial cells called "nurse cells," and also macrophages and dendritic cells that express both classes of MHC molecules.) The precursors that failed to rearrange their TCR V-region genes productively, or those that have rearranged these genes but have TCR that are unable to associate with self MHC molecules, cannot bind to the MHC-presenting cells and do not receive signals from these cells. Notably, the precursors that enter the thymus have already initiated their apoptotic programs. The positive signals that these precursors receive from the MHC-presenting cells rescue them from apoptosis. If they do not associate with self MHC, they inevitably die "**by neglect**."

The precursors that are able to associate with self MHC and therefore do receive a positive, rescuing signal, are further screened on the basis of the **affinity** of their TCR. Since essentially all peptides that are present in the thymus are self peptides, the precursors that would bind to self MHC with a high affinity are likely

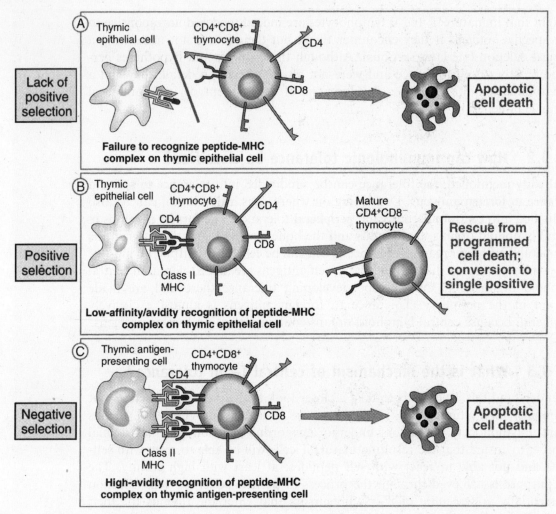

Fig. 9-1. Selection of T-cell precursors in the thymus. In order to mature and successfully exit the thymus, T cells must be able to express TCRs able to associate with self MHC, but not with self peptides. (A) T-cell precursors whose TCR is unable to bind self MHC die because of the lack of positive selection. (B) T-cell precursors whose TCR is able to establish a low-affinity interaction with self MHC (presumably having high affinity to foreign peptides) are rescued from programmed cell death by positive selection signals. (C) T-cell precursors whose TCR binds self MHC with high affinity or avidity (presumably being specific to self peptides) die, because they receive "killing" (negative selection) signals. From Abbas, K., A.H. Lichtman, and J.S. Pober, *Cellular and Molecular Immunology*, fourth edition, W.B. Saunders, 2000, p. 154 (their Fig. 7-20).

to be those that recognize self peptides in the context of self MHC. These autoreactive or potentially autoreactive precursors of T lymphocytes are eliminated through the mechanism of **negative selection**. The molecular mechanism of this process is still not known, but it is thought that in the case the affinity of the TCR–self MHC interaction is high, the precursor receives a negative, deleterious signal from the MHC-presenting cell. Like the death by neglect, the death of negatively selected cells is thought to be apoptotic. The efficiency of negative selection in the thymus is amazingly high, as judged from the fact that normally, very few if any autoreactive T lymphocytes can be found in the peripheral lymphoid organs and in

CHAPTER 9 Immunologic Tolerance

the tissues. If some autoreactive T cells still escape the thymic deletion, they are made tolerant in the periphery (see later sections).

9.4 How was the thymic selection demonstrated experimentally?

The experimental demonstration of the clonal deletion of self-reactive T cells in the thymus relied on the transgenic murine model. The idea behind the experiment, which was conceived and performed by the laboratories of P. Kisielow and H. von Boehmer in the late 1980s, was as follows. If autoreactive T cells are really deleted, and if maturing thymocytes are exposed to an antigen that is ubiquitously expressed both in the thymus and in the periphery, the animal will essentially lack peripheral T cells. Kisielow and von Boehmer made transgenic mice that expressed an MHC Class I-restricted TCR specific to the so-called **H-Y antigen**. The latter is a protein encoded by a locus of the Y chromosome and, as such, is expressed exclusively in males. The tissue distribution of the H-Y antigen is very broad, so that virtually all cells of the male organism express it. Kisielow and von Boehmer observed that in female mice that were transgenic for the anti-H-Y TCR, the development of T-lymphocytes was normal. In the male transgenic littermates, however, one could find large numbers of maturing T-lymphocyte precursors in the thymic cortex, but no $CD8^+$ MHC Class I-restricted precursors in the medulla and no mature $CD8^+$ MHC Class I-restricted T lymphocytes in the periphery. These data strongly implied that the T-lymphocyte precursors that recognize self peptides in the context of the self MHC are not allowed to mature, being instead deleted.

9.5 Are autoreactive B lymphocytes also deleted in the central lymphoid organs?

Not necessarily. While a large fraction of self-reactive B lymphocytes are rendered tolerant in the bone marrow, their fate is not always physical deletion. The possible outcomes of self antigen recognition by immature B lymphocytes in the bone marrow were studied in detail in the late 1980s and early 1990s. Goodnow designed a transgenic murine model where the mice transgenic for hen egg lysozyme (HEL) were crossed with the mice transgenic for anti-HEL antibody. In the resulting **double transgenic mice**, virtually all of the immature B lymphocytes expressed the anti-HEL antibody, and HEL was either expressed on the surface of the murine cells or secreted (depending on the promoter). If the latter was expressed on the cell surface, most of the immature B lymphocytes were deleted. If, however, HEL was secreted, most of the transgenic B lymphocytes remained alive. Instead of being deleted, they entered the state of **anergy**, or functional quiescence. This meant that these B lymphocytes expressed a reduced number of surface IgM, and neither proliferated nor secreted antibodies when adoptively transferred or cultured with the antigen or polyclonal activators *in vitro*. Subsequent experiments showed that the affinity and avidity of the self-reactive antibody is a major factor determining the fate of the lymphocyte that expresses it. If the affinity and avidity were high (for example, when HEL was expressed on the cell membrane at high density, causing the antibody cross-linking), the B cells that carried the transgenic anti-HEL antibody were deleted. If they were low (for example, when HEL was secreted), they entered the state of anergy.

9.6 Is the negative selection against autoreactive B lymphocytes in the bone marrow as ruthless as the negative selection against autoreactive T lymphocytes in the thymus?

No. In fact, unlike precursors of T lymphocytes, immature B lymphocytes that recognize self antigens in the bone barrow have one more chance to change the specificity of their antibodies. The recently discovered process that allows them to do it is called **receptor editing**. Essentially, receptor editing means re-activation of RAG-1 and RAG-2 genes and the enzymes that they code, so that the immature B lymphocyte becomes able to launch another antibody V-region gene rearrangement (see Chapter 4). This "re-rearrangement" occurs usually in the V_κ locus: the "old" $V_\kappa J_\kappa$ unit is deleted and replaced by a "new" $V_\kappa J_\kappa$ (or, rarely, $V_\lambda J_\lambda$). Receptor editing was observed both in transgenic and in nontransgenic murine B lymphocytes, as well as in human B lymphocytes. Very recent data indicate that TCR V-region genes can also be "edited," but rather rarely.

9.7 How do peripheral tolerance mechanisms work?

We should begin this part of the discussion with peripheral T-cell tolerance. During the last decade, it has been firmly established that T lymphocytes can be readily made tolerant in the periphery if they encounter their specific antigens (peptide–MHC complexes) in the absence of co-stimulators. Schwartz *et al.* worked with a $CD4^+$ T-cell clone that recognized a known peptide in the context of an MHC Class II molecule. If the clone was exposed to the peptide presented *in vitro* by the MHC-compatible antigen-presenting cells, it responded by vigorous proliferation. If, however, the same clone was exposed to the peptide–MHC complex inserted into liposomes (tiny microscopic spheres that are formed by an artificial lipid bilayer), no proliferative response was observed. Further, if the clone was first exposed to the peptide–MHC complex embedded in liposomes and then to the same peptide–MHC complex presented to it by the MHC-compatible antigen-presenting cells, it still failed to proliferate, i.e., was apparently rendered tolerant. These data were interpreted to mean that the recognition of the specific TCR ligand (i.e., peptide plus MHC) alone triggers T-cell tolerance rather than T-cell activation. This explanation fit into the hypothesis of Bretscher and Cohn (1970) that the activation requires two signals (one from antigen and one from co-stimulator), while the tolerance can be induced if only the "signal one" is delivered.

CHAPTER 9 Immunologic Tolerance

9.8 How was it shown that it is the lack of the "signal two," i.e., of the co-stimulatory molecules, that causes tolerance?

It was directly shown using the same liposome model. The T-cell clone was exposed to liposomes in which the peptide–MHC complex had been embedded, and, during the exposure, stimulated with an anti-CD28 antibody (see Chapter 7). Under these conditions, a strong proliferative response was detected, indicating that the engagement of CD28 by the antibody (which, in this case, mimicked the co-stimulator molecule B7) delivered the "second signal" necessary for the T-cell activation.

9.9 What is the fate of the T cells that are made tolerant because of the lack of co-stimulatory signal?

It is generally considered that the majority of T lymphocytes that are made tolerant in the periphery because of the lack of "signal two" are not deleted, but rather enter the state of anergy. We will discuss molecular mechanisms of anergy and deletion in later sections of this chapter.

9.10 Is the lack of co-stimulation the only mechanism of peripheral T-cell tolerance induction?

No. There exist two other ways to induce T-cell tolerance in the periphery. In addition to the tolerance due to a lack of co-stimulation, T cells may be made tolerant in the periphery if they are: (a) exposed to a persisting antigen and co-stimulators and activated repeatedly (in this case they are deleted, and the phenomenon is called activation-induced cell death); and/or (b) exposed to a peptide that has an altered (mutated) amino acid sequence at the site of the TCR contact. Finally, T lymphocytes can be rendered tolerant in the periphery because of the action of other T lymphocytes.

9.11 How does the activation-induced cell death work?

Activation-induced cell death means, essentially that recently activated T lymphocytes, when exposed to high concentrations of the same antigen undergo rapid activation, followed by apoptosis. It can be observed, for example, when an anti-CD3 antibody polyclonally activates peripheral blood T lymphocytes. This antibody mimics the activator signal triggered by antigen and transmitted from the TCR to the CD3 molecule. If the anti-CD3 antibody is added to a T-cell culture, it triggers a strong proliferative response that peaks at days 3–4 of culturing. If the same anti-CD3 antibody is added at that time (i.e., at the peak of the proliferative response), the lymphocytes rapidly die by apoptosis. The same phenomenon is observed when T-cell clones specific to particular antigens are activated by their antigens and then, at the peak of their proliferative response, exposed to high concentrations of the antigen again. The phenomenon of activation-induced cell death is believed to be accountable for elimination of T-cell clones specific to those self antigens that are not expressed in the thymus, but are abundantly expressed in the periphery.

9.12 How does the tolerance induced by altered peptides work?

This phenomenon was discovered when T-lymphocyte clones with known peptide–MHC specificity were exposed to peptides with altered amino acids at the region of the peptide–TCR contact. As expected, these clones did not respond to the altered peptides but, somewhat unexpectedly, they lost the ability to respond to their original antigen, i.e., they acquired tolerance. The peptides in which the amino acid was changed at the region of the peptide–MHC contact were called **altered peptide ligands** or **peptide antagonists**. In contrast to activation-induced cell death, the tolerance caused by altered peptide ligands manifests itself as clonal anergy rather than deletion. Until recently, the significance of this phenomenon was not appreciated enough, but it received closer attention after it was discovered that peptide antagonists can prevent an autoimmune disease called experimental autoimmune encephalomyelitis (see later).

9.13 What is known about the molecular mechanisms of the activation-induced T-cell death?

It is generally agreed that the T lymphocyte clonal deletion results from unwinding of the apoptotic programs, and is a result of the Fas–Fas ligand interaction. When T lymphocytes are activated, they simultaneously express Fas and Fas ligand. The interaction between these two molecules on the same cell or on two neighboring cells may lead to apoptosis because it initiates the production of active caspases, similar to the process described in Chapter 8 (Section 8.14). During repeated stimulation of previously activated T lymphocytes with large concentrations of the same antigen, unusually large amounts of a cytokine called **interleukin-2 (IL-2)** are produced. Although IL-2 was discovered as a growth factor for activated T lymphocytes, in large concentrations it significantly increases the susceptibility of the cells to Fas–Fas ligand-mediated apoptosis.

9.14 Is the mechanism of death of T-cell precursors that do not receive positive selection signals the same as the mechanism of the activation-induced T-cell death in the periphery?

No. Recent studies have firmly established that although the "death by neglect" and the activation-induced T-cell death are both apoptotic, they are mediated by different molecular mechanisms. When a T-cell precursor dies from the lack of

CHAPTER 9 Immunologic Tolerance

positive selection, it does not up-regulate the expression of Fas and Fas ligand. In this case, the primary mediator of the cell death is **cytochrome c** – a protein that is involved in oxidative phosphorylation and is normally stored in mitochondria. If the T-cell precursor is not rescued from apoptosis by positive selection signals, it allows the cytochrome c to leak from the mitochondria. When in the cytoplasm, the cytochrome c binds another protein, called **Apaf-1** ("**ap**optosis **a**ctivating **f**actor 1). The complex of cytochrome c and Apaf-1 binds to, and activates an inactive cytoplasmic precursor of an enzyme called **caspase-9**. When this enzyme is activated, it in turn activates a cascade of the so-called effector caspases. These cleave a variety of protein substrates on their aspartic acid residues and thus cause the fragmentation of the nucleus, the so-called "blebbing" of the cellular membrane and, eventually, the death of the cell. On the contrary, the activation-induced T-cell death in the periphery involves up-regulation of Fas and Fas ligand and the binding of Fas to Fas ligand on the same or neighboring cell. This leads to **clustering** of the Fas molecules. The cytoplasmic domains of three identical clustered Fas molecules bound by a special adapter protein called **FADD** (**F**as-**a**ssociated **d**eath **d**omain; see also Section 8.14) activate an inactive cytoplasmic precursor of an enzyme called **caspase-8**, which activates a number of effector caspases, eventually leading to the apoptotic death.

9.15 What is known about the molecular mechanism of the reversion of apoptosis?

It has been established that a variety of stimuli that reverse or prevent apoptosis (for example, antigen stimulation, co-stimulation, moderate concentrations of growth factors like IL-2) prevent apoptosis by up-regulating molecules that belong to the so-called **Bcl family**. The two most known representatives of this family, called **Bcl-2** and **Bcl-X**, inhibit apoptosis by blocking the release of cytochrome c from mitochondria, and by binding to Apaf-1 and inhibiting the activation of caspase-9. A different mechanism is employed for prevention or inhibition of the activation-induced cell death. This mechanism involves a protein called **FLIP**. (The name of this protein is an abbreviation of "**FLICE**-**i**nhibiting **p**rotein"; "FLICE" stands for "FADD-like ICE"; and "ICE" means "interleukin-1-converting protein." These names are historical, and reflect earlier findings of the role of caspases in the conversion of an inactive precursor of the cytokine interleukin-1 (IL-1) into its active form.) FLIP binds to caspase-8, inhibiting its binding to Fas-associated protein complex, thus preventing the development of apoptosis. It has been shown that naive T lymphocytes express large amounts of FLIP, but in the presence of high concentrations of IL-2, the expression of this protein is shut down. This explains why naive T cells can be activated by antigen and co-stimulators without undergoing apoptosis: although they do up-regulate Fas, they are resistant to Fas-mediated apoptosis, because FLIP efficiently turns off the entire Fas signaling pathway. Recently activated T cells, however, are very sensitive to Fas-mediated apoptosis because IL-2 down-regulates FLIP.

9.16 What is known about the molecular mechanism of anergy?

The exact molecular mechanism of clonal anergy as an outcome of deficient co-stimulation or exposure to peptide antagonists is rather poorly defined. Perhaps the best way to characterize the status of anergic T and B lymphocytes in molecular terms is to say that they have a partial or selective defect in their intracellular signaling. In T-cell clones that are rendered anergic, engagement of TCR does activate some PTKs, but not others. In anergic B cells from double transgenic mice described in Section 9.5, the density of surface IgM (but not IgD) is dramatically reduced and the kinetics of calcium flux in response to surface IgM triggering differs from that in normal B cells. Also, the activity of some PTK and protein phosphatases appeared to differ from that in nonanergic B lymphocytes.

9.17 What is known about peripheral B-cell tolerance?

B lymphocytes that are reactive with self antigens and fail to edit their antibody receptors can escape central tolerance. These mature autoreactive B cells can be made tolerant in the periphery, the principal reason being **the absence of specific T cell help**. The cellular and molecular mechanism of this phenomenon was addressed in a modified HEL-anti-HEL double transgenic system (see Section 9.5), where HEL-specific B lymphocytes were transferred to mice transgenic for HEL in its circulating form. In such mice, HEL-specific T cells are deleted, because HEL is present in the thymus. The fate of the "T cell-helpless" HEL-specific B lymphocytes in this system is long-lasting anergy. The anergic B cells show a partial or selective defect in cellular signaling (see Section 9.16). For example, they fail to activate syk and, although they are able to respond to the antigen by a rise in $[Ca^{2+}]_i$, they are unable to maintain the high levels of Ca in the way that functional B cells do. Also, importantly, anergic B cells lose the ability to migrate into lymphoid follicles of the spleen and lymph nodes. This phenomenon of **follicular exclusion** is thought to play a significant role in the prevention of antibody responses to self antigens.

9.18 What can make foreign antigens tolerogenic?

For a foreign antigen to induce tolerance, several factors are important. First of all, the form of antigen itself determines whether the tolerance to this antigen will be established. For example, purified protein antigens injected into the bloodstream often induce tolerance; the same antigens, however, induce vigorous immune response if they are emulsified in oily substances called **adjuvants**, and then injected intradermally or subcutaneously. If a protein is injected into the bloodstream without adjuvants and then subcutaneously with adjuvants, the response will not develop, i.e., the first encounter with the antigen will result in true immunologic tolerance. Thus, the same antigen can exist in an immunogenic

CHAPTER 9 Immunologic Tolerance

form (i.e., induce an immune response) and in a tolerogenic form (i.e., induce tolerance).

The exact cellular and molecular mechanisms responsible for the induction and maintenance of immunologic tolerance to foreign antigens are not yet completely elucidated. One very important factor that influences the development of tolerance is known, however, and that is the lack of up-regulation of co-stimulatory molecules in the tissues that encounter the foreign antigen. It is generally thought that nonspecific, innate defense mechanisms such as local inflammation (see next chapters) must first influence on the local antigen-presenting cells, making them up-regulate co-stimulatory molecules such as B7-1 and B7-2. The reason why the addition of adjuvants makes antigens immunogenic rather than tolerogenic is perhaps that they induce local inflammation, and hence up-regulate co-stimulatory molecules in local antigen-presenting cells.

9.19 What is the consequence of the encounter of a foreign peptide presented by an antigen-presenting cell with little or no co-stimulators?

If co-stimulatory molecules are not present, the T lymphocyte that recognizes the foreign peptide will not receive "signal 2," and will therefore enter the state of anergy. A different outcome, however, is possible if an antigen-presenting cell expresses small amounts of B7-1 and/or B7-2. In this case, the "signal 2" will be delivered, but it will be most likely **inhibitory** for the responding T lymphocyte. The reason for this inhibition is that if the amount of B7 is limiting, the T-cell structure that is likely to bind B7 molecule(s) will be not CD28, but the inhibitory molecule **CTLA-4**, which has a higher affinity to B7 molecules than CD28. It is currently believed that CTLA-4 plays a significant role in the maintenance of peripheral T-cell tolerance to both self and foreign antigens. It is not clear however, why – and under what circumstances – do T lymphocytes preferentially use either CD28 or CTLA-4.

9.20 Other than the form in which antigens are administered, what else determines their immuno- or tolerogenicity?

The route of antigen entry is also an important factor. As has already been mentioned, systemic administration of an antigen (into the bloodstream rather than locally, under the skin or in the skin) tends to be tolerogenic. It has also been established that oral administration of some protein antigens leads to lasting tolerance. This phenomenon of **oral tolerance** is thought to play a major role in the prevention of immune responses to food antigens, and is believed to be related to the overproduction of an inhibitory cytokine TGF-β (see next chapter). Not all proteins induce oral tolerance however. It remains to be investigated why some proteins – for example protein components of the polio vaccine – induce very strong immune responses and long-lasting immune memory instead of tolerance.

9.21 How do other lymphocytes influence the development of immune response or tolerance?

As early as the late 1960s, immunologists knew that some T lymphocytes tended to inhibit immune responses mediated by other T and B lymphocytes. In the early 1970s, there appeared a notion that certain T lymphocytes belong to a separate class or lineage of the so-called **suppressor cells**. The evidence in favor of the existence of T lymphocytes with properties of suppressors was based mostly on the adoptive transfer technique. For example, if a mouse is immunized with a large dose of antigen, and its splenic cells are then adaptively transferred into a naïve recipient, such a recipient will not respond to any dose of the antigen at any time. During the 1970s and 1980s, many attempts were made to obtain pure populations of suppressor cells, but these were unsuccessful. Likewise, investigators failed to purify soluble factors that were at some time believed to mediate the specific immunosuppressive function of suppressor cells, or to clone the genes that code for such factors. At present, it is thought that "suppressor T lymphocytes" do not exist as a separate class or lineage. Rather, some T_h cells can produce cytokines that inhibit certain immune responses, thus exerting suppressor functions. For example, the previously mentioned TGF-β profoundly inhibits most immune responses, inducing a kind of immunologic tolerance. As will be seen in Chapter 10, the T_h cells that produce such cytokines as IL-4 and IL-13 (T_h2 cells) dramatically inhibit immune responses of the T_h cells that produce such cytokines as interferon (IFN) γ and interleukin-2 (T_h1 cells), and vice versa. However, T_h1 cells strongly stimulate the responses of T cells that belong to their "kind," and so do the T_h2 lymphocytes. Based on these factors, it makes more sense to talk about **immunoregulatory** T lymphocytes instead of "suppressor T lymphocytes."

9.22 Can B lymphocytes be immunoregulatory?

This question is still unresolved. B lymphocytes are not immunoregulatory in that they do not produce inhibitory cytokines. However, some immunologists believe that B cells can, and do, regulate immune responses through production of **anti-idiotype antibodies**. As mentioned in Chapters 4 and 7, idiotype is a marker of the antibody or TCR V-region genes utilized by the given clone of lymphocytes. Since an idiotype may be unique, not shared with other lymphocyte clones, it can be taken by the immune system as being a foreign antigen, triggering the production of an anti-idiotype antibody. The existence of anti-idiotype antibodies has been demonstrated experimentally. The epitope on the antibody molecule that is physically bound by an anti-idiotype antibody is called an idiotope. (In this regard, an idiotype can be viewed as a sum of idiotopes.) Some idiotopes actually overlap with, or even are identical to, the sites of complementary interaction between the antigen and the antibody to which they are specific. In this case, the anti-idiotype antibody can compete with the antigen, and may modify or regulate the response to this antigen. In other cases, when the site of the anti-idiotype antibody binding does not overlap with the site of antigen–antibody interaction, the anti-idiotype antibody may not have any influence on the response to the antigen. Yet, in any

CHAPTER 9 Immunologic Tolerance

case, the anti-idiotype antibody itself has its own idiotype and can trigger the production of anti-anti-idiotype antibodies.

In 1974, Niels Kai Jerne (the author of the "natural selection" theory of immunity mentioned in Chapter 1) offered the so-called "**network hypothesis**." According to this hypothesis, all lymphocytes are interconnected through idiotype–anti-idiotype interactions. Whenever an antibody or a TCR is made, the anti-idiotype response immediately follows, being itself immediately followed by an anti-anti-idiotype response, and so on. In the absence of antigen, the net result of the work of this "network" of idiotype–anti-idiotype interactions is immunologic homeostasis, i.e., such a state that not one single lymphocyte clone is expanded and its antibody or TCR overproduced. The role of antigen, according to this hypothesis, is to shift this balance, causing not only an expansion of the specific clone but also a disturbance in the entire network of idiotype-anti-idiotype interactions. The network hypothesis, however, remains largely an elegant intellectual construction that lacks a substantial experimental support, in part due to the fact that idiotypes and anti-idiotypes are extremely diverse. This in turn makes it an enormously difficult task to trace the regulatory influence of any particular idiotype–anti-idiotype interaction.

Questions

REVIEW QUESTIONS

1. Self-reactive lymphocytes occasionally fail to recognize or respond to some self antigens in the periphery, but neither die nor become anergic. This phenomenon is called clonal ignorance and is a common situation during the encounter with sequestered self antigens, i.e. the antigens of the central nervous system or the eye. What, if any, is the difference between the clonal ignorance and the immunologic tolerance?
2. How would the positive thymic selection work in mice that are transgenic for Bcl-2 and/or BclX?
3. There are two T-cell precursors, A and B, both recognizing the same peptide X and restricted to H-2K^d. The precursor A binds the peptide X with a K_d of 10^{-5} M; the precursor B with a K_d of 10^{-7} M. Which of the two precursors is more likely to be subject to negative selection in the thymus of an H-2^d mouse?
4. In mice, staphylococcal enterotoxin B acts as a superantigen, recognized by all T lymphocytes that express the TCR coded by a $V_\beta 17a$ gene. In the mid-1980s, P. Marrack and J. Kappler showed that peptides that mimic parts of this superantigen are presented to thymic precursors by the I-E molecule. How many mature T cells that utilize $V_\beta 17a$ do you expect to find in the periphery, if: (a) the mice express I-E; and (b) the mice do not express I-E (actually, some strains that do not express I-E molecules do exist)?

5. In the mid-1980s, D. Nemazee's laboratory performed an experiment similar to that of Goodnow with double transgenic mice, except that instead of HEL and anti-HEL, they used H-2K and anti-H-2K transgenes. How do you imagine the fate of B lymphocytes in Nemazee's system?

6. In a follow-up to Schwartz's experiment, the T-cell clone was exposed to its specific peptide–MHC complex expressed on chemically fixed antigen-presenting cells. Such an exposure rendered the T lymphocytes tolerant. If, however, the clone was presented with its antigen by a mixture of fixed and freshly isolated antigen-presenting cells, the tolerance was not induced. Explain this finding.

7. In the late 1980s, P. Lipsky's laboratory developed a tissue culture system where B lymphocytes are stimulated by T lymphocytes that have been seeded into plastic wells precoated with an anti-CD3 antibody. Some T cell lines, however, cannot replace the normal T lymphocytes in this system, because when they are seeded into anti-CD3-precoated wells, they show signs of apoptosis after a few hours. What phenomenon does this remind you of?

8. What feature of altered peptide ligands is consistent with the idea that immunologic tolerance develops exclusively after the antigen recognition?

9. Name three consecutive events that follow the leakage of cytochrome c from mitochondria during the development of "death by neglect," and three consecutive events that follow the up-regulation of Fas and Fas ligand during activation-induced cell death.

10. What will be the phenotype of mice transgenic for FLIP?

11. Why is the phenomenon of follicular exclusion beneficial?

12. What defects would you expect to see in the knockout mice that lack the CTLA-4 gene?

13. As was mentioned in Chapter 3, antibodies that belong to the IgA class are produced mostly in the mucosal compartment of the immune system. On the other hand, it is known that TGF-β is a factor that promotes the antibody class switch to IgA. Offer a hypothesis that would link these two phenomena.

14. How would you summarize the current opinion about suppressor T cells?

15. What is immune homeostasis (in the terms of Jerne's "network hypothesis")?

MATCHING

Direction: Match each item in Column A with the one in Column B to which it is most closely associated. Each item in Column B can be used only once.

Column A

1. Neonatal
2. TCR affinity
3. Females with anti-H-Y TCR transgene
4. Extracellular HEL
5. Receptor editing
6. Anti-CD28 antibody
7. Caspase-8
8. Caspase-9
9. TGF-β
10. Network hypothesis

Column B

A. T cells intact
B. Anti-idiotype regulation
C. Mimics signal 2
D. Mediates "death by neglect"
E. Important for oral tolerance
F. Medawar's experiment
G. Activation-induced cell death
H. B cells anergic
I. V_κ rearrangement
J. Crucial for negative selection

CHAPTER 9 Immunologic Tolerance

 ## Answers to Questions

REVIEW QUESTIONS

1. The clonal ignorance may mean that the recognition simply did not occur, while the clonal tolerance absolutely requires previous antigen recognition.
2. It will not work at all, because in such mice the "death by neglect" will not be possible (the transgene products will prevent it).
3. The precursor B, because it has a higher affinity to the peptide X.
4. (a) Zero or close to zero; (b) normal numbers.
5. They will likely be deleted, because H-2K antigens are expressed on the surface of all nucleated cells and can cross-link the transgenic antibodies.
6. The freshly isolated antigen-presenting cells provide the necessary co-stimulators.
7. Activation-induced cell death.
8. The substitution exclusively in the region of physical peptide-TCR contact.
9. In "death by neglect," cytochrome c binds to Apaf-1, the complex of the two activates a precursor of caspase-9, and the latter triggers effector caspases. In activation-induced cell death, Fas molecules cluster together, FADD binds the three clustered Fas cytoplasmic domains, and this activates a precursor of caspase-8.
10. Their lymphocytes will be resistant to Fas-mediated apoptosis, but the "death by neglect" (e.g., during positive selection in the thymus) will proceed normally.
11. Because it ensures that autoreactive B cells that received T cell help (e.g., if some autoreactive T cells escaped negative selection) do not undergo proliferation, differentiation, and especially class switch and affinity maturation in the germinal centers.
12. Their T lymphocytes will be constantly activated in the periphery, which may lead to autoimmune disorders.
13. Perhaps mucosal T cells are specialized in the production of TGF-β.
14. They do not exist as a separate lineage; helper T cells may exert suppressor functions towards other helper T cells or antigen-presenting cells.
15. The lack of "disturbance" in the idiotype–anti-idiotype network that does not allow any particular idiotype bearer (clone) to expand; it is temporarily abolished during an antigen invasion.

MATCHING

1, F; 2, J; 3, A; 4, H; 5, I; 6, C; 7, G; 8, D; 9, E; 10, B

Cytokines

Introduction

Cytokines are **soluble mediators** that play an important role in immunity. In previous chapters, we mentioned that cytokines produced by activated T lymphocytes direct antibody class switch (Chapter 3, 4, and 8) and T helper cell differentiation (Chapter 2), thus affecting the effector phase of immunity and regulating the immune response (Chapter 9). The role of cytokines is much broader however; these polypeptide, hormone-like substances act in both innate and acquired immunity, as well as in the maturation of lymphocytes and other hemopoietic cells from their progenitors. Cytokines influence on a wide variety of cells that do or do not belong to the immune system.

In our discussion, we will analyze some common features that all cytokine have. Instead of giving an exhaustive definition of a cytokine, we will try to convey the concept of a cytokine through the properties of these molecules, taking a "cytokine is what cytokine does" approach. We will focus on the role that cytokines play in innate and acquired immunity, and especially on how the production of cytokines affects an outcome of the immune reaction. Also, we will pay special attention to the structure of individual cytokines, their interaction with specific receptors, and the transduction of the signals that are delivered upon this interaction. Throughout the chapter, we will highlight the most important avenues of the clinical use of cytokines.

Discussion

A. COMMON PROPERTIES OF CYTOKINES

10.1 What are the origins of the term "cytokine?"

During the late 1960s and 1970s, immunologists discovered that leukocytes derived from recently immunized mice and cultured *in vitro* secrete substances

CHAPTER 10 Cytokines

that affect the biological properties of other lymphocytes. For example, the supernatant of such leukocyte cultures strongly enhances the response of thymocytes to submitogenic concentrations of polyclonal T-cell activator. Supernatants of cultured immune leukocytes are also capable of stimulating antibody production by B lymphocytes, thus partially replacing helper T lymphocytes. Migration of T and B lymphocytes observed *in vitro* was also affected. In 1976, it was shown that polyclonally activated T lymphocytes secrete a factor that sustains the survival and growth of other activated T lymphocytes. Although each of the groups of experimenters coined a name for their "factor," investigators agreed on a collective name for all these factors, and called them **interleukins** – factors that are used for a "communication" between different leukocytes. This name appeared to be convenient for the "factors'" nomenclature. The "factor" that stimulated thymocytes became known as interleukin-1 (IL-1), the one that sustained the survival of activated T cells as interleukin-2 (IL-2), etc. Since interleukins were believed to be produced by activated lymphocytes and affect other lymphocytes, they were also called **lymphokines**. It became clear, however, that not all of the "factors" fit this definition. IL-1 was found to be actually produced by adherent monocytes rather than lymphocytes. Because of that, IL-1 and other monocyte/macrophage-derived interleukins were called **monokines**. However, some interleukins (for example, IL-1 or γ-interferon) can be produced by more than one type of cell. This made the investigators agree that a broader name is needed for these "factors," and the term **cytokines** appeared the best in this regard. However, the name "interleukins" was also retained, because it appeared to be convenient for the use in cytokine nomenclature.

10.2 Since cytokines are obviously so numerous and diverse, what properties unite all of them in a single group of substances?

Such common properties indeed exist. All cytokines are **polypeptides**. Although structurally diverse, they always contain regions that mediate their binding to their specific **receptors**. The binding of cytokines to their receptors is characterized by a **high affinity**: the K_d of such binding is usually less than 10^{-9} M, but sometimes as small as 10^{-12} M. This means that very small amounts of cytokines are enough to saturate their receptors and to initialize the signaling through the latter. The expression of these receptors is **regulated** by external signals, which makes the responses to cytokines subject to regulation by these signals (see later).

Cytokines are usually **secreted** by the cells that make them; they do not remain expressed on the cell surface, and neither are they stored in cellular depots. **The secretion of a cytokine is always a brief and self-limited event**, triggered by either a specific (antigen) or a nonspecific stimulus (e.g., invasion of lipopolysaccharide expressing Gram-negative bacteria). The action of cytokines is often **pleiotropic** and **redundant**. Pleiotropy means that one cytokine performs more than one function, acting on a variety of cell types. Redundancy means that almost always, the function of one particular cytokine is "backed" by the identical activity that can be performed by one or more other cytokines. In other words, multiple cytokines exert similar functional effects (which explains why knockout mice deficient for any one cytokine may be phenotypically normal). Finally, one other common property of cytokines is their ability to induce transcription of new genes, and to serve as growth factors, i.e., to stimulate the proliferation of their target cells.

10.3 What is the current working classification of cytokines?

All cytokines are classified into the following three categories:

1. **Mediators and regulators of innate immunity**. These cytokines are produced mostly by macrophages activated by lipopolysaccharide (LPS) produced by Gram-negative bacteria, or viral products such as double-stranded RNA. These cytokines can also be secreted by macrophages activated by antigen-stimulated T lymphocytes; this illustrates a connection between the innate and the adaptive immunity. The major function of these cytokines is to induce early inflammatory reactions to infectious agents. The examples of cytokines that belong to this group are **TNF**, **IL-1**, **Type I interferons**, **IL-6**, **IL-10**, **IL-12**, **IL-15**, and **chemokines**.

2. **Mediators and regulators of adaptive immunity**. These cytokines are produced mainly by activated T lymphocytes in response to specific antigens. Some of these cytokines act primarily on other lymphocyte populations, regulating their growth and differentiation; other cytokines of this group affect mostly nonlymphoid effector cells (macrophages and inflammatory leukocytes). The examples of cytokines that belong to this group are **IL-2**, **IL-4**, **interferon γ**, **IL-5**, **lymphotoxin**, **TGF-β**, **IL-13**, **IL-16**, **IL-17**, and **migration inhibiting factor (MIF)**.

3. **Stimulators of hemopoiesis**. These cytokines are produced by bone marrow stromal cells, leukocytes, and other cells and act as growth and differentiation factors of leukocyte precursors of various lineages. The examples of cytokines that belong to this group are **stem cell factor** (also known as "**steel factor**" or **c-kit ligand**), **IL-3**, **IL-7**, **granulocyte-monocyte colony-stimulating factor (GM-CSF)**, **IL-9**, and **IL-11**.

10.4 Is there a classification of cytokine receptors?

Yes. The current classification of cytokine receptors divides them into five distinct groups:

1. **Type I cytokine receptors**, or **hemopoietin receptors**. Their "signature" feature is the presence of one or more copies of an N-proximal domain that contains two conserved pairs of cysteine residues, and a membrane-proximal sequence WSXWS, where W is tryptophane, S is serine, and X is any amino acid residue. The name "hemopoietin receptors" comes from

CHAPTER 10 Cytokines

the fact that these receptors serve for binding of hemopoietins, or cytokines that stimulate hemopoiesis (see Section 10.3; M-CSF and stem cell factor are exceptions). They are also the receptors for IL-2, IL-4, IL-5, IL-6, IL-12, IL-13, and some noncytokine hormones (growth hormone and prolactin).

2. **Type II cytokine receptors**. These are similar to type I cytokine receptors in that they have two extracellular domains with conserved cysteines, but they do not have the WSXWS motif. The type 2 cytokine receptors serve for the binding of interferons alpha, beta and gamma, and of IL-10.

3. **Cytokine receptors that belong to the immunoglobulin superfamily**. These receptors contain extracellular immunoglobulin-like domains (see Chapter 3) and serve for the binding of IL-1, M-CSF, and stem cell factor.

4. **TNF receptors**. These are receptors with cysteine-rich extracellular domains; as mentioned in Chapters 7 and 8, they have cytoplasmic domains that bind signaling proteins collectively called TRAF. The receptors of this family bind TNF, CD40 ligand, Fas ligand, lymphotoxin, and nerve growth factor.

5. **Serpentines**, or receptors with seven transmembrane alpha-helical domains, have a peculiar structure: they span the cell membrane back and forth seven times (thus resembling the coils of a serpent, hence the name). These receptors bind chemokines.

10.5 What is the subunit composition of these receptors?

Only TNF receptors consist of one polypeptide chain. Other cytokine receptors consist of unique ligand-binding chains and one or more signal-transducing chains. The latter can be unique for the particular cytokine receptor, but also can be (and often are) shared with other cytokine receptors. For example, the IL-2 receptor consists of the alpha, beta, and gamma chains, and its gamma chain is identical to the gamma chain of the IL-4 receptor (-R) and of the IL-15R. The IL-5R and the GM-CSFR share their beta subunits, and the IL-6R and the IL-11R- the subunit called gp130.

10.6 How are the signals generated by cytokine binding transduced inside the cell?

Cytokines employ a number of signal transduction pathways, of which the best studied and perhaps the most important is the so-called **JAK-STAT pathway**, used by type I and type II cytokine receptors. Other pathways include signaling through GTP-binding proteins similar to described in Chapter 7 (chemokine receptors), signaling through TRAF or "death domains" (TNFR family, see Chapters 8 and 9), signaling through receptor-associated tyrosine kinases (hemopoietins), and signaling through a protein called Toll (IL-1R, IL-18R) (more about the Toll pathway in Chapter 11).

10.7 What is the JAK-STAT signal transduction pathway?

This pathway was discovered and dissected in the late 1980s and early 1990s, during the analysis of interferon signaling (see later sections), but is currently thought to be principally similar in all cells that express type I and type II cytokine receptors. **JAK** stands for "**Janus kinases**" – tyrosine kinases named after the two-headed Roman god Janus, because these molecules have two kinase domains. (JAK also stands for "just another kinase," the name given to this family of molecules because for several years their ubiquitous expression was known, but their function was not.) **STAT** stands for "**signal transducers and activators of transcription**" – transcription factors activated by JAKs and capable of up-regulation of gene transcription. Four individual JAK molecules, called JAK-1, -2, -3, and tyk-2, and seven individual STAT molecules, called STAT-1, -2, -3, -4, -5a, -5b, and -6, were identified in mammals. The current working hypothesis states that the unique amino acid sequences in the different cytokine receptors provide the scaffolding for specifically binding, and thereby activating, different **combinations** of JAKs and STATs.

The transduction of the signal triggered by the binding of a cytokine to its specific receptor begins with the activation of a JAK enzyme, which is a large (120–140 kDa) molecule loosely attached to the cytoplasmic domain of a cytokine type I or type II receptor. Such activation becomes possible when **two subunits of a cytokine receptor, each containing an attached JAK**, are **brought together** as a result of the cytokine binding. When this juxtaposition happens, the two JAKs mutually phosphorylate (transphosphorylation), and they also phosphorylate tyrosine residues of the receptor subunits to which they are attached. This, in turn, allows inactive cytosolic STAT proteins to bind to these newly appearing phosphotyrosine residues. One monomeric STAT (mol. wt. 80–90 kDa) binds to each of the two phosphorylated JAKs. After this, the bound STAT proteins immediately become phosphorylated by the receptor-associated JAKs, after which the two STAT molecules bind to each other, forming a dimer, and dissociate from the receptor. The STAT dimers are then **translocated** to the nucleus, where they bind to DNA sequences of the promoter regions of cytokine-responsive genes and activate gene transcription. Interestingly, once a STAT dimer dissociates from the receptor, new STAT monomers can immediately bind to the vacated phosphotyrosine residues of the receptor–JAK complex, be phosphorylated, dimerize, dissociate, and be translocated. Thus, the high-affinity binding of one cytokine molecule to its specific receptor initiates a long series of sequential phosphorylations, dimerizations, and nuclear translocations of STAT molecules.

10.8 Is the JAK-STAT signal transduction pathway the only one that can be activated by the binding of cytokines to cytokine type I and type II receptors?

No. Cytokines that bind to these receptors may, in addition to the JAK-STAT pathway, activate other pathways of signal transduction. For example, IL-2 can activate the Ras-MAP pathway described in Chapter 7.

CHAPTER 10   Cytokines

10.9   Are negative signals also transduced upon cytokine–cytokine receptor interaction?

Yes. Recently, a group of proteins collectively called suppressors of cytokine signaling (SOCS) were identified. The eight known members of this group of proteins exert a negative effect on cytokine signaling. In addition, SHP-1 phosphatase (described in Chapter 8) can also be engaged during cytokine signaling, and its activation can provide a negative feedback on the cytokine signal transduction similar to the suppression of B lymphocyte activation described in Chapter 8. Finally, a group of proteins was recently discovered, where individual members can selectively inhibit individual STATs. These proteins were called PIAS (for "protein inhibitors of individual STATs"). Their biological role remains unknown.

B. CYTOKINES THAT MEDIATE AND REGULATE INNATE IMMUNITY

In this and the following sections we will list the cytokines that mediate and regulate innate immunity, as given in Section 10.3.

TUMOR NECROSIS FACTOR (TNF)

10.10   What are the principal functions of TNF?

The name "tumor necrosis factor" (TNF) is actually a misnomer. This cytokine was discovered in the early 1980s as a serum factor produced in mice treated with LPS. A similar factor was isolated from cultures of LPS-activated murine and human macrophages, macrophage-derived cell lines, and some leukemia cell lines. This factor was called TNF because it showed a cytotoxic and cytostatic effect of a number of malignantly transformed cell lines and caused necrosis of tumors *in vivo*. Later, however, it became apparent that the anti-tumor effect of TNF is caused by very high concentrations of this agent, while smaller, physiologic concentrations of this cytokine cause other effects. **The principal physiologic effect of TNF is to mediate an acute inflammatory response to Gram-negative bacteria and some other infectious microorganisms**. Large concentrations of TNF cause pathologic effects that include tissue necrosis and other phenomena (see later sections).

10.11   What is the cellular source and the target of TNF?

The principal **source** of TNF is an **LPS- or enterotoxin-activated macrophage**. *In vivo*, macrophages become activated by LPS and enterotoxin during acute and chronic infections by Gram-negative bacteria that express these substances. In addition, TNF can be produced by antigen-activated T lymphocytes, natural killer (NK) cells, and mast cells (see Chapter 2). The **target** of TNF is virtually **any cell**, because all mammalian cell types examined express TNF receptors.

10.12 What is the structure of TNF and its receptor(s)?

TNF is initially synthesized as a transmembrane protein that belongs to the so-called **type II proteins**. This means that, unlike in most transmembrane (type I) proteins), its N-terminus is cytoplasmic and its C-terminus is extracellular. It is expressed as a homotrimer, in which each of the three subunits has a molecular weight of 25 kDa. After the homotrimer is expressed, a membrane-bound enzyme cleaves from each of its subunits a 17-kDa fragment; the three cleaved fragments assemble together, forming a **pyramid-like secreted 51-kDa homotrimer**. The receptor-binding sites are at the base of the pyramid; this allows more than one TNF receptor to be bound by one TNF molecule simultaneously.

There are two types of TNF receptor, the so-called **TNF-RI** and **TNF-RII**. The TNF-RI is a 55-kDa protein, to which TNF binds with the K_d of 10^{-9} M. The TNF-RII is a 75-kDa protein, to which the cytokine binds with the K_d of 5×10^{-10} M. Both receptors are expressed ubiquitously, but there is a difference in the way they transduce the signal upon the TNF binding. Most biologic effects of TNF are mediated through its binding to TNF-RI, which is followed by recruitment of an adapter protein, **c-E10**. This protein, when recruited, activates intracellular **caspases** and leads to **apoptosis** (see Chapters 7–9). Alternatively, the binding of TNF to TNF-RI can recruit an adapter protein, **Grb2**. This protein simultaneously binds to another cytoplasmic mediator protein called **SOS**, leading to the activation of **c-Raf-1 kinase pathway** and eventually to the up-regulation of gene transcription by activated transcription factors. It is not clear what factors play the decisive role in the balance between cellular activation triggered by the Grb2-SOS-Raf pathway and apoptosis triggered by c-E10-caspase pathway. The binding of TNF to TNF-RII leads to the attachment of TRAF proteins to the cytoplasmic portion of the receptor, to the recruitment of RIP adapter protein, and to activation of transcription factors similar to that induced by CD40L-CD40 interaction (see Chapter 8).

10.13 What are the physiologic effects of TNF?

As we mentioned in Section 10.10, TNF is a cytokine whose major physiologic effect is **to induce and promote inflammatory reactions**. The latter involve **recruitment** of the so-called "**inflammatory leukocytes**" (neutrophils and monocytes) to the site of infection, and activation of these cells. Briefly, the inflammatory reactions stimulated by TNF can be described as follows. The binding of TNF to its receptor on the **endothelial cells** makes these **express adhesion molecules** (which we already mentioned in Chapters 7 and 8). Among these molecules, there are structures (called "addressins") that are able to bind selectins and other receptors expressed by leukocytes. Because of the interaction between addressins and their counterparts on leukocytes, the latter are **retained** on the surface of endothelial cells and eventually **migrate** from bloodstream to the tissue at the site of infection. In parallel, TNF stimulates endothelial cells, and also macrophages, to secrete **chemokines** (see about them in later chapters), which further increase the migration of leukocytes from blood to tissue through chemotaxis. TNF also stimulates the secretion of **IL-1** by macrophages (see later). Finally, TNF **induces apoptosis** in some cells that it binds; the biological significance of this effect is unknown.

10.14 What are the pathologic effects of TNF?

The injurious effects of TNF develop after a severe bacterial infection, when this cytokine is secreted in **large amounts** in response to a strong stimulus. Under these conditions, TNF enters the bloodstream and functions at long distances, as an **endocrine** factor. The principal **systemic actions** of TNF are the following: (1) its binding to its receptor expressed on neurons of the hypothalamus (a region in the brain) triggers production of prostaglandins, resulting in **fever**; (2) its binding to hepatocytes (liver cells) stimulates the latter to produce amyloid A and other proteins of the so-called "**acute phase response**" of the liver to inflammation, as well as fibrinogen; (3) TNF induces **cachexia**, a rapid loss of muscle and fat tissue, partially due to the inhibition of appetite and partially because TNF inhibits enzymes that control the utilization of circulating lipids by tissues; (4) at very large concentrations, TNF inhibits cardiac muscle contractility and vascular smooth muscle tone, resulting in the rapid **loss of blood pressure** (shock); (5) also at very large concentrations, TNF causes **disseminated intravascular coagulation** (or intravascular thrombosis) by shifting the normal balance between procoagulant and anticoagulant factors secreted by endothelial cells; (6) TNF may cause a profound **decrease in blood glucose** concentration, in part because of the overutilization of glucose by muscle and in part because of the liver failure. The combination of the blood pressure loss, disseminated intravascular coagulation, fever, and severe hypoglycemia is indicative of the **septic shock**, and may result in death.

10.15 Can septic shock be diagnosed and treated?

Yes. The most informative parameter indicative of septic shock is the serum concentration of TNF. It is now possible to measure it in patients, because many standard diagnostic kits that include monoclonal anti-TNF antibodies are available. Unfortunately, these antibodies are not efficient as therapeutic devices, perhaps because of the redundancy of cytokines (see above). The treatment of septic shock includes intensive antibiotic therapy and detoxication therapy (infusions of glucose, polyvinylpyrrolidone, balanced salt solutions, diuretics, etc.)

IL-1

10.16 What is the main function of IL-1?

Although IL-1 was originally discovered as a factor that enhanced the proliferation of thymocytes (see Section 10.1), it is now clear that its main function is **similar to that of TNF**, i.e., to induce and promote the inflammatory reaction in response to Gram-negative bacteria and some other infectious microorganisms.

10.17 What is the cellular source and the target of IL-1?

The major cellular **source** of IL-1 is **the same as for TNF**, i.e., the activated macrophage. Unlike TNF, however, IL-1 can be also produced in smaller quantities by a rather broad range of cells that includes neutrophils, epithelial cells (especially keratinocytes), and endothelial cells.

10.18 What is the structure of IL-1 and its receptor(s)?

There are **two forms of IL-1**, called **IL-1α** and **IL-2β**, coded by separate genes and showing only 30% of the structural homology but, nevertheless, binding to the same IL-1 receptors. Both of them are synthesized as 33-kDa precursors and are then enzymatically cleaved, resulting in the formation of the active 17-kDa protein. (IL-1α, but not IL-1β, can be active in its 33-kDa form as well). The enzyme that cleaves IL-1β is called "**i**nterleukin-1 **c**onverting **e**nzyme" (**ICE**); we mentioned it in Chapter 9 as being historically the first enzyme belonging to the family of apoptosis-inducing caspases. Most of the IL-1 found in the circulation is IL-1β.

There are **two forms of IL-1 receptor** (IL-1R), called **IL-1R type 1** and **IL-R1 type 2**. The IL-1R type 1 is expressed ubiquitously and actually mediates the biological responses to IL-1. The IL-1R type 2 is expressed exclusively on B lymphocytes and is nonfunctional. It may be a "decoy" that inhibits the biological effect of IL-1, competing for the cytokine with the IL-1R type 1. The cytoplasmic portion of the IL-1R type 1 is homologous to the *Drosophila* surface protein called **Toll**. As we will detail in Chapter 11, mammalian homologs of Toll play a significant role in the innate as well as adaptive immunity. The binding of IL-1 to IL-1R type 1 activates a signaling pathway similar to the one activated by CD40L-CD40 interaction (see Chapter 8).

10.19 What are the physiologic and pathologic effects of IL-1?

They are very similar to those of TNF. The two differences in the effects of the two cytokines are that: (1) IL-1 does not induce apoptotic death of cells even at high concentrations; and (2) IL-1 does not lead to septic shock, even although at high concentrations it can induce hepatic acute phase responses, fever, intravascular thrombosis, loss of appetite, hypoglycemia, and other metabolic disturbances.

Mononuclear phagocytes are capable of producing the so-called **IL-1 receptor antagonist** (IL-Ra), which is structurally similar to IL-1 and able to bind the same receptors, but is biologically inactive. IL-Ra is thought to be a natural antagonist of IL-1 and a means of regulation of the cytokine's biological effects.

TYPE I INTERFERONS

10.20 Which cytokines belong to the category of type I interferons, and what is their principal function?

This category includes interferons alpha (IFN-α) and beta (IFN-β). The principal function of these cytokines is to mediate the early innate immune response to viral infections.

CHAPTER 10 Cytokines

10.21 What is the cellular source and the target of IFN-α and IFN-β?

IFN-α, which is, actually, a family of molecules (see next section), is produced by a subset of **mononuclear phagocytes**, and it is also called **leukocyte interferon**. IFN-β is a single substance produced by a variety of cell types, most notably **fibroblasts**, and is also called fibroblast interferon. The target of type I interferons is, essentially, any cell. The production of both type I interferons is strongly **activated by viral infection**. Experimentally, it can be also activated by double-stranded RNA, which is thought to mimic viral infection.

10.22 What is the structure of type I interferons and their receptors?

The IFN-α family includes approximately 20 species of molecules, each coded by a separate gene. These molecules consist of five individual alpha-helical subunits, of which four are held closely together and one is "loose." The IFN-β is a single polypeptide coded by one single gene. There is very little structural homology between the IFN-β and any of the 20 representatives of the IFN-α family. In spite of this pronounced structural difference, **all 20 molecules of the IFN-α family, as well as the IFN-β, bind to one single type I interferon receptor**. This receptor is expressed ubiquitously. It is a heterodimer of two polypeptide subunits, sometimes called INFAR-1 and INFAR-2, each of which has a binding site for a type I interferon molecule. As discussed in Section 10.7, simultaneous binding of the two interferon receptor subunits by two type I interferon molecules activates the receptor-bound JAKs and triggers the JAK-STAT pathway of intracellular signal transduction.

10.23 What is the physiologic effect of type I interferons?

Type I interferons were known for a long time to be capable of "**inducing an anti-viral state**" in target cells. By inducing the expression of target cell enzymes such as $2',5'$-oligoadenylate synthetase (OAS) upon binding to their receptors, type I interferons **inhibit viral replication**. Type I interferons **enhance the expression of MHC Class I antigens** by target cells, thus facilitating their recognition by the MHC Class I-restricted $CD8^+$ cytotoxic T lymphocytes and providing a link between innate and adaptive immunity to viruses. These two effects are performed in both paracrine fashion (i.e., a neighboring cell serves as the target for the produced cytokine), or in an autocrine fashion (the same cell that secretes the cytokine is its target). Type I interferons also **enhance the expression of IL-12 receptor**, thus promoting the T_h1 pathway of T helper differentiation (see later sections). Finally, these cytokines inhibit the proliferation of many cell types; the biologic significance of this effect is not quite understood.

IL-6

10.24 What are the main properties of IL-6?

This cytokine functions in innate as well as in adaptive immunity. Its role in innate immunity is to **stimulate the synthesis of acute phase proteins by hepatocytes**, similar to the effect of TNF at large concentrations (see Section 10.14). By stimulating the acute phase response, IL-6 contributes into the body's systemic inflammatory reaction. In adaptive immunity, IL-6 serves as a **growth factor for cells of the B-lymphocyte lineage**, especially terminally differentiated plasma cells. IL-6 is also capable of promoting growth of plasmacytomas (or myelomas) and hybridomas.

IL-6 is produced by mononuclear phagocytes, endothelial cells, fibroblasts, and other cells, in response to microbial infections and to other cytokines (especially IL-1 and TNF). Some activated T lymphocytes also produce IL-6. The cells responding to IL-6 are the cells that produce it (macrophages, endothelial cells, and fibroblasts), and also differentiated antibody-producing B lymphocytes and plasma cells. Structurally, IL-6 is a homodimer, consisting of two subunits with globular domains. Its receptor is a two-subunit structure; one of the subunits binds IL-6 and the other one (called **gp130**) transduces the signal by activating the JAK-STAT signaling pathway.

IL-10

10.25 What are the main properties of IL-10?

IL-10 is a major **inhibitor of activated macrophages**. It is produced by activated macrophages, and thus serves as a negative feedback in macrophage activation. By inhibiting macrophage responses, IL-10 down-regulates and terminates many reactions of innate immunity that involve these cells. This cytokine inhibits the production of macrophage-derived TNF and IL-12, thus suppressing inflammatory reactions as well as the T_h1 pathway of T helper cell differentiation. IL-10 also **down-regulates the expression of Class II MHC molecules and co-stimulatory molecules** on the affected macrophages. Finally, IL-10 plays an as yet not very clearly defined role in adapted immunity, because it **enhances the proliferation of B lymphocytes**. In combination with IL-2, IL-10 stimulates the differentiation of CD40-stimulated B lymphocytes into plasma cells (see Chapter 8 and later sections of this chapter). IL-10 is a polypeptide with four α-helical domains that binds to a typical type 2 cytokine receptor and activates the JAK-STAT signaling pathway.

IL-12

10.26 What are the main properties of IL-12?

IL-12 is **a principal mediator of the early innate responses to intracellular microbes**, and a key inducer of cellular immunity to these microbes. It is a cytokine that serves as a very good illustration of a link between innate and

CHAPTER 10 Cytokines

adaptive immunity. The two principal **sources** of IL-12 are activated **macrophages** and **dendritic cells**. Structurally, IL-12 is a heterodimer that consists of a 35-kDa subunit (p35) and a 40-kDa subunit (p40). Interestingly, the p40 subunit of this cytokine is homologous to the receptor of another cytokine, namely of the IL-6R. The receptor for IL-12 is a heterodimer composed of β1 and β2 subunits and a typical activator of the JAK-STAT signaling pathway. Cytokines, notably interferons, strongly up-regulate the β2 subunit of the receptor. The latter (but not the former) subunit serves as a signal transducer. The stimuli for IL-12 production include LPS, viral infection, and infection by intracellular bacteria (for example, *Listeria monocytogenes* and *M. tuberculosis*). Activated helper T lymphocytes can induce macrophages and dendritic cells to synthesize IL-12, mainly through the CD40L-CD40 interaction. Interferon gamma (IFN-γ, see later) also stimulates IL-12 production.

The **biological role** of IL-12 is, essentially, **to initiate a series of responses involving macrophages, NK cells, and T lymphocytes**. IL-12 is **a potent stimulator of the T_h1 pathway of helper T cell differentiation** (see later sections). It is also a potent stimulator of IFN-γ by NK cells and T lymphocytes, and an enhancer of the cytolytic functions of activated NK cells and $CD8^+$ cytolytic T lymphocytes.

IL-15 AND IL-18

10.27 What are the main properties of IL-15 and IL-18?

These cytokines are involved in innate immunity; both of them are strong **activators of NK cells**. IL-15 is structurally homologous to IL-2, and may serve as its functional analogue early in the response to infectious agents, when reactions of adaptive immunity are still not developed. IL-18 is structurally homologous to IL-1 and signals through a Toll-like receptor, but functionally is homologous to IL-12 rather than to IL-1, and shows a strong synergism with IL-12.

CHEMOKINES

10.28 What are chemokines, and what are their main functions?

The name "chemokine" is a fusion of the terms "cyto**kine**" and "**chemo**taxis," the latter meaning the movement of cells toward a chemical stimulus or attractant. Just as the name indicates, chemokines promote leukocyte **movement** and regulate the migration of leukocytes from blood to tissues. Obviously, if they possess such an ability, they play an important role in inflammatory reactions. However, recent studies showed that chemokines also play a role in **lymphocyte homing** in secondary lymphoid organs, as well as in the development of T_h1 and T_h2 pathways of T helper cell differentiation, in the function of B lymphocytes, dendritic cells, in the viral infection of lymphocytes, in angiogenesis, and metastatic processes. No wonder that chemokines and their receptors have been a major focus of an extensive research.

10.29 What is the source of chemokines, their structure, and receptor(s)?

Chemokines are produced by **leukocytes** and by **several types of tissue cells**, notably by endothelial cells, epithelial cells, and fibroblasts. The **stimuli** for their production include **microbial infections**, TNF, IL-1, and **also antigen-activated T cells**. Thus, chemokines provide yet another important link between innate and adaptive immunity. While most chemokines require the above stimuli, some are produced **constitutively**; these latter are believed to play an especially significant role in the lymphocyte homing. (For example, a chemokine has been recently identified that controls the migration of murine B lymphocytes into the follicles of secondary lymphoid organs.) Chemokines may be secreted, but they also may be displayed to leukocytes while being associated with heparan sulfate proteoglycans on the surface of endothelial cells. In this case, their high local concentration facilitates the leukocyte migration very strongly. Chemokines also **control the nature of the inflammatory infiltrate**, because different chemokines act on different leukocytes; for example, the chemokine called IL-8 preferentially recruits neutrophils, while the chemokine called eotaxin preferentially recruits eosinophils.

The large family of chemokines includes more than 50 individual molecules; they are 8- to 12-kDa polypeptides that contain two internal disulfide loops. Chemokines are usually divided into two subfamilies. The first of these subfamilies is called CC, and includes chemokines in which the two terminal cysteines ("C") are adjacent. The second subfamily, CXC, includes chemokines in which the two terminal cysteines are separated by one (varying) amino acid. The receptors for CC cytokines do not bind CXC chemokines, and vice versa. There are as many as ten known receptors for the CC, and six known receptors for the CXC chemokines. These receptors show overlapping specificity for chemokines within each family, and are expressed on leukocytes. T lymphocytes show the largest number of distinct chemokine receptors on their surface. Chemokine receptors consist of seven α-helical domains and signal as GTP exchange proteins (see Chapter 7). Certain chemokine receptors, notably CCR5 and CXCR4, act as **co-receptors for HIV** (see Chapter 18).

C. CYTOKINES THAT MEDIATE AND REGULATE ADAPTIVE IMMUNITY

IL-2

10.30 What are the principal functions of IL-2?

As was mentioned at the beginning of this chapter, IL-2 was discovered in 1976 as a "T cell growth factor" (TCGF). In those early studies, IL-2 was characterized as a polypeptide that was released into the culture medium by polyclonally activated T lymphocytes; it acted on other activated T lymphocytes, supporting their survival and proliferation in culture. Currently, IL-2 is viewed as **a major growth factor for antigen-activated T lymphocytes that is responsible for T-cell clonal expansion after antigen recognition**. However, it is now clear that IL-2 is a typical pleiotropic

CHAPTER 10 Cytokines

cytokine that exerts a number of effects on a variety of cell types. In particular, IL-2 stimulates the growth and differentiation of B lymphocytes, activates NK cells, and potentiates apoptotic death of antigen-activated T lymphocytes.

10.31 What is the source of IL-2, and what is its structure?

IL-2 is produced by **activated T lymphocytes**, **mostly CD4$^+$ T cells** and, **in smaller amounts, by CD8$^+$ T cells**. Activation of T cells by antigens and co-stimulators (or polyclonal activators like plant lectins PHA and ConA) is the necessary prerequisite for the production of IL-2. As was mentioned in Chapter 9, T lymphocytes that recognize antigen in the absence of co-stimulators are rendered tolerant and do not produce IL-2. The IL-2 production becomes possible after the transcriptional activation of the IL-2 gene as a result of antigen- and co-stimulator-triggered signal transduction (see Chapter 7). IL-2 production is transient, its peak occurring 8–12 hours after activation.

IL-2 in its mature secreted form is a 14-kDa to 17-kDa glycoprotein with a globular tertiary structure. Like all cytokines that interact with type I cytokine receptors, it contains four α-helices that contact the chains of its specific receptor.

10.32 What is the structure of the IL-2 receptor (IL-2R)?

The receptor consists of **three non-covalently associated transmembrane proteins called α, β, and γ subunits**. The α subunit has a molecular weight of 55 kDa, does not formally belong to the type 1 cytokine receptors, and is expressed exclusively on activated T lymphocytes. It was the first of the three subunits to be identified by a monoclonal antibody called "anti-Tac." This subunit binds IL-2, but the binding of the cytokine to this subunit alone has no biologic effects since the α subunit does not transduce any signals. The affinity of the IL-2–IL-2Rα interaction is very low. The β chain of the IL-2R is expressed at low levels on resting T lymphocytes and on NK cells in association with the IL-2R γ chain. The latter is also called "common (or "shared," or "promiscuous") γ chain" (γc), because this chain is also a component of other cytokine receptors (notably IL-4, IL-7, and IL-15). The affinity of the binding of IL-2 to the complex of IL-2Rβ and IL-2Rγc ($K_d \sim 10^{-9}$ M) is somewhat higher than the affinity of the cytokine's binding to the IL-2Rα alone, and the complex of IL-2Rβ and IL-2Rγc does transduce activating signals. (Because of that, IL-2 can activate NK cells as well as IL-R2γc-expressing B cells at relatively high concentrations.) **When cells are activated by antigen and co-stimulators, the IL-2Rα is rapidly expressed**; this leads to the formation of the complex of all three subunits (IL-2Rα, IL-2Rβ, and IL-2Rγc). The binding of IL-2 to the three-subunit complex is characterized by very high affinity ($K_d \sim 10^{-11}$ M), allowing the activated T cells to respond to small concentrations of the cytokine. The transduction of activator signals after the cytokine binding involves the JAK-STAT pathway (see Section 10.7), and also the PLCγ- and the Ras signaling pathways (see Chapter 7). Chronic T-cell stimulation leads to shedding of IL-2Rα; when this subunit is detected in the serum, it is indicative of a strong antigenic stimulation, and is used as a diagnostic marker in transplantation clinics.

10.33 What are the biologic effects of IL-2?

As already mentioned, these include the promotion of the antigen-activated T cells' clonal expansion; the activation of NK cells; the promotion of the B cells' proliferation and antibody production, and the potentiation of the activated T cells' apoptotic death. A brief examination of the mechanisms of these effects is worthwhile at this point.

(a) **Stimulation of activated T cells**. The principal mechanism of this effect is thought to be mediated by proteins called **cyclin D2** and **cyclin E**. These proteins, whose concentration rises upon the binding of IL-2, activate a number of cyclin-dependent kinases. The latter phosphorylate and activate a variety of cellular proteins that stimulate transition from the G_1 to the S phase of the cell cycle. On the other hand, the binding of IL-2 reduces the concentration of a cellular protein called **p27**. This latter protein inhibits cyclin-dependent kinases. IL-2 thus lifts the suppression of cell cycle mediated by p27. The cytokine binding also **increases the concentration of the anti-apoptotic protein Bcl-2** (see Chapter 9), thus increasing the T-cell survival. IL-2 also **promotes the secretion of other T-cell-derived cytokines** (notably IFN-γ and IL-4). All these effects are mostly autocrine (i.e., the same cell that made IL-2 responds to it), but they may be also paracrine (i.e., the adjacent T cells also may respond).

(b) **Activation of NK cells**. This effect is mediated through the IL-2Rβ–IL2Rγc complex and, because of the properties of these subunits described in Section 10.32, can be seen only at high concentrations of IL-2. As was mentioned previously, IL-15, which is structurally homologous to IL-2, can mimic this effect. NK cells activated by IL-2 or IL-15 are called **LAK** (from "**l**ymphokine-**a**ctivated **k**iller cells"). The appearance of LAK serves as a diagnostic trait indicative of strong antigen stimulation, e.g., in transplantation clinics during acute transplant rejection.

(c) **Promotion of B-cell proliferation and antibody production**. Since B lymphocytes do not express IL-2R, it is likely that IL-2 mediates this effect through some molecule that serves as its mimic. The signal is likely to be transduced through the γ subunit of IL-4R, which B cells express. Little is known about the molecular mechanism of this effect; notably, the moderate activation of antibody production by IL-2 is **not accompanied by class switch**. IL-10 acts synergistically with IL-2 to promote the antibody production in CD40-activated B lymphocytes.

(d) **Potentiation of the activated T cells' apoptotic death**. As was mentioned in Chapter 9, IL-2 acts as a powerful promoter of the activation-induced cell death, because **it inhibits the anti-apoptotic protein FLIP** (see Section 9.15). It may be that if an immune response is persistent and T cells are exposed to increasing quantities of IL-2, this pro-apoptotic property of IL-2 takes over its pro-survival property described in (a). Knockout mice lacking IL-2 or IL-2R develop a systemic autoimmune disorder, indicating that normally, IL-2 eliminates autoimmune T-cell clones that are repeatedly stimulated by self antigens.

CHAPTER 10 Cytokines

IL-4

10.34 What are the main functions of IL-4?

IL-4 was discovered as a factor produced by activated T cells and capable of enhancing the proliferation of B lymphocytes stimulated by antibodies to surface IgM. Initially, IL-4 was called "B cell growth factor-1" (BCGF-1) or "B cell stimulatory factor-1" (BSF-1). As it often happened with cytokines, it became soon apparent that IL-4 is a pleiotropic cytokine, and that the activity associated with it at the time of its original discovery was not its major function. Currently, it is considered that **the major function of IL-4 is to stimulate the development of T_h2 cells from naive $CD4^+$ helper T lymphocytes, and to direct antibody class switch to the production of IgE.**

10.35 What is the source, the target, and the structure of IL-4?

IL-4 is produced predominantly by **activated $CD4^+$ helper T lymphocytes** that belong to the **T_h2 helper cell subset**. Activated mast cells and basophils can also produce some IL-4. The target of IL-4 are activated T cells, B cells, mast cells, and basophils. IL-4 is a typical member of the four α-helical cytokine family (see above).

10.36 What is the structure of the IL-4 receptor, and how does it transduce signals?

The IL-4 receptor (IL-4R) is a heterodimer that consists of the cytokine-binding α chain (structurally distinct from the IL-2Rα) and (in lymphocytes) the signal-transducing γc chain, which is shared with the IL-2R. Nonlymphoid cells express the IL-4R α chain in association with a different protein, which is also a component of the receptor for IL-13 (see later). IL-4R transduces its signal through the JAK-STAT pathway and the pathway that involves the so-called **insulin response substrate 2 (IRS-2)**. Most biologic properties of IL-4 are associated with the JAK-STAT pathway, and, in particular, with the STAT6 protein, which seems to be uniquely activated by IL-4.

10.37 What are the biologic effects of IL-4?

The most important biologic effect of IL-4 is **to promote IgE- and mast cell/eosinophil-mediated reactions, and to suppress macrophage-mediated reactions**. This effect is mediated through promotion of the T_h2 pathway of T helper cell differentiation (discussed later). In addition, IL-4 serves as a growth factor of B lymphocytes and acts in strong synergy with anti-CD40 antibodies or CD40 ligand to stimulate B-cell growth, antibody production, and class switch to IgE in B lymphocytes cultured *in vitro*. This is presumed to reflect the *in-vivo* role of this cytokine in B-cell differentiation.

IL-5

10.38 What are the principal function and properties of IL-5?

IL-5 is an activator of eosinophils that serves as the link between T-cell activation and eosinophilic inflammation. The principal **sources** of IL-5 are the T_h2 subset of activated $CD4^+$ **T lymphocytes** and activated **mast cells**. The IL-5 receptor (IL-5R) is a typical type 1 cytokine receptor. Its 150-kDa signaling subunit is shared with receptors for IL-3 and GMCSF, and is sometimes called the common beta chain. Signaling through IL-5R is mediated through the JAK-STAT pathway.

IL-5 stimulates the growth and differentiation of eosinophils (see Chapter 2), and activates mature eosinophils. The latter express Fc receptors that can bind IgE antibodies, which feature prominently in immediate hypersensitivity reactions (discussed in later chapters). IL-5 acts in concert with IL-4, which is also produced by the T_h2 cells and stimulates the class switch to IgE. Eosinophilic inflammation is a consequence of the immediate hypersensitivity reactions, and is thought to play a major role in the protection against helminthes and arthropods (see later chapters). In addition, IL-5 stimulates the proliferation of B lymphocytes and the production of IgA antibodies.

IFN-γ

10.39 What are the principal functions of IFN-γ?

IFN-γ (also called immune interferon or type II interferon) **is the principal activator of macrophages and the "signature cytokine" of the T_h1 subset of helper T lymphocytes**. IFN-γ possesses some anti-viral activity but, unlike in type I interferons, this activity is not very strong. Instead, IFN-γ functions mainly in the effector phase of immune reactions, and serves as a critical link between innate and adaptive immunity.

10.40 What is the source of IFN-γ?

IFN-γ is produced by **NK cells, $CD4^+$ T_h1 cells, and $CD8^+$ cells**. NK cells produce IFN-γ in response to an unknown microbial product or IL-12. T cells produce IFN-γ in response to antigen recognition, and this production is enhanced by IL-12.

10.41 What is the structure of IFN-γ and its receptor?

IFN-γ is a homodimeric protein containing two 21-kDa to 24-kDa subunits. Its receptor (IFN-γR) is genetically and structurally distinct from the receptor of type I interferons, and consists of two structurally similar subunits, one of which binds the cytokine and the other which transduces the signal. The IFN-γR

CHAPTER 10 Cytokines

belongs to the type II cytokine receptor family and transduces the signal through the JAK-STAT pathway.

10.42 What are the biologic effects of IFN-γ?

IFN-γ is a pleiotropic cytokine that exerts the following biologic effects: it **activates macrophages; enhances the expression of Class I and Class II MHC molecules and co-stimulators on antigen-presenting cells; promotes the differentiation of naive $CD4^+$ T cells to the T_h1 subset; inhibits the proliferation of T_h2 cells; promotes class switch to IgG subclasses (in mice; notably IgG2a); inhibits class switch to IgE; activates neutrophils; and enhances the cytolytic activity of NK cells**.

The mechanisms of these effects may be briefly outlined.

- **Macrophage activation**. This effect is mediated through the STAT-1 protein component of the JAK-STAT pathway, which serves as a transcriptional activator of the genes that control production of biologically active substances by macrophages. These substances include reactive oxygen intermediates and reactive nitrogen intermediates (to be discussed in more detail in later chapters).
- **MHC and co-stimulator up-regulation**. This effect is also mediated through STAT-1, which serves as a transcriptional activator of MHC and co-stimulator genes, as well as genes that code for several important proteins involved in antigen processing (TAP, LMP-2 and LMP-7 components of proteasome, HLA-DM, and others; see Chapter 6).
- **Promotion of the T_h1 subset and inhibition of the T_h2 subset**. This effect is thought to be indirect; it is perhaps mediated through the production of IL-12 by macrophages activated by IFN-γ. IL-12 is stimulated through transcriptional activation by STAT-1. IL-12 directly activates the differentiation of naive ("T_h0") $CD4^+$ cells into T_h1 cells and inhibits the proliferation of T_h2 cells (see Section 10.26).
- **Promotion of IgG subclass switch and inhibition of IgE class switch**. This effect is thought to be mediated through the mechanism of switch recombination (see Chapter 4).
- **Activation of neutrophils and enhancement of NK-mediated cytolytic activity**. These effects are thought to be mediated through the JAK-STAT pathway, similarly to the effect of IFN-γ on macrophages.

TGF-β

10.43 What are the main properties of TGF-β?

As the name indicates, transforming growth factor-beta (TGF-β) was originally discovered as a cytokine that conferred on normal cells some properties of malignantly transformed cells. In particular, cells treated with TGF-β acquired the ability to survive in culture media with low percentage of serum or in semisolid agar. Subsequently, it was shown that TGF-β is actually a family of three polypeptides designated TGF-β1, TGF-β2, and TGF-β3. TGF-α is structurally related to these molecules, but its function, as well as the functions of TGF-β2 and TGF-β3, is not related to the immune system. Operationally, TGF-β1 in the context of immunology can be called TGF-β.

The principal functions of TGF-β are **to inhibit the proliferation of T lymphocytes and the activation of macrophages, and to enhance the production of IgA**. Because of its strong negative effect on T cells and macrophages, TGF-β is sometimes called an "anti-cytokine." TGF-β is also **able to inhibit the recruitment of inflammatory leukocytes** (mostly neutrophils) to the site of inflammation, and to **stimulate the synthesis of extracellular matrix proteins** like collagen, thus promoting tissue repair. The source of TGF-β is mostly a subset of antigen-activated T helper cells (sometimes called "T_h3" to stress on their unique suppressor properties, distinguishing them from both T_h1 and T_h2). In addition, LPS-activated macrophages and other cells produce TGF-β. TGF-β is a homodimer that is secreted after proteolytic cleavage of an inactive precursor. There exist two distinct TGF-β receptors (TGF-βRI and TGF-βRII), both of which signal through a serine-threonine kinase domain.

LYMPHOTOXIN (LT)

10.44 What are the main properties of LT?

LT is a cytokine that is very similar to TNF in regard to its structure and biological effect; because of its close similarity to TNF, the latter is sometimes called TNFα and the LT is called TNFβ. However, while TNF is produced by LPS-activated macrophages, **LT is produced by some antigen-activated T lymphocytes**. Since the quantities of LT made by T cells are much smaller than the quantities of TNF made by macrophages, LT never exerts any systemic effects, but acts like a local promoter of inflammation. Studies in knockout mice showed that LT also has an interesting role in the **development of lymphoid organs**. In particular, mice lacking LTβ (a membrane-associated form of LT) show various defects in the architectonics of their lymph nodes, Peyer's patches, and spleen.

IL-13

10.45 What are the main properties of IL-13?

IL-13 is a cytokine structurally similar to IL-4 and is **produced by T_h2 cells and some epithelial cells**. Its receptor is expressed on macrophages. IL-13 **suppresses macrophage activation**, acting in concert with IL-4, IL-10, and TGF-β and antagonizing IFN-γ.

CHAPTER 10 Cytokines

IL-16, IL-17, AND MIF

10.46 What are the main properties of IL-16, IL-17, and MIF?

Although these cytokines have been identified as distinct molecules, their functions remain unclear. IL-16 is implicated in the chemotaxis of eosinophils, IL-17 is thought to mimic pro-inflammatory effects of TNF and LT, and MIF (as the name indicates) inhibits the migration of macrophages.

D. CYTOKINES THAT STIMULATE HEMOPOIESIS

STEM CELL FACTOR (C-KIT LIGAND)

10.47 What are the main properties of stem cell factor?

Stem cell factor, or c-kit ligand, is a cytokine **produced by bone marrow stem cells** and **acting on bone marrow stem cells, immature thymocytes, and mast cells**. Stem cell factor promotes the survival of these cells and stimulates their proliferation. It is thought that stem cell factor also makes bone marrow stem cells receptive to other hemopoietins. The name "c-kit ligand" originates from the fact that this cytokine binds to a cell surface-expressed tyrosine kinase coded by a cellular proto-oncogene c-*kit*. Stem cell factor is also known as "steel factor" because the mutant mouse strain that does not have its secreted form is called steel. In these mutants, the development of mast cells in mucosal tissues is impaired, but all other pathways of hemopoiesis are normal, suggesting that membrane-bound form of stem cell factor is of a greater importance for hemopoiesis. Elimination of both forms of stem cell factor by gene knockout is lethal.

IL-3

10.48 What are the main properties of IL-3?

In the mouse, IL-3 is a cytokine **produced by $CD4^+$ T lymphocytes** and **acting on immature bone marrow progenitors of various lineages**. Precursors of all known blood cells have been shown to be affected by IL-3. Because of that, and because IL-3 and other hemopoietins are often assayed for their ability to stimulate the formation of colonies in culture, IL-3 is also called "multilineage cytokine" or "multi-CSF" (where CSF stands for "colony-stimulating factor"). IL-3 is a member of the four-α-helical cytokines. Its receptor is a heterodimer with the signaling subunit transmitting the signal through the JAK-STAT pathway. In the human, IL-3 has been biochemically characterized and its gene has been cloned, but its function remains unclear.

IL-7

10.49 What are the main properties of IL-7?

IL-7 is a four-α-helical cytokine that is **produced by bone marrow stromal cells** and **acting on immature bone marrow progenitors committed to B- and T- lymphocyte lineages**. This cytokine seems to be crucial for lymphopoiesis because knockout mice lacking IL-7 have decreased numbers of B and T lymphocytes. The receptor for IL-7 signals through JAK-STAT and involves the JAK-3 kinase. Interestingly, IL-7 influences also on mature T lymphocytes cultured *in vitro*, mimicking IL-2. It is unclear whether it possesses such an activity *in vivo*.

GM-CSF, M-CSF, AND G-CSF

10.50 What are the main properties of GM-CSF, M-CSF, and G-CSF?

GM-CSF stands for "**g**ranulocyte-**m**acrophage **c**olony-**s**timulating **f**actor," M-CSF- "**m**onocyte **c**olony-**s**timulating **f**actor," and G-CSF- "**g**ranulocyte **c**olony-**s**timulating **f**actor." **These three cytokines are produced by activated T cells, macrophages, endothelial cells, and bone marrow stromal cells**. They act on the respective bone marrow progenitors to increase production of inflammatory leukocytes. GM-CSF was also shown to activate macrophages and to promote the differentiation and maturation of dendritic cells (see Chapter 2). These cytokines are successfully used in bone marrow transplantation clinics to stimulate the growth and maturation of the progenitors in marrow grafts.

IL-9 AND IL-11

10.51 What are the main properties of IL-9 and IL-11?

These two cytokines are **produced by bone marrow stromal cells**, and their exact role in hematopoiesis is unclear. IL-9 stimulates the growth of some T-cell lines, as well as mast cell progenitors, *in vitro*. IL-11 is known to stimulate megakaryocytopoiesis (maturation of megakaryocytes, the immediate precursors of platelets), and is used to treat patients with platelet deficiencies.

Questions

REVIEW QUESTIONS

1. Of the two terms coined by early immunologists to refer to cytokines, "interleukins" and "lymphokines," why does the first one stay and the second not?
2. Why is it beneficial that the production of most cytokines is a brief and self-limited event?
3. Which of the two terms – "type I cytokine receptors" and "hemopoietin receptors" – would you consider being a better one? Why?
4. In the literature of the 1980s and early 1990s one can often find the term "promiscuous subunit," referring to a cytokine receptor subunit. Explain the term, and give examples.

CHAPTER 10 Cytokines

5. Would signal transduction through the JAK-STAT pathway be possible if the density of the cell membrane increased to such an extent that lateral movement of transmembrane proteins were no longer permitted?
6. How would T-cell-dependent and T-cell-independent immune responses be affected in mice transgenic for SHP-1? In mice that are knockout for SHP-1?
7. What is the advantage of the pyramidal shape of TNF? What other immunologically important molecules have this shape?
8. What facts about the patient, and what symptoms should you consider, when you think about prescribing a test for circulating TNF?
9. Will the functions of IL-1 be affected in mice that are knockout for the ICE gene? If not, why not?
10. Is it, in your opinion, easy or difficult to obtain an anti-IFN-α monoclonal antibody with cytokine-neutralizing properties?
11. It has been known since perhaps the late 1970s that hybridoma clones grow better when peritoneal macrophages are used as a "feeder layer." How would we explain this situation now, knowing so much more about cytokines than immunologists knew then?
12. What is the difference between the CC and the CXC families of chemokines?
13. If freshly isolated human peripheral blood lymphocytes are incubated with IL-2, they will show a proliferative response, but only if IL-2 is present at large concentrations and the cultures are kept for at least 7 days. Explain this finding.
14. What is the connection between IL-2 and cyclins?
15. You have two cultures of B lymphocytes with identical starting B-cell numbers. One of them is supplemented with recombinant CD40 ligand and IL-4, and the other one with recombinant CD40 ligand, IL-2, and IL-10. In which of them would you expect to find more antibodies? In which of them would you expect to find a larger concentration of IgE?
16. How does IFN-γ regulate the expression of MHC molecules, and what are the consequences of this regulation compared to type I interferons?
17. Immunologists say that the production of TGF-β is one of the means that tumor cells use to avoid immune surveillance. Explain this idea.
18. Why is IL-3 called a "multilineage cytokine?"
19. In what way are IL-4, the secreted form of stem cell factor, and IL-9 redundant?

MATCHING

Direction: Match each item in Column A with the one in Column B to which it is most closely associated. Each item in Column B can be used only once.

Column A
1. Redundancy
2. Hemopoietin receptors
3. Serpentines
4. Juxtaposition of two JAKs
5. SOCS
6. Type II protein
7. Amyloid A

Column B
A. Precedes the binding of STAT
B. TGF-β
C. Shares a subunit with IL-13R
D. Acts on T- and B-cell precursors
E. Eosinophils
F. CCR5
G. Is up-regulated by IL-2

8. INFAR-2
9. gp130
10. Down-regulates co-stimulators
11. Production triggered by *M. tuberculosis*
12. Acts as a co-receptor for HIV
13. Binding of IL-2 to the three-subunit complex
14. Bcl-2
15. IL-4R
16. T_h1
17. Activated by IL-5
18. IFN-γ
19. Enhances collagen build-up.
20. Steel factor
21. IL-7
22. Stimulates hemopoiesis *in vivo*

H. Inhibits class switch to IgE
I. GM-CSF
J. Cytokines back-up each other
K. Bind IL-2, IL-3, IL-4, prolactin
L. Inhibit cytokine signaling
M. High concentration of TNF
N. Is inhibited by IL-4
O. Bound by IFN-α and IFN-β
P. Binds IL-6
Q. $K_d \sim 10^{-11}$ M
R. Span membrane seven times
S. Same as c-*kit* ligand
T. TNFR
U. IL-12
V. IL-10

Answers to Questions

REVIEW QUESTIONS

1. Because some cytokines are either not produced by lymphocytes, or do not act on lymphocytes. On the other hand, the production or reception of most cytokines involves leukocytes, and the abbreviation IL, followed by a number, is convenient for nomenclature.
2. Because for cytokines to exert their physiologic effect, small concentrations are sufficient, and the timing of the release is important. Large concentrations of cytokines produce pathologic rather than physiologic effects (e.g., TNF).
3. The first, because some hemopoietins (e.g., GM-CSF) do not bind to these receptors.
4. It indicates that signal-transducing subunits can be shared by several cytokine receptors, e.g., IL-Rγc is shared by IL-2, IL-4, IL-7, and IL-15.
5. No, because to initiate the signaling, two cytokine receptor subunits must be juxtaposed; for this to happen, some lateral movement is needed.
6. In transgenics, both types of responses will be suppressed; in knockouts, they will be enhanced.
7. Three binding sites can bind three identical TNF receptors simultaneously. Other examples are CD40 ligand and Fas ligand.
8. It is important to establish whether the patient was infected by Gram-negative bacteria, and to consider such symptoms as fever, loss of blood pressure, cachexia, hypoglycemia, and signs of abnormal blood clotting.
9. No, because other cytokines (notably, TNF) will mask the defect.
10. Very difficult, because there are more than 20 different molecular species of IFN-α.

CHAPTER 10 Cytokines

11. Perhaps peritoneal macrophages are in contact with Gram-negative bacteria and/or substances that mimic LPS; because of that, they produce IL-6.
12. They are structurally different and bind each to its own receptor.
13. Because unless T cells are activated by antigen and co-stimulators (or polyclonal activators like PHA), they express only low-affinity IL-2R (IL-2Rβ and γc subunits). Low-affinity interactions require higher concentrations of the ligand, and more time, to produce an effect.
14. IL-2 signaling enhances their intracellular concentration.
15. There will be more antibodies in the second one, but more IgE antibodies in the first one.
16. By activating the transcription of MHC genes through STAT-1. This leads to enhanced expression of both MHC Class I and MHC Class II molecules, while type I interferons up-regulate Class I and down-regulate Class II.
17. TGF-β is known to inhibit the proliferation of activated T cells, thus, possibly, suppressing local anti-tumor immune responses.
18. Because it affects all lineages of hemopoiesis (at least in mice).
19. Both of them activate mast cell progenitors.

MATCHING

1, J; 2, K; 3, R; 4, A; 5, L; 6, T; 7, M; 8, O; 9, P; 10, V; 11, U; 12, F; 13, Q; 14, G; 15, C; 16, N; 17, E; 18, H; 19, B; 20, S; 21, D; 22, I

CHAPTER 11

Innate Immunity

Introduction

Innate immunity is a complex series of reactions that provide **the first line of defense against infections**. As the name indicates, the mechanisms of innate immunity exist before an encounter with infectious agents, but these agents rapidly activate them. As we mentioned in Chapter 1, innate immunity is also called nonspecific or nonadaptive. "Nonspecific" means that the reactions of innate immunity do not discern between various infectious agents the way reactions of specific immunity discern between antigens. **Diverse cellular clones that express clonospecific receptors are not characteristic for the innate immune system**. Nonetheless, cells of the innate immune system (neutrophils, macrophages, NK cells) do express surface receptors, and these receptors as well as other molecules of the innate immune system (e.g., chemokines or complement proteins) are able to tell infectious microorganisms from the tissues of the host. "Nonadaptive" means that the innate immune system does not develop immune memory and secondary responses, thus not "adapting" to developing a faster, stronger, and qualitatively better response to a next encounter with the same invader the way that the adaptive immune system does. Yet, responses generated by the innate immune system are rapid and in many cases strong enough to eliminate infectious agents on the first encounter with them.

One important feature of innate immunity is that it is not just a "back-up" for adaptive immunity. Phylogenetically, innate immunity is much older: many mechanisms of innate immunity function in invertebrate animals, including arthropods and worms. Some of these mechanisms can be traced in plants. In higher (jawed) vertebrates, adaptive immunity is intertwined with specific immunity. The latter greatly enhances the former; moreover, the latter cannot exist without the former. The specific immune system used nonspecific mechanisms of innate immunity to develop effector reactions against antigens. For example, B lymphocytes (part of adaptive immunity) produce antibodies, and these molecules, after they bind their specific antigen, bind and activate complement (part of innate immunity). T lymphocytes (part of adaptive immunity) produce cytokines that

CHAPTER 11 Innate Immunity

activate macrophages, NK cells, and inflammatory leukocytes (parts of innate immunity). Adaptive immunity focuses the mechanisms of innate immunity at the sites of antigen invasion and uses them against specifically recognized antigens, thus making them even more efficient.

In the discussion, we will go through some basic features of innate immunity. We will analyze its cellular and molecular components and their functions. Special attention will be given to receptors that are expressed on cellular mediators of nonspecific immunity (neutrophils, macrophages, NK cells). We will see how these receptors, although not clonally distributed and not nearly as diverse as specific antigen receptors, recognize certain molecular patterns that "warn" the organism about infection. We will dissect the transduction of signals through these receptors and analyze the mechanics of the activation of cells that express them. Finally, we will describe the effector phase of innate immunity and go through examples that illustrate the significance of these effector reactions for protection against infectious microorganisms.

Discussion

11.1 What are the components of the innate immune system?

The innate immune system consists of three principal components:

- **Physical barriers**. These include the skin and the mucosal surfaces of the respiratory and gastrointestinal tract. Their role in innate immunity is to guard against free entrance of potentially or actually infectious microorganisms into the systems of the host.
- **Circulating cells** that specialize in recognizing and eliminating microorganisms. These include leukocytes (in particular, neutrophils and macrophages), and some T and B lymphocytes with limited diversity of their antigen receptors.
- **Circulating proteins** that exert microbicidal effects. These include complement system, as well as a wide variety of other antimicrobial proteins (lysozyme, collectin, pentraxin, coagulation factors, etc.)

11.2 Specific immune responses include recognition, activation, and effector phases; does this also apply to innate immune responses?

Yes. Reactions of innate immunity include all of the above phases, although the basic features of these phases are distinct from the features observed during adaptive immune responses.

11.3 What are the basic features of innate immune recognition?

Similarly to what is characteristic for adaptive immune system, innate immunity uses specialized **receptors** in order to recognize its target structures. However, receptors expressed on cellular components of innate immune system recognize structures that are characteristic of **microbial pathogens** and are not present on mammalian cells. This is a fundamental difference from the recognition of antigens by the adaptive immune system. For example, T cells can recognize peptides derived from bacterial or viral proteins, but also from mammalian proteins (including those made in the host), or synthetic peptides. B cells can recognize virtually any structure that can be bound by antibody V-regions. The microbe-recognizing receptors of innate immune system are **not clonally distributed**. Again, this is a fundamental difference with antigen receptors of lymphocytes, where each lymphocyte clone produces a receptor with unique specificity. The receptors of innate immune system are **encoded in the germline**. In contrast, as was discussed in Chapters 4 and 7, genes that encode antibodies and TCR undergo somatic rearrangement. Because of the germline coding, the receptors of innate immune system possess a limited repertoire of specificities. Receptors of the innate immune system recognize certain **molecular patterns**. Such patterns can be found in the DNA, proteins, lipids, and carbohydrates made by certain classes of pathogenic microorganisms. For example, double-stranded RNA is peculiar to replicating viruses; unmethylated CpG DNA sequences, LPS, or mannose-rich oligosaccharides are characteristic of many bacteria, etc. (see later). On the other hand, these patterns are completely lacking in mammalian tissues. Notably, the molecular patterns recognized by the innate immune system play a crucial role in the survival of microorganisms and, unlike antigens recognized by the adaptive immune system, are conserved (do not mutate). This makes innate immune responses very efficient.

11.4 What is remarkable about the physical barriers and their role in innate immunity?

One obvious characteristic of the above-mentioned physical barriers is their integrity, which does not allow microbes to penetrate freely inside the host. However, the epithelia of the skin, and gastrointestinal and respiratory tracts also produce substances that actively kill microorganisms. The best known of these "natural antibiotics" are peptides called **defensins**. These are 29–34 amino acid long, cysteine-rich peptides that are present in the skin of many species. Their synthesis is greatly enhanced by locally produced TNF and IL-1; the latter can be produced by keratinocytes of the skin epithelium (see Chapter 10). The epithelium of the intestine secretes substances similar to defensins, which are called **cryptocidins**.

Yet another important feature of the natural barriers is that they contain **lymphocytes** with receptors presumably specific to abundantly expressed microbial components. These lymphocytes include **intraepithelial T lymphocytes**, **B-1 lymphocytes**, and **mast cells**. We have already mentioned these cells in the respective chapters about adaptive immunity. Nonetheless, they can be also viewed as com-

CHAPTER 11 Innate Immunity

ponents of innate immunity because of their location and function. Intraepithelial T lymphocytes are located in the epidermal layer of the skin, interspersed between epithelial and dendritic (Langerhans) cells. In the mouse, many of these cells express γδ TCR (Chapter 7), and in both mouse and human, these cells possess a limited TCR repertoire. B-1 lymphocytes (Chapter 2) are located predominantly on the immediate outside of the gastrointestinal and respiratory barriers – in the peritoneal and pleural cavities. They, too, express antigen receptors with limited repertoire of specificities and are thought to recognize "molecular patterns" characteristic of common microorganisms. It has been hypothesized that the target of B-1 lymphocytes are microorganisms that colonize the gastrointestinal and respiratory tracts early in life and stay there as commensals (e.g., *E. coli*, possibly *Haemophilus influenzae*, etc.). Mast cells contain granules that can be released upon recognition of microbes, which causes inflammation.

11.5 What is the principal role of phagocytes (neutrophils and macrophages) in natural immunity, and what are their main properties?

As the term itself explains, the principal function of these cells is to phagocytose, i.e., to "eat," engulf and digest, microbial particles. Neutrophils and macrophages both are able to perform this function; it is more intense in macrophages, however. Neutrophils are weaker phagocytes, but the advantage of having them is that they mediate very early inflammatory responses (hours after infection). We have already given a brief description of **neutrophils** in Chapter 2. They are small cells (12–15 μm in diameter), with numerous ciliary projections, segmented nuclei, and granules in their cytoplasm. There are two kinds of neutrophil granules: the so-called "**specific granules**," filled with degradative enzymes like lysozyme, collagenase, and elastase, and the so-called "**azurophilic granules**," which are in fact lysosomes. These granules are not stained strongly by either acidic or basic dyes, hence the name of the cell. Neutrophils are produced in the bone marrow, this process being stimulated by G-CSF (see Chapter 10). In adult humans, about 6×10^{11} neutrophils are made each day, and the life span of a circulating neutrophil is only about 6 hours. If the cell is not recruited into a site of inflammation, it undergoes a programmed cell death and is engulfed and digested by macrophages in the spleen or liver.

Macrophages are perhaps phylogenetically the oldest mediators of innate immunity. Cells resembling macrophages can be identified in *Drosophila*, in worms, and even in plants. We have given a description of a macrophage in Chapter 2. As mentioned there, macrophages are descendants of bone marrow-derived circulating cells called monocytes. These cells differentiate into macrophages after they migrate from blood into tissues, and the resulting macrophages have been in the past given different names depending of the tissue where they reside (e.g., Kupffer cells in the liver, microglial cells in the nervous system, etc.). One important difference between a macrophage and a neutrophil is that the former has a much longer life span: macrophages can live for many days and even weeks after they move into a site of inflammation. While neutrophils dominate inflammatory sites early after infection, macrophages are the predominant cell type at these sites later, 1–2 days after the infection. Interestingly, the differentiation from monocytes into macrophages is not terminal: differentiated macrophages still retain the ability to undergo mitotic division at inflammatory sites.

11.6 How do neutrophils and macrophages recognize microbial particles?

As was already mentioned, they do it with the help of germline-encoded, nonclonal surface receptors. The predominant type of receptors that are expressed on neutrophils and macrophages and used in the recognition of microbial agents are the so-called **seven transmembrane alpha-helical receptors** or **serpentines** (see Chapter 10). As was discussed in Chapter 10, these receptors bind **chemokines**. Serpentines expressed on neutrophils and macrophages bind predominantly chemokines of the C-X-C family, and especially IL-8. Other serpentine receptors of these cells bind **short peptides containing N-formylmethionyl residues.** This feature allows neutrophils and macrophages to detect and respond to bacteria and not to mammalian cells, because the former always contain membrane proteins that are initiated by N-formylmethionyl residues. Yet other receptors of neutrophils and macrophages bind peptides derived from complement proteins (e.g., C5a—see later); these peptides are generated when complement is activated. Finally, a number of neutrophil and macrophage receptors bind **lipid mediators of inflammation**. These include platelet-activating factor (PAF), prostaglandin E (PGE), and leukotriene B_4 (LTB_4). Binding of ligands to these receptors initiates signal transduction that ultimately results in such cellular responses as migration, activation of the so-called respiratory burst, production of microbicidal substances, and phagocytosis.

11.7 How do neutrophil and macrophage seven transmembrane alpha-helical receptors transduce the signal inside the cell?

The major signal-transducing system employed by these receptors is the system of trimeric **GTP-binding proteins**. These proteins are attached to the cytoplasmic portion of serpentine receptors, and consist of three subunits, $G\alpha$, $G\beta$, and $G\gamma$. In resting neutrophils and macrophages, $G\alpha$ subunits of the trimer contain attached GDP. When a ligand binds a serpentine receptor, GDP is exchanged for GTP, and the trimer dissociates into $G\alpha$·GTP and the $G\beta$-$G\gamma$ dimer. Both of these entities act as allosteric activators of a membrane enzyme called phospholipase C-β (**PLC-β**). Recall from Chapters 7 and 8 that T and B lymphocytes employ a different isoform of PLC, called PLC-γ; both PLC-γ and PLC-β catalyze the breakdown of membrane phospholipid and convert phosphatidylinositol biphosphate (PIP2) into inositol triphosphate (IP3) and diacylglycerol (DAG). The appearance of IP3 and DAG leads to a rapid increase in the concentration of intracellular free ionized calcium ($[Ca^{2+}]_i$) and to activation of protein kinase C (PKC) (Chapters 7 and 8). Activated PKC and calcium-calmodulin complexes strongly promote actin–myosin cytoskeletal changes, leading to cell spreading

CHAPTER 11 Innate Immunity

(which allows migration through endothelial layers), and the growth of pseudopodia (which allows phagocytosis). PKC also enhances the avidity of leukocyte integrins (e.g., LFA-1) to their ligands expressed on the endothelium, thus further enhancing the cell migration towards the site of inflammatory response.

11.8 What other receptors, besides serpentines, do phagocytes express, and what is their role in innate immunity?

In addition to serpentine receptors, neutrophils and macrophages express two other kinds of receptors that play important roles in innate immunity, namely, **receptors for opsonins** and the **receptor for LPS**.

11.9 What are opsonins and receptors for opsonins?

Opsonins is a collective name of substances or particles that **enhance phagocytosis** upon their binding to their receptors expressed on phagocytes. Bacterial cells, components of their cell walls, and also various plasma proteins can serve as opsonins. The plasma proteins can be classified into **specific opsonins** (antibodies) and **nonspecific opsonins** (components of the complement system, and a number of other plasma molecules like fibronectin, fibrinogen, mannose-binding lectin, and C-reactive protein). The group of opsonin receptors thus includes several kinds of receptors for components of bacterial cell wall, and several kinds of receptors that bind specific and nonspecific plasma opsonins. Among the opsonin receptors that bind components of bacterial cell wall, the most important ones are the **mannose receptor**; several different **scavenger receptors** (these bind mostly modified low-density lipoproteins expressed by a variety of bacteria), and **integrins** (for example, Mac-1 or CD11bCD18 – the molecule that serves as a complement receptor as well as the receptor for a number of microbial proteins). Among the receptors for specific plasma opsonins (antibodies) the most important is **FcγRI**, which serves as a high-affinity receptor for the Fc fragment of IgG antibodies. Among the receptors for nonspecific opsonins, the most important are **receptors for the C3 fragments**, which we will describe in more detail during the discussion about the complement system, and also receptors for the above-mentioned nonspecific plasma proteins. Binding of opsonins to their receptors enhances phagocytosis; specific opsonins – IgG antibodies – are the strongest enhancers of phagocytosis, but the role of nonspecific opsonins is also important because they appear early after infection, when specific antibodies are not yet made.

11.10 What is the receptor for LPS, and how does it function?

The receptor that binds bacterial LPS is called **CD14** (Fig. 11-1), and is synthesized and expressed by macrophages. This protein can also be shed from the macrophage surface and attach to surfaces of other cells; in this case, the cells that

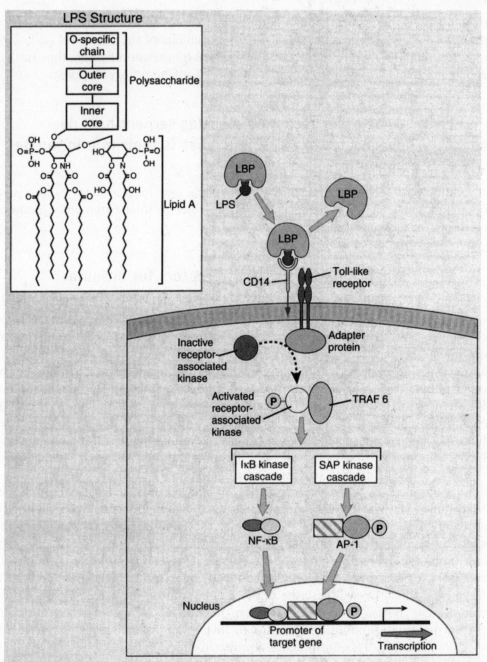

Signaling by lipopolysaccharide (LPS).
The receptors and signal transduction pathway by which LPS (structure shown in inset) activates the transcription factors NF-κB and AP-1 are shown. IκB, inhibitor of NF-κB; SAP kinase, stress-activated protein kinase; TRAF, tumor necrosis factor (TNF) receptor–associated factor.

Fig. 11-1. LPS signaling system. The diagram shows the receptors and signal transduction pathway by which LPS (structure shown in inset) activates the transcription factors. From Abbas, K., A.H. Lichtman, and J.S. Pober, *Cellular and Molecular Immunology*, fourth edition, W.B. Saunders, 2000, p. 278 (their figure in Box 12-1).

passively harbor CD14 on their membranes, acquire the ability to respond to LPS, but only to relatively large concentrations of this substance. Macrophages, on the other hand, respond to as little as 10 pg/ml of LPS. The actual entity that is bound by CD14 is the conserved **lipid moiety** of the bacterial LPS (the polysaccharide moiety is highly variable and is a major antigen of bacterial capsules, recognized and responded to by specific antibodies). The lipid portion of LPS represents a typical "molecular pattern" recognized by macrophages through CD14 (see Section 11.3).

The reception and response to LPS has some important features. First, LPS expressed on the bacterial cell wall, or LPS that is shed from the bacteria and circulates in body fluids, cannot be bound by CD14. For the binding to occur, the circulating LPS must be "delivered" to CD14 by a plasma protein called **LBP** (from **l**ipopolysaccharide-**b**inding **p**rotein). Second, CD14 does not transduce the signal upon the binding of the LPS–LBP complex. A protein called mammalian **Toll-like receptor protein 4** performs this function. Thus, macrophages have a three-component "LPS-sensing system" (Fig. 11.1), which allows them to bind bacterial LPS and generate responses to it.

11.11 How does the Toll-like receptor transduce the LPS signal, and what responses does it generate?

As was mentioned in Chapter 10, Toll-like structures are known to be used for signal transduction by some cytokine receptors, e.g., IL-1R. Toll is a protein that was first identified in the fruit fly, *Drosophila melanogaster*. In *Drosophila*, Toll plays an important role during development, and several additional Toll-like proteins serve in the fly's defense against bacteria and fungi. In *Drosophila*, Toll signals to activate a transcription factor called Dorsal, which is homologous to NF-κB. In mammals, the binding of ligands (e.g. LPS–LPB complex) to structures associated with Toll-like receptors (e.g., CD14) promotes the attachment of an inactive receptor-associated kinase to an adapter protein (see Fig. 11-1). The kinase becomes thus activated and phosphorylates itself, as well as **TRAF6**, a representative of the TNF receptor-associated factor (TRAF) family (see Chapters 7–10). TRAF6 initiates the cascade of events leading to the activation of transcriptional factors NF-κB and AP-1 (see Chapter 7). We have already discussed one of the principal effects of LPS recognition, i.e., **the production of TNF**, in Chapter 10. This effect is possible because the transcription of TNF gene is activated by the LPS signal through NF-κB, AP-1 and, possibly, other transcription factors. The other principal effect of LPS signaling, which is called **respiratory burst**, will be discussed in later sections of this chapter.

11.12 Are there special receptors that recognize unmethylated CpG DNA sequences?

It is not known so far. Macrophages are known to respond to these sequences, which are much more frequent in bacterial than in mammalian DNA, by producing IL-12 and type I interferons. The exact mechanism of the recognition or "sensing" these sequences has not been elucidated, however.

11.13 How are leukocytes recruited to the sites of infection during innate immune responses?

The process of their recruitment is stimulated by cytokines, and depends on the interaction of the leukocytes' **homing receptors** with the ligands of these receptors expressed on endothelial cells. The crucial cytokines that control this reaction are **TNF** and **chemokines**. TNF especially strongly stimulates endothelial cells to express molecules that mediate the preferential attachment of different kinds of leukocytes. The earliest endothelial molecule to be expressed is **E-selectin**, which appears within just 1–2 hours after exposure of endothelial cells to TNF. Within 6–12 hours after exposure, these cells begin to express **ICAM-1** (a ligand for an integrin molecule LFA-1, see Chapter 7, and a complement receptor Mac-1), and **VCAM-1** (the ligand for an integrin molecule called VLA-4). Integrins expressed on leukocytes interact with these newly expressed endothelial ligands, which leads to a stable attachment of leukocytes to endothelium. The so-called "**multistep model of leukocyte recruitment**" presumes that this attachment is preceded by brief stages of "tethering" and "rolling." These stages are followed by the interaction of serpentine receptors expressed on leukocytes with chemokines bound to endothelial surface heparan sulfate glucosoaminoglycans. In response to chemokines, leukocytes **rearrange their cytoskeleton**, spread from a spherical to a flattened shape, and become more **motile**. They stop "rolling" over the endothelium, but instead slowly move towards interendothelial cell junctions and penetrate through the fibrin or fibronectin scaffold that is formed in these junctions, leaving blood vessels and ending up in the infected tissue. Recently, a new adhesion molecule called **PECAM-1** (from **p**latelet-**e**ndothelial **c**ell **a**dhesion **m**olecule), or **CD31**, has been implicated in this process.

11.14 How do phagocytic cells actually phagocytose?

Phagocytosis (Fig. 11-2) is a **cytoskeleton-dependent** process that includes engulfing of large (>0.5 mm in diameter) particles and destroying them by means of **proteolysis** and "poisoning" with the so-called "**reactive oxygen intermediates**" (**ROI**). A phagocyte attaches to its "victim" particle with the help of its various surface receptors, forms a cup-like membrane projection around it, and has this projection "zip up" on the particle. The result of such an engulfing is the so-called **phagosome** – a vesicle that contains the engulfed particle that is surrounded by the "pinched off" piece of cellular membrane. The next step is **fusion** of this phagosome with a **lysosome**, which results in the formation of a phagolysosome. After the phagolysosome is formed, the process of destruction of the engulfed particle begins. Interestingly, the receptors that attach to the particle that is subject to phagocytosis also signal (mostly through GTP-exchange proteins) for turning on the processes of proteolysis and ROI formation.

CHAPTER 11 Innate Immunity

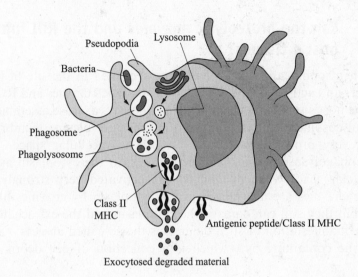

Fig. 11-2. Phagocytosis. A schematic diagram showing phagocytosis and processing of the phagocytosed material.

11.15 How does the proteolysis of engulfed particles occur?

Once the phagolysosome is formed, their content becomes exposed to lysosomal proteolytic enzymes. Neutrophil granules are especially rich in **elastase**, an enzyme that can digest elastin and a variety of microbial proteins. Macrophages use mostly other proteolytic enzymes, and can present the resulting peptides to $CD4^+$ T cells via their MHC Class II molecules (see Chapter 6).

11.16 How are ROI formed, and what do they do?

These entities are formed as **a result of the catalytic conversion of molecular oxygen to oxyhalide free radicals**. Neutrophils and macrophages have two enzymatic systems that form free radicals: the **oxidase** system and the **inducible nitric oxide synthase (iNOS)** system. The first is operating with the help of the large multisubunit enzyme oxidase, which in activated phagocytes is assembled in plasma membranes and in the phagolysosome membrane. Oxidase reduces molecular oxygen into superoxide radicals, which are then converted into hydrogen peroxide with the help of superoxide dismutase. Hydrogen peroxide reacts with halide ions (this reaction being catalyzed by the enzyme myeloperoxidase) to form hypohalous acids. The latter compounds are extremely toxic to bacteria. The iNOS is a cytosolic enzyme that is absent in resting phagocytes but rapidly induced in response to LPS, especially in the presence of IFN-γ. iNOS converts arginine into citrulline; this reaction is accompanied by release of nitric oxide (NO). The latter diffuses into phagolysosomes, combines with hydrogen peroxide or superoxide, and forms peroxynitrite radicals, which are also extremely toxic to bacteria.

11.17 Can the proteolytic enzymes and the ROI injure the host's tissues?

Yes. When neutrophils and macrophages are recruited to the site of infection in large numbers and activated strongly, their proteolytic enzymes and ROI can, and do, injure the host's tissues. As mentioned previously, oxidase assembles not only in the phagolysosome membrane but also in the outer plasma membrane, which makes ROI very prone to penetrate into the extracellular space. Proteolytic enzymes of phagolysosomes also can penetrate into the cytoplasm as well as in the extracellular space when phagocytes are activated very strongly. To some extent, these enzymes are neutralized in the cytoplasm by enzymes like **alpha-1-antitrypsin**, but they still can damage the cytoplasm and the extracellular matrix. One extreme manifestation of such damage is the so-called **abscess** – a pocket of liquefied tissue containing pus, which is a collection of cell debris and dying neutrophils.

11.18 If pus contains mostly neutrophils, does it mean that macrophages do not injure host tissues?

The tissue injury inflicted by macrophages is "milder" than that of neutrophils, for a number of reasons. First, as already mentioned, macrophages have little or no elastase. For proteolysis, they employ other enzymes, mostly **metalloproteinases**, which are less injurious, easier to neutralize, and of a narrower spectrum of activity. Second, activated macrophages produce factors that participate in the **repair of tissue damage** ("tissue remodeling"). Among such factors are **platelet-derived growth factor** (**PDGF**), which stimulates the proliferation of fibroblasts, **TGF-β**, which in turn stimulates the growth of extracellular fibrous matrix (Chapter 10), and other factors. Therefore, in the settings of prolonged activation, macrophages mediate tissue **fibrosis** (formation of scars) rather than abscess formation.

11.19 What is the role of natural killer (NK) cells in innate immunity?

As was discussed in Chapter 2, NK cells are bone marrow-derived cells that resemble lymphocytes (and sometimes are called "large granular lymphocytes"), but do not make clonally distributed antigen receptors. These cells are able to release their preformed cytotoxic granules and kill cells infected by viruses or by some intracellular bacteria. In addition, NK cells are powerful producers of cytokines, especially IFN-γ.

11.20 How do NK cells recognize their targets?

NK cells express three kinds of receptors that serve for the recognition of their targets. The first of them is used to recognize cells coated by antibodies,

CHAPTER 11 Innate Immunity

and is actually an Fc receptor (see Chapter 3) called **FcγRIII** or **CD16**. It is a low-affinity receptor that binds IgG1 and IgG3 antibodies in the human (IgG1 and IgG2a antibodies in the mouse). The binding of an NK cell to the target cell via the CD16–IgG interaction leads to the release of the NK toxic granules and the killing of the target cell. This process is called **antibody-dependent cell-mediated cytotoxicity (ADCC)**, and will be discussed in more detail in the chapter about effector mechanisms of humoral immunity. Remarkably, CD16 has a signaling subunit that is homologous to the TCR ζ chain and contains ITAMs that are phosphorylated upon ligand binding, similarly to what happens during T-cell activation (Chapter 7). The end result of CD16-mediated activation of NK cells is, however, the release of their granules and not the clonal expansion or differentiation, as in T cells. The second type of NK-recognizing receptors is used for the binding to infected cells in the absence of antibody. Such binding leads to the activation of the NK cell, and therefore receptors of this second type are called activatory receptors. They have not been identified as distinct molecules, although studies show that some molecules not unique to NK but, rather, shared between NK and T cells (e.g., **CD2** and **integrins**) can play this role. The third kind of NK receptors are called **inhibitory receptors**. These molecules are represented by three types (see below) and recognize particular alleles of **MHC molecules**, especially HLA-A and HLA-B alleles in humans. The binding of these receptors to MHC molecules inhibits the NK cells, and prevents them from releasing their toxic granules. If, however, an NK cell approaches a cell of the host and binds to it through integrins or other surface receptors, but does not engage the inhibitory receptor because of the lack or down-regulation of the MHC molecule, this NK cell would release its granules and kill the host cell. This function of NK cells is very important because in fact, many viruses down-regulate MHC molecules on the cells that they infect. Such a down-regulation makes $CD8^+$ T-cell-mediated killing of the targets difficult; therefore, NK serve as a crucial alternative in this case.

11.21 What three types of NK inhibitory receptors exist, and why are they needed?

Two of these types have been recently identified in humans, and one in rodents. The first type is called the **KIR family** (from **k**iller **i**nhibitory **r**eceptor), and is a group of molecules that belong to the immunoglobulin superfamily. These molecules signal through immunoreceptor tyrosine-based inhibitory motifs (**ITIMs**) of their cytoplasmic portions, and recruit tyrosine phosphatases, similarly to the FcγRII signaling described in Chapter 8. The second type of NK inhibitory receptors is represented by the so-called **CD92-NKG2 heterodimers**. In these heterodimers, CD92 is an invariant protein and NKG2 is a variable lectin. They are thought to signal through ITIMs as well. The third type of NK inhibitory receptors was identified exclusively in rodents and is a lectin called **Ly49**. Interestingly, a subset of KIR family contains ITAMs instead of ITIMs and functions as activatory receptors.

11.22 What is the mechanism used by NK cells to kill target cells?

When NK cells attach to their target and are activated, they release their preformed **cytotoxic granules**. The granules contain a wide variety of proteins, of which two are most important for the killing of target cells. One of them is called **perforin**. This large, multi-subunit protein acts like a drill, making "holes," or pores, in the plasma membrane of the target cell. The other is called **granzyme**; this smaller protein penetrates through the pores made by perforin, and causes apoptosis of the target cell. This mechanism is actually almost identical to the mechanism of killing used by $CD8^+$ cytolytic T lymphocytes (CTL) (see Chapter 12). However, $CD8^+$ CTL need a rather long time (approximately 48 hours) to synthesize granules after the contact with their target, while NK cells synthesize granules in advance. Therefore, the nonspecific killing mediated by NK cells is much more **rapid**. The cytolytic function of NK is greatly **potentiated by cytokines** of innate immunity, especially by IL-15 and IL-12. Activated NK cells produce large amounts of IFN-γ, which activates macrophages and stimulates their production of IL-15 and IL-12; thus, there exists a positive feedback in the effector function of NK cells.

11.23 What are tumor targets of NK cells, and what is the significance of these cells' anti-tumor effect?

NK cells indeed can lyse a number of tumor cell lines *in vitro*, this lysis being a classical laboratory test of NK activity. This lysis perhaps occurs because many tumor cell lines either down-regulate their MHC molecules, or express aberrant forms of these molecules. In neither case can the inhibitory NK receptors be engaged, and the cytotoxic granules are released by NK cells upon their binding to tumor cells. However, there is doubt that NK possess a strong anti-tumor effect *in vivo*. Tumor-associated inflammatory infiltrates do not usually have many NK cells. The only *in-vivo* setting where NK cells are really abundant and thought actively to kill their target is the so-called graft-versus-host disease (GvHD). This situation will be discussed in detail in the chapter about transplantation immunity.

11.24 What is complement, and what is its role in innate immune reactions?

Complement will be discussed in detail in the chapter dedicated to the effector mechanisms of humoral immunity. Here, we will just briefly highlight some most basic characteristics of the complement system.

Complement is a complex mixture of proteins that have functions of **zymogens**, i.e., they show no enzymatic activity when they are just synthesized, but acquire such activity when they are modified (e.g., cleaved) by other proteins. Components of the complement system form "cascades" of zymogens, where cleavage of one of the components generates products that activate another component. These cascades can be triggered in two different ways. One of these ways is called the **classical**

CHAPTER 11   Innate Immunity

pathway of complement activation; it starts when one of the complement proteins, called C1, binds to an IgM or IgG antibody bound to an antigen. The other way to initiate complement "cascades" is called the **alternative pathway** of complement activation. It starts when a derivative of one of the complement components, called C3b, binds directly to the microbial cell wall. When complement is activated by either of the pathways, a variety of products are formed, some of which serve as agents of inflammation, recruiting neutrophils and macrophages to the sites of infection. Others act as opsonins, binding complement receptors on neutrophils and macrophages and enhancing phagocytosis. Finally, proteins called C5, C6, C7, C8, and C9 form a multimeric complex called "**membrane attack complex**" (**MAC**). This complex acts like a drill, boring holes in microbial cell walls and killing bacterial and other cells.

11.25   What other circulating proteins are important for innate immunity?

The so-called **collectin** family of molecules plays an important role in innate immune responses. The name "collectin" originates from words "**col**lagen" and "**lectin**", and is given to these molecules because they are proteins that have a collagen domain and a lectin (i.e., carbohydrate-binding) moiety. One interesting representative of collectins is **mannose-binding lectin** (**MBL**) – a plasma protein that functions as an opsonin. MBL is structurally similar to a component of the complement system called C1q. This protein is able to bind to microbial cell wall glycoproteins that contain mannose and fucose. It can also bind to C1q receptor and also to two other complement proteins, C1s and C1r, and thus to trigger the classical pathway of complement activation. Through its binding to yet another complement component, C3, MBL can bypass both classical and alternative pathways of complement activation and to trigger the so-called lectin pathway of complement activation.

Another important circulating protein that plays a role in innate immunity is called **C-reactive protein** (**CRP**). The name of this protein originates from its ability to bind capsules of pneumococcal bacteria. It belongs to the so-called pentraxin family of plasma proteins. This protein enhances phagocytosis by binding to C1q receptors or directly to Fcγ receptors of phagocytes. It can also trigger the classical pathway of complement activation by binding to C1q.

Yet another important family of circulating proteins that participate in innate immunity is the family of coagulation factors (see physiology and pathology textbooks for details).

11.26   How exactly are the above-described components of innate immunity involved in the stimulation of adaptive immunity?

The innate immunity serves as the **initial warning signal** for the adaptive immune system to mount a specific response. In more precise terms, it means that molecules produced or up-regulated during innate immune responses, such as co-stimulators, cytokines, and complement breakdown products, function as **second signals** for lymphocyte activation. We have already discussed some of these second signals. Recall the role of B7-1 and B7-2 (the molecules that are up-regulated by local inflammatory reactions) in T-cell activation or the role of C3d in B-cell activation, or the role of IL-12 in regulating T_h1 and T_h2 T-cell subsets. Co-stimulators like B7-1 and B7-2, cytokines like IL-12, and complement products like C3d are products of innate immunity. Up-regulation of co-stimulators by local inflammatory reactions is the mechanism of a well-known effect of adjuvants – derivatives from microbial cells that trigger local inflammatory reactions and strongly enhance the immunogenicity of antigens injected in the same site (usually together with an adjuvant under the skin).

Innate immunity does not merely enhance specific immune responses; it also plays an important role in "customizing" the latter, determining the nature and character of specific immune responses. For example, macrophages that engulf bacteria up-regulate co-stimulators and produce cytokines that activate T-cell adaptive responses. Components of complement that bind circulating bacteria can selectively activate B-cell responses. Finally, interactions between innate and adaptive immunity are bidirectional. Activated lymphocytes produce cytokines that can inhibit or enhance the function of inflammatory leukocytes.

Questions

REVIEW QUESTIONS

1. Why is innate immunity called "the first line of defense against infection?"
2. Unlike adaptive immunity, innate immunity is not known to attack self tissues. Does this mean that innate immunity is unable to distinguish them from foreign tissues?
3. What is the role of keratinocytes in innate immunity?
4. The so-called "natural antibodies" are predominantly low-affinity IgM antibodies that are widely cross-reactive, binding epitopes on a number of substances including bacterial polysaccharides. The appearance of these antibodies is not associated with deliberate immunization. In the early 1990s, it was shown that injections of distilled water in a certain anatomical location of mice greatly reduce the titers of "natural antibodies." What is this location, and what components of innate immunity are affected?
5. How would you prove that inflammatory infiltrates consist mostly of neutrophils early in the response, and become rich in macrophages later? What are the reasons for this change?
6. What feature of neutrophil and macrophage cytoskeleton is especially important for recruitment of these cells from circulation into tissues?
7. How can the "division of labor" between innate and adaptive immunity be demonstrated in the case of responses to bacterial lipopolysaccharide?
8. What exactly do chemokines do to facilitate the penetration of leukocytes through the endothelial layer?

CHAPTER 11 Innate Immunity

9. Why is the tissue injury mediated by macrophages "milder" than the one mediated by neutrophils?
10. Name three differences between the killing of targets by NK cells and the killing of targets by $CD8^+$ CTL.
11. What can be said about the anti-tumor effect of NK cells?
12. Can complement be activated in genetically deficient mice that have no lymphocytes?
13. While MBL and CRP are both opsonins, what are the differences in the way they opsonize microbes?
14. How do factors of innate immunity serve as second signals for the activation of adaptive immunity?
15. A rabbit is immunized with bovine serum albumin intravenously. Another rabbit is immunized with the same protein emulsified in mineral oil that contains a suspension of killed *M. tuberculosis*, subcutaneously. In which case would you anticipate anti-BSA T lymphocytes to be activated, and why?

MATCHING

Direction: Match each item in Column A with the one in Column B to which it is most closely associated. Each item in Column B can be used only once.

Column A

1. Molecular patterns
2. Cryptocidins
3. Azurophilic granules
4. *N*-Formylmethionyl residues
5. Scavenger receptors
6. Mammalian Toll-like protein 4
7. VCAM-1
8. Metalloproteinase
9. iNOS
10. CD16
11. KIR family
12. C1

Column B

A. Bind low-density lipoproteins
B. Are lysosomes
C. Contain ITIMs
D. Absent in mammalian tissues
E. Binds to IgM bound to antigen
F. Produced in the intestine
G. Ligand for VLA-4
H. Contains homologue of zeta chain
I. On bacterial peptides
J. Produces NO
K. Transduces LPS signal
L. Participates in phagocytosis

 Answers to Questions

REVIEW QUESTIONS

1. Because its components are ready to attack microbes when the responses of adapted immunity have not yet developed.

CHAPTER 11   Innate Immunity

2. No, innate immunity can distinguish self from foreign if this foreign is of microbial origin. It does not attack self tissues because they have no molecular patterns peculiar to microbes.
3. They produce IL-1 and enhance the effect of defensins.
4. The peritoneal cavity. The cells that will be affected are B-1 lymphocytes.
5. Microscopically, neutrophils can be identified as cells with nuclei that consist of several lobes, while macrophages are mononuclear. The reason is because neutrophils are recruited into tissues by chemokines and other early signals, but have a much shorter life span than macrophages.
6. Actin and myosin of their cytoskeleton can be activated by calcium-calmodulin complexes after transduction of signals from their surface receptors.
7. Variable polysaccharide moieties of LPS are targets of antibodies, while invariant lipid moieties are recognized by the LPS-sensing system of macrophages.
8. They bind to their receptors on leukocytes and stop them from rolling over the endothelial layer.
9. Because macrophages do not employ elastase; their proteolytic enzymes, metalloproteinases, do little damage to extracellular matrix. Besides, macrophages produce PDGF and TGF-β, which promote tissue remodeling.
10. NK recognize a variety of targets with down-regulated MHC, while T cells recognize targets that present specific antigens; NK cells have preformed granules, while T cells need to produce granules after contact with their antigens; NK cells kill if inhibitory receptors are not engaged, while cytolytic T cells do not employ inhibitory receptors for the killing of targets.
11. It has little significance *in vivo*, because tumor infiltrates usually contain few NK cells.
12. Yes, through the alternative and lectin pathways.
13. MBL binds C1q, C1s, and C1r, while CRP binds C1q or FcγR.
14. They up-regulate co-stimulators on T cells (B7-1, B7-2), B cells (C3d-CD21), and enhance the production of cytokines and complement breakdown products.
15. In the second rabbit, because mycobacteria derivatives act as adjuvants.

MATCHING

1, D; 2, F; 3, B; 4, I; 5, A; 6, K; 7, G; 8, L; 9, J; 10, H; 11, C; 12, E

CHAPTER 12

Effector Mechanisms of Cell-Mediated Immunity

Introduction

Previous chapters were dedicated to mechanisms of antigen recognition and activation of immune cells. It is obvious, however, that the **effector phase** of immune reactions needs a special attention because during this phase, potentially or actually dangerous antigens are purged from the organism. In this chapter, we will dissect the effector phase of cell-mediated immunity.

Historically, adaptive immunity was classified into **humoral** and **cell-mediated immunity (CMI)**. While antibodies mediate the former, cells, more precisely antigen-activated T lymphocytes, macrophages, and other leukocytes, as the name tells, mediate the latter. T cells play a central role in these reactions. They can recognize antigens and "execute the sentence" on these antigens themselves. That happens when the peptides generated from intracellular antigens end up in the cytosol, and are presented to $CD8^+$ cytolytic T lymphocytes (CTL) in the context of MHC Class I molecules (see Chapter 6). The typical intracellular antigens that generate cytosolic peptides are viral and tumor-associated antigens. The effector phase of CMI in this case is the direct killing (cytolysis) of the cell that harbors the antigen. In the case when antigens are captured and internalized by professional antigen-presenting cells (dendritic cells, macrophages, or B lymphocytes) and the peptides generated from these antigens end up in the cell's acidic vesicular compartment, these peptides are recognized by CD4$^+$ helper T lymphocytes (T$_h$) in the context of MHC Class II molecules (Chapter 6). The effector phase of CMI in this case is different. T$_h$ still play a central role in the effector reactions because they direct these reactions to, and focus them on specific antigens. However, it is not the T lymphocytes themselves, but macrophages and other leukocytes, activated by antigen-specific CD4$^+$ helper lymphocytes, that "execute the sentence" on the antigens.

As we shall discuss in this chapter, the effector reactions directed by CD4$^+$ helper T lymphocytes greatly depend on the exact "background" that is created, essentially, by innate immunity developing upon the encounter of a particular antigen. Many microbial antigens activate the mechanisms of innate immunity described in Chapter 11; these mechanisms involve the production of a crucial cytokine, IL-12, by macrophages and dendritic cells. IL-12 stimulates the differentiation of T$_h$ into the subset called **T$_h$1**, which produces mostly IFN-γ and strongly activates macrophages (Chapters 7, 9, 10, and 11). In other cases, however, due to mechanisms that are not yet fully elucidated, T$_h$ lymphocytes differentiate into the subset called **T$_h$2**. This subset produces mostly IL-4 (Chapters 7, 9, and 10), and strongly inhibits the activation of macrophages, as well as the expansion of the T$_h$1 subset. It activates a reaction, mediated predominantly by eosinophils and mast cells, which may be of significance in the defense against helminthes and some arthropods.

The knowledge of factors that influence on the differentiation of naïve T$_h$ (sometimes referred to as T$_h$0) into either T$_h$1 or T$_h$2 is crucial for our understanding of the effector phase of CMI. Throughout the following discussion, we will try to highlight important features of both subsets, and ponder on the interaction of adaptive and innate immune mechanisms in their differentiation and function.

Discussion

12.1 How was it first demonstrated that cells can mediate the effector phase of immunity?

As was mentioned in Chapter 2, Eli Metchnikoff observed in the 1880s that cells could engulf foreign particles. This was perhaps the first demonstration of the cell-mediated effector phase of innate immunity. In the 1950s, G. Mackaness proved that cells could also mediate the effector phase of adaptive immunity. Mackaness's experiment was performed on mice immunized with an intracellular bacterium *Listeria monocytogenes*, and employed the technique of adoptive transfer (see Chapters 1 and 8). Mice were immunized (or "primed") with doses of *L. monocytogenes* that were not lethal, but conferred immunity to subsequent injections of lethal doses of the bacterium. Mackaness discovered that **cells, but not sera, of the primed mice could adoptively transfer this immunity**. In later experiments, it was shown that the protection of mice against *L. monocytogenes* is mediated by macrophages, but depends on T lymphocytes; when mice are primed, their T cells "instruct" their macrophages to develop an attack on the bacterium.

CHAPTER 12 Cell-Mediated Immunity

12.2 Why is the *Listeria* experiment of Mackaness so significant?

Mackaness's experiment actually revealed that protection against infectious agents could be conferred through **T lymphocyte-mediated macrophage activation**. We now know that this is one of the most common effector mechanisms of CMI. It is also a good illustration to the point that innate and adaptive immunity in higher vertebrates are intertwined and interdependent. In CMI against *L. monocytogenes* as well as other intracellular (or phagocytosed) antigens, the specificity of the response is due to T lymphocytes (which are cellular components of adaptive immunity). The effector functions are provided by macrophages (cellular components of innate immunity). The communication between the two cell types is provided mostly by cytokines (molecules that function in both innate and adaptive immunity). Using cytokines, T cells stimulate the function and focus the activity of macrophages (or, broader, phagocytes), thus "converting" the latter into agents of adaptive immunity.

12.3 What other observations supported Mackaness's experiment?

There exists a common pattern of organisms' responses to infectious agents or their derivatives, and to some chemicals. This pattern was called the **delayed-type hypersensitivity** (**DTH**). This reaction develops when a person or an animal is immunized ("**sensitized**") with a microorganism and then immunized again (or "**challenged**") by injecting the same antigen intradermally. In this person or animal, typical signs of local inflammation (swelling, redness, local raising of the temperature, and pain) accompanied by the **induration** (hardening) of the tissue around the injection site, develop in 24–48 hours after the "challenge." DTH is widely used in clinics as a standard test to establish whether a person has been infected or previously vaccinated with *M. tuberculosis*. As far as its mechanism is concerned, however, it is a typical case of T-cell-mediated macrophage activation, very similar to the phenomenon observed by Mackaness in his *Listeria* experiment.

12.4 What are the mechanisms of DTH?

The process of DTH can be divided into a number of steps. The first of them is called **induction** and corresponds to "sensitization," when naïve T lymphocytes first encounter the antigen in peripheral lymphoid organs. The specific T-cell clones recognize their antigens, are activated, proliferate, and differentiate into effector T cells. The next step in the development of DTH takes place after the "challenge", and is the **migration** of the differentiated effector T lymphocytes, as well as other leukocytes, to the site of the challenge. We have already discussed some of the mechanisms of leukocyte migration in Chapter 11, and will stop at mechanisms peculiar to T-lymphocyte migration in later sections of this chapter. Finally, the third step in the development of DTH is **macrophage activation**.

The latter is actually responsible for the elimination of the antigen and for the remodeling of the infected and possibly injured tissue.

12.5 What features of the induction step of the DTH are important for its further course?

Before T cells recognize them, the antigens that attack the organism (in particular, microbial agents) are either phagocytosed by macrophages, or are picked up by Langerhans cells (see Chapter 2). Both cell types transport the antigens to regional lymph nodes, and Langerhans cells differentiate into mature dendritic cells on their way there (Chapter 2). Since both macrophages and dendritic cells are professional antigen-presenting cells that express Class II MHC molecules, they can (and do) present antigenic peptides to Class II-restricted $CD4^+$ T_h. Therefore, the T cells that participate in the induction of DTH are naive T_h (T_h0) lymphocytes. However, microbial antigens always elicit innate immune responses when they invade (see Chapter 11). Factors produced during these early innate responses, especially cytokines like IL-12, IL-18 and a variety of other factors whose role is less well elucidated, influence on the differentiation of the effector T_h cells into T_h1 or T_h2. If the net effect of the innate immunity "background" is T_h1 differentiation, the DTH response may predominate.

12.6 What features of leukocyte migration after the "challenge" are most important for further development of the DTH reaction?

To develop efficient responses against antigens, leukocytes must "locate" them, i.e., move from the circulation into the tissues where antigens have entered. The effector T lymphocytes must **migrate** from the sites of their differentiation (i.e., local draining lymph nodes) to the circulation, and then into the tissue site of antigen invasion. In order to move towards the antigen and cross the endothelial barriers, leukocytes express and use a number of surface receptors mentioned in Chapter 11. These receptors bind to their counter-structures expressed on endothelial cells and in the tissues. We have given some details of the migration of neutrophils and macrophages across the endothelial barrier in Chapter 11. As far as the effector T cells are concerned, one major factor that helps them to proceed successfully from the lymph node to the circulation and then to tissues is the down-regulation of some **selectins** and the up-regulation of other **adhesion molecules**. In particular, **naive** T lymphocytes express high levels of the homing receptor **L-selectin** (or CD62L). This molecule binds to sulfated glucosoaminoglycans expressed on high endothelial venules of lymph nodes, or their analogues expressed in Peyer's patches and elsewhere in the secondary lymphoid organs (see Chapter 2). Once a T cell encounters its specific antigen and is activated in the presence of co-stimulators, it rapidly down-regulates L-selectin and instead up-regulates the expression of a series of other adhesion molecules. Among the latter, there are **ligands for E-selectin** and **P-selectin,** as well as integrins **LFA-1** and

CHAPTER 12 Cell-Mediated Immunity

VLA-4, and a molecule called **CD44**. E-selectin and P-selectin, as well as counter-structures for LFA-1, VLA-4, and CD44, are expressed at high levels on the endothelial cells outside of the secondary lymphoid organs, and in peripheral tissues. The interaction between the above structures helps effector T cells to adhere to the endothelium of blood capillaries and to penetrate through it, as well as to be retained in tissues at the site of antigen invasion. For example, the ligands ("addressins") that bind VLA-4 and CD44 are such common components of the extracellular matrix of the connective tissue as fibronectin and hyaluronidate, respectively.

12.7 Which factors are the most essential for the development of full-blown macrophage activation during the DTH reaction?

Macrophage activation is a rather vague term. It refers to the ability of a macrophage to perform certain activities that cannot be performed by the immediate precursors of tissue macrophages, i.e., resting circulating monocytes. For example, a macrophage will be called "activated" if it phagocytoses, or produces ROI (Chapter 11), or performs some other activity that can be tested *in vitro*. In this regard, macrophages may be activated in some respect and not activated in another. Yet, regardless of what is considered "activation," macrophages are activated, principally, by **two signals**. One of them is a direct **contact** of the macrophage with an activated T_h cell; this contact involves an interaction between **CD40L** expressed on the activated T_h and **CD40** expressed on the macrophage. The other signal is IFN-γ. As was discussed in Chapter 8, B lymphocytes are similarly activated by two signals, one of which is CD40L–CD40 interaction, and the other a set of cytokines. CD40L knockout mice were shown to have a profound defect in the development of DTH reaction against phagocytosed microorganisms. IFN-γ knockout mice develop a similar defect. A somewhat less dramatic, but also pronounced defect in macrophage activation develops in mice with the knockout of the IL-12 gene.

12.8 What events actually take place at the site of antigen invasion during macrophage activation in the DTH reaction?

T-cell-induced macrophage activation at the site of antigen invasion usually manifests in **phagocytosis** and **respiratory burst**. We have already dissected the molecular mechanisms of both of these phenomena in Chapter 11. Here, let us re-emphasize that the activation of macrophages is a "double-edged sword:" it can lead to the destruction of antigen, but it can also cause tissue damage. We have already mentioned in Chapter 11 that macrophages are not as prone to cause the formation of abscesses, as are activated neutrophils, but instead are capable of inducing **fibrosis**. One of the extreme manifestations of macrophage-related tissue damages not mentioned in Chapter 11 is the so-called **granulomatous inflammation** (Fig. 12-1). This happens when macrophages are chronically activated and fail to eradicate the microbes that resist digestion in phagolysosomes and the toxic action of ROI. In this case, the effector T_h1 lymphocytes continue to produce cytokines in response to persisting antigen, and the macrophages, accumulated in large numbers, undergo certain changes in response to these persistent cytokine signals. These changes include an increase in the cytoplasm and cytoplasmic organelles. The macrophages that undergo this change begin to look like epithelial cells, and are sometimes called "**epithelioid cells**." The epithelioid cells often fuse and form multinucleate "**giant cells**." The formation of granulomatous tissue that contains epithelioid and giant cells often takes place in the lung during tuberculosis, when a DTH-like T-cell-induced macrophage activation is a typical, but not very efficient response. It is

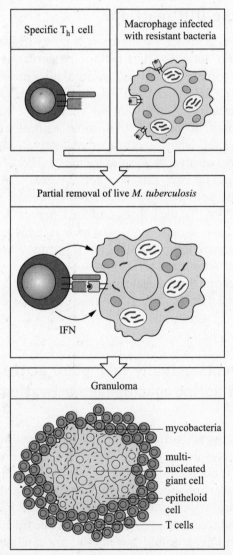

Fig. 12-1. Granulomatous inflammation. The diagram shows how continuous stimulation of macrophages by T_h1 cells leads to the development of granulomas.

CHAPTER 12 Cell-Mediated Immunity

accompanied by tissue fibrosis and a gradual replacement of the alveoli by tumor-like granulomatous tissue, which results in profound impairment of respiration. The granulomatous inflammation during tuberculosis is an example of tissue injury that has very little to do with the primary cause of tuberculosis (*M. tuberculosis*), but rather with unsuccessful, chronic activation of the immune system.

12.9 What are the consequences of T_h2 involvement in CMI?

If T_h2 cells mediate the effector phase of immunity, the consequence is mostly B-lymphocyte activation and/or a peculiar reaction called "**immediate hypersensitivity**." In both cases, antibodies will serve as actual effector molecules. Because of that, it is more logical to discuss the consequences of T_h2 involvement in the next chapter, which will be dedicated to the effector phase of humoral immunity. If $CD8^+$, MHC Class I-restricted CTL are involved, the effector phase of CMI will be, essentially, characterized by the **direct lysis** of antigen-presenting target cells. We will discuss mechanisms of the CTL reaction in next sections.

12.10 What are the "rules of engagement" of CTL?

As was discussed in Chapter 2, CTL are a subset of T lymphocytes that are capable of synthesizing **cytolytic granules** and, by releasing these granules in the vicinity of a target cell, kill the target cell directly. CTL recognize antigens presented by the MHC molecules of the target cell. Most CTL are $CD8^+$, MHC Class I-restricted T lymphocytes, although a minority of them are $CD4^+$ MHC Class II-restricted T lymphocytes. The $CD8^+$ MHC Class I-restricted CTL can recognize peptides presented by MHC Class I molecules on the surface of any nucleated cell, because all nucleated cells in higher vertebrates express these molecules (see Chapters 5 and 6). Therefore, the $CD8^+$ Class I-restricted CTL can recognize and kill an enormously wide range of cellular targets. The peptides that they recognize are generated in the cytosol, mostly from viral or tumor-associated antigens. However, the $CD8^+$ Class I-restricted CTL are also involved in cytotoxic responses against allografts (see later chapters) and against those microorganisms that are normally phagocytosed, but escape the phagolysosome and are processed in the cytosol of the phagocytosing cell.

12.11 At what stage of their development do T lymphocytes actually become CTL?

When T-cell precursors mature in the thymus, they undergo a stage when they express both CD4 and CD8 molecules. When the maturing precursors contact peptides presented to them by MHC molecules, they lose one of the two accessory molecules at random. The cells that happen to be specific to peptides presented in the context of MHC Class I molecules and to lose CD4 (and to retain CD8), are able to survive if the affinity of their TCR interaction with the peptides is moderate (see Chapter 9). Importantly, the resulting $CD4^-$, $CD8^+$ T lymphocytes can be considered mature, and yet they are still not functional CTL, because they contain no cytolytic granules. These cells, sometimes called "**pre-CTL**," exit the thymus, and populate the secondary lymphoid organs. Their final progression into CTL requires two signals; one of these is the contact with the specific antigen, but the nature of the other signal needs special consideration.

12.12 What is the second signal for the development of CTL?

The nature of the second signal necessary for the induction of CTL activity has been very elusive. Obviously, since most pre-CTL are MHC Class I-restricted, they are able to kill many types of cells that are not known to express co-stimulatory molecules like B7-1 and B7-2. Yet, at least three ways in which pre-CTL can obtain the second signal seem to be possible. First, since many cells that are infected by viruses or affected by the process of malignancy are apoptotic, they are phagocytosed by the resident macrophages or other professional antigen-presenting cells. The cytosolic peptides of the infected or malignantly transformed cells may be presented to T cells by the MHC Class I molecules of the antigen-presenting cell, and this same cell will simultaneously provide co-stimulators for the T cells' CD28 molecule. The phenomenon where a cell generates cytosolic peptides from foreign antigens, but is then ingested by an antigen-presenting cell that actually presents the peptides and co-stimulators is called **cross-priming**. Another possibility is that for pre-CTL, the second signal is delivered not through the contact of accessory molecules with co-stimulators, but by cytokines produced by T_h in the vicinity of the CTL–target cell interaction. The exact cytokines that may play such a substitute role, and the exact sequence of events in this scenario have not been identified however. Finally, it may be that there is an interaction between CD40L on activated T_h and CD40 on professional antigen-presenting cells (e.g., macrophages, which do express CD40) and the production of cytokines by T_h in the vicinity of the interaction of developing CTL and target cells. As a result of CD40L–CD40 interaction and cytokine production, target cells may begin to express co-stimulatory molecules that are recognized by the developing CTL as second signal.

12.13 What happens during the transition of pre-CTL into CTL?

The developing CTL undergo a strong, extensive **clonal expansion** and **differentiation**. It is noteworthy that the extent of antigen-induced pre-CTL's proliferation is much greater than that of antigen-induced T_h proliferation. For example, in mice at the peak of viral infection, a 50,000- to 100,000-fold increase in the number of $CD8^+$ T cells specific to the antigens of the virus can be detected. Because of such a tremendous proliferation, up to one-third of the $CD8^+$ cells of the spleen in infected mice is specific to the antigens of that virus. A similar extent of proliferation can be observed in humans infected with Epstein–Barr virus or HIV, when up

CHAPTER 12 Cell-Mediated Immunity

to 10% of the circulating $CD8^+$ cells are specific to the virus during the acute phase of infection. The differentiation of pre-CTL into CTL involves the synthesis of characteristic membrane-bound granules that contain a number of proteins, including perforin and granzyme (see below). It also includes the transcriptional activation of cytokine genes and secretion of **cytokines**, notably IFN-γ, LT, and TNF (see Chapter 10).

12.14 What are the mechanisms of the CTL-mediated killing?

The most important features of this killing are the following: the killing is strictly **antigen-specific**, **contact-mediated**, and **directed towards the target**. Because of these features, CTL kill target cells that express the same MHC Class I-associated antigen that triggered the activation of pre-CTL, and do not kill neighboring uninfected (or untransformed) cells that do not express this antigen. Moreover, CTL themselves are not injured by the toxic granules that they release. The process of CTL-mediated killing consists of: antigen **recognition**; activation of the CTL and **delivery of the "lethal hit,"** and **release** of the CTL.

12.15 What happens during the recognition step?

The recognition of the target includes the interaction between the following molecules: (1) **TCR** of the recognizing CTL and the **peptide presented by the MHC Class I molecule** on the target cell; (2) the **CD8** molecule on the CTL and the MHC Class I molecule on the target cell; and (3) the **LFA-1** molecule on the CTL and the **ICAM-1** molecule in the target cell. Microscopically, the interacting CTL and their targets look like "conjugates" of cells. Antibodies to either of the six above-mentioned molecules can abrogate the conjugate formation. The recognition leads to **clustering** of TCR and generates biochemical signals that activate the CTL.

12.16 How does the activated CTL deliver the "lethal hit?"

Within a few minutes of the CTL recognizing its target, the latter undergoes changes that lead to its apoptotic death within 2–6 hours. Remarkably, the CTL may (and usually does) detach from the target, but the apoptotic dying of the target continues (hence the term "delivery of the lethal hit"). The central moment of the delivery of the "lethal hit" is the **directed** release or the delivery of cytotoxic granules to the target cell recognized by CTL (Fig. 12-2). Immediately after recognition of the antigen of the target, the activated CTL reorganizes its cytoskeleton in such a way that its microtubule-organizing center comes to proximity with the target. The cytotoxic granules become concentrated in the same area; the granule membranes fuse with the CTL membrane and are exocytosed onto the surface of the target cell. The process of cytoskeleton reorganization and granule fusion is thought to be mediated by TCR-triggered activation of such GTP-binding proteins as Rho and Rab.

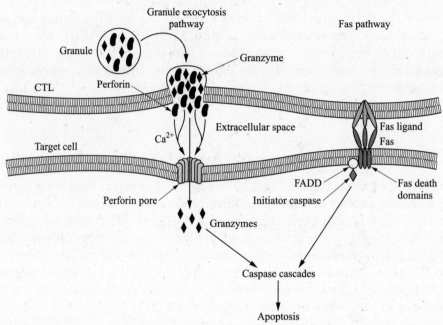

Fig. 12-2. Molecular mechanisms of "lethal hit." Proposed model of killing target cells by a mechanism that includes the formation of perforin pore and the activation of apoptosis via Fas ligand-Fas pathway. See text for details.

12.17 Why, and how, do the cytotoxic granules kill the target cell?

The cytotoxic granules contain, among other proteins, enzymes called **perforin** and **granzymes**. Perforin is a "drill," a pore-making protein that is present in the granules in its inactive monomeric form. Once the granules are released, the perforin monomer comes into contact with high extracellular concentrations of calcium (about 1–2 mM) and polymerizes. The resulting polymer forms an **ion channel** in the lipid bilayer of the target cell, that is permeable to water and ions (including calcium). If a sufficient number of perforin channels are formed, the cell becomes unable to exclude water and ions and dies. Interestingly, this mechanism is very similar to the action of the MAC (Chapter 11; see also Chapter 13), and perforin is structurally homologous to the ninth component of complement (C9), which is the principal constituent of the MAC. **Granzymes** are serine proteases, the most important one being granzyme B. These proteins enter the target cell through channels made by perforin, and cleave a variety of proteins on serine and aspartic acid residues. As a result of this cleavage, some intracellular enzymes are activated, among them **effector caspases** (see Chapters 7–9) that mediate the process of apoptosis. Some of these caspases activate the target cell **nucleases**, which degrade the target cell DNA that is especially important for the eradication of viral infection.

In addition to the perforin–granzyme-mediated killing, CTL use another mechanism of cytolysis; this alternative mechanism involves the interaction between **Fas and Fas ligand** (Chapters 7–9).

CHAPTER 12 Cell-Mediated Immunity

12.18 Why is the CTL not injured by its own cytotoxic granules?

The answer to this question is still unclear, although part of the reason is that the CTL detaches from the target cell almost immediately after the granule exocytosis.

Questions

REVIEW QUESTIONS

1. In 1942, K. Landsteiner and M. Chase showed that the so-called contact hypersensitivity (a reaction to some chemicals that manifests similarly to DTH) can be adoptively transferred. It was later admitted that their experiment was in fact very similar to the experiment of Mackaness. What exactly was adoptively transferred? Describe the way you see Landsteiner's and Chase's experiment.
2. Based on what you learned from Chapter 11 and this chapter, explain why DTH is accompanied by induration.
3. A knockout mouse is infected with *L. monocytogenes*, but develops no DTH. Name at least three genes that may be inactivated in this mouse.
4. Will CD44 transgenic mice show an abnormal pattern of T-lymphocyte migration? If yes, what will be this pattern?
5. How would you comment on the idea to treat tuberculosis patients with IFN-γ? With IL-4?
6. Why is the defect in macrophage activation more pronounced in CD40L-deficient mice than in IL-12-deficient mice?
7. In patients with deficient T-cell responses, intradermal injections of common antigens (e.g., a fungus, *Candida albicans*) does not elicit DTH. Clinicians called this phenomenon anergy. What is the difference between this anergy and the anergy described in Chapter 9?
8. Chromium ions have the capacity to penetrate cells through membrane ion channels, but it is difficult for them to exit an intact cell. How would you use this feature to assay CTL activity?
9. Name two reasons why CTL response is more efficient against tumors than T_h response.
10. The differentiation of pre-CTL into CTL requires contact with antigen. For how long should this contact be?
11. Design an experiment proving that, at the height of EBV infection, up to 10% of circulating $CD8^+$ cells are specific to the virus.
12. How would you prevent the delivery of a lethal hit other than by blocking surface molecules with antibodies?

13. Why is the combination of perforin and granzyme B a more efficient weapon against viruses than just perforin?
14. The efficiency of cytolysis is often plotted against the ratio of effector to target cells (on the X-axis). How would you imagine the curve to look?
15. What is common between the killing of target cells by CTL and the killing of antibody-bound cells by complement?

MATCHING

Direction: Match each item in Column A with the one in Column B to which it is most closely associated. Each item in Column B can be used only once.

Column A

1. Macrophages
2. $T_h 1$
3. Requires "sensitization"
4. P-selectin
5. Granulomatous inflammation
6. Pre-CTL
7. Synthesized after antigen contact
8. IFN-γ
9. GTP-binding proteins
10. Calcium ions

Column B

A. Ligand expressed on activated T_h
B. Needed for macrophage activation
C. Cytolytic granules
D. "Execute sentence" in DTH
E. Polymerize perforin
F. Activated by IL-12
G. DTH
H. Important for granule fusion
I. Unsuccessful DTH
J. $CD8^+$ cells that exit the thymus

Answers to Questions

REVIEW QUESTIONS

1. Lymphoid cells (most likely, spleen cells) from a mouse that was exposed to chemicals were adoptively transferred. The experiment must have included sensitization with chemicals, irradiation of a naïve recipient, adoptive transfer of splenic cells from the sensitized animal into the irradiated recipient, and challenge of the recipient with the same chemicals. Local inflammation accompanied by induration must have been observed in recipients of cells, but not the sera of the sensitized animals.
2. Because of the changes in endothelial layers induced by chemokines and other mediators of inflammation, many components of plasma are lost from blood into the tissues. Among these components is fibrinogen, a precursor of insoluble fibrin. The cleavage of fibrinogen into fibrin and the accumulation of fibrin in tissues cause their hardening (induration).
3. CD40L, IL-12, IFN-γ (can be also IL-18, CD40, and other genes).

CHAPTER 12 Cell-Mediated Immunity

4. They may not develop any abnormalities because CD44 binds hyaluronidate, which is not expressed in lymph nodes but rather in the extracellular matrix. Therefore, the constitutively expressed CD44 will not find its ligand until the cells exit from the lymph node into tissues.
5. IFN-γ may be beneficial because it might enhance the functions of macrophages and accelerate the clearance of *M. tuberculosis*. IL-4 may also be beneficial, especially during the granulomatous inflammation, because it might decrease the accumulation of macrophages and the formation of granulomas. Both cytokines must be used with caution however, because of their pleiotropic effects.
6. Because of the redundancy of cytokines. There might exist "back-up" cytokines that would take over the function of IL-12.
7. This anergy is general, while the anergy that results from the lack of co-stimulation or recognition of altered peptides (Chapter 9) is strictly clonal.
8. Load cells with a radioactive isotope of chromium (e.g., ^{51}Cr), wash them by pelleting and resuspending several times, and incubate them with CTL in a culture tray. After several days, measure ^{51}Cr in the culture medium. If cells are lysed, the counts in the medium will be high.
9. First, malignant transformation can affect any cell, and most T_h can only recognize MHC Class II-expressing cells. Second, CTL, unlike T_h, can destroy the nucleus in their targets (because granzyme B from cytolytic granules activates effector caspases and these activate cellular nucleases). Thus, CTL are more likely to destroy not only cells but also genes that are affected by the cancerous process.
10. Contact should be for long enough to allow the biosynthesis of new proteins and the assembly of cytolytic granules, i.e., at least 18–20 hours.
11. Prepare a virus lysate, and coat polystyrene dishes with it. Draw several milliliters of the patient's blood, isolate $CD8^+$ cells, count their number per 1 milliliter of the medium, then incubate the $CD8^+$ cells in the dishes coated with the viral lysate, and re-count $CD8^+$ cells. The number should decrease by 10%, which is possible to detect if cells from several replica cultures are counted.
12. By paralysis of the cytoskeleton (e.g., by adding colchicine), or by removing calcium from the extracellular solution (e.g., by adding EDTA or EGTA).
13. Because granzyme B can destroy DNA and perforin cannot. Without perforin, however, granzyme B will not be able to penetrate the cell.
14. The curve will go up at first, because the more CTL attach to the target, the more pores will be formed and the more efficient the lysis will be. However, it will plateau at the maximal number of CTL that are able to form conjugates with one target.
15. Polymerization of a protein that possesses features of the ninth component of complement (C9).

MATCHING

1, D; 2, F; 3, G; 4, A; 5, I; 6, J; 7, C; 8, B; 9, H; 10, E

CHAPTER 13

Effector Mechanisms of Humoral Immunity

Introduction

As was mentioned in the introductory chapter, humoral immunity can be adoptively transferred by the cell-free liquid portion of the blood (serum) and **is mediated by** soluble **antibody molecules** that float in biological fluids (Latin, *humori*). However, as we also mentioned, antibody binding by itself does not destroy the antigen. The effector phase of humoral immunity therefore includes a variety of mechanisms that eventually lead to the destruction of antigens bound by antibodies. As in the case of cellular immunity, the effector mechanisms of humoral immunity are based on the interaction between adaptive and innate immunity. For example, specific antibodies (agents of adaptive immunity) coat microbial particles and, through the interaction with Fc receptors, activate the destruction of these particles by phagocytes, natural killer cells, or eosinophils (agents of innate immunity). Throughout this chapter, we will illustrate how the interaction between innate and adaptive immunity serves the goal of ridding the body of potentially or actually dangerous antigens.

The effector phase of humoral immunity also serves as a bright illustration to the concept of specialization of immune responses (Chapter 1). Indeed, the availability of various antibody heavy chain isotypes provides a means to "customize" the effector phase of humoral immunity. Different isotypes mediate different effector functions: for example, IgG antibodies serve best as specific opsonins, IgM and some subclasses of IgG antibodies are the best in complement binding, and IgE are the only type of antibodies that can mediate immediate hypersensitivity, etc. As we discuss the effector phase of humoral immunity, we will try to strengthen this point by providing examples of particular humoral immune responses.

Previously, we have introduced the concept that antibodies may or may not be protective against infectious agents and other pathogens (Chapter 3), and that

CHAPTER 13 Humoral Immunity

immunologic reactions can injure normal tissues instead of defending the body from pathogens (Chapters 11 and 12). In this chapter, we will re-emphasize these concepts and show that, along with their protective role, certain antibodies can participate in a pathologic reaction called "immediate hypersensitivity" and thus be harmful to the organism.

We will begin the discussion by analyzing such effector mechanisms of humoral immunity as antigen neutralization and antibody-mediated opsonization and phagocytosis. The discussion will continue by the dissection of complement system and the analysis of different complement pathways. We will then proceed to characterization of antibody responses in certain local anatomic sites. At the end of the chapter, we will discuss mechanisms and consequences of immediate hypersensitivity and tissue injury that can appear as a result of the effector phase of humoral immunity.

Discussion

13.1 How do antibodies perform their "neutralizing" function?

Antibodies can neutralize (i.e., abrogate the function of) bacterial toxins as well as whole bacteria or viruses. The neutralization can be achieved merely through the binding of antibodies to these entities, or to cellular receptors used by these entities for their binding. For example, antibodies specific to antigens of bacterial pili (small appendages that are used by bacterial cells to attach to eukaryotic cells) can prevent the bacteria from attachment and thus to abrogate their function. Antibodies specific to cellular receptors used by viruses for their attachment to cells can block the viral entry into the cells. In the case when antibodies bind certain structures on microorganisms, they can neutralize these microorganisms by **steric hindrance**, i.e., directly competing with a receptor or a ligand engaged in the interaction between the microorganism and a host's cell. Alternatively, antibodies may have an **allosteric effect** on microbial particles or toxins. A few antibody molecules can bind to a microbial particle or toxin and induce conformational changes that prevent the interaction with the receptor for this microbe or toxin.

An important point about antibody-mediated neutralization is that it is a reaction that requires only two components: an antigen and an antibody. No "third party" is needed, no involvement of antibody constant region is necessary, and the antibody heavy chain isotype (which is determined by amino acid sequences in constant region; see Chapter 3) does not matter. What does matter, however, is the antibody **affinity**. High-affinity antibodies perform the neutralizing function better than low-affinity antibodies. Since IgG antibodies on average have higher affinity than antibodies of other classes (because they tend to accumulate more mutations in their V-regions; see Chapter 4), these antibodies serve best as neutralizing agents.

13.2 How efficient is the antibody-mediated neutralization of clinically important human pathogens?

Many antibodies produced against these pathogens are efficient, and many prophylactic vaccines have proven to be useful exactly because they stimulate the production of efficient neutralizing antibodies. However, this effector function of humoral immunity has one major limitation. Pathogenic microorganisms are capable of extensive antigenic variation, i.e., they use molecular mechanisms constantly to modify their surface antigens, including those that are targets for neutralizing antibodies. We will discuss these mechanisms in more detail in later chapters.

13.3 How does antibody-mediated opsonization and phagocytosis work?

We have already given the definition of opsonins, including specific opsonins (antibodies) in Chapter 12. Recall that opsonins are substances that enhance functions of activated phagocytes. Specific antibodies enhance these functions because as they bind their antigens, phagocytes (macrophages and neutrophils) attach to their constant regions via their Fc receptors. The process of coating microbial particles with antibodies that would serve as opsonins is sometimes referred to as opsonization. Unlike neutralization, the function of opsonins strictly depends on the ability of their constant regions to interact with Fc receptors. Therefore, **antibodies of certain isotypes are better opsonins than antibodies of other isotypes**. Generally, antibodies of the IgG class serve as strong specific opsonins. In humans, IgG1 and IgG3 and in mice IgG2a and IgG2b are the best opsonins. Microbial particles can also be opsonized by a product of complement activation called C3b, which binds to its receptor expressed on leukocytes. When constant regions of antigen-bound antibodies or the C3b bound to a microbial particle are sequentially bound by phagocytes via Fc or C3b receptors, the particles are engulfed, intensively digested, and subjected to ROI by the receptor-expressing phagocytes. The functions of phagocytes are thought to be stimulated by positive signals that Fc and C3b receptors transmit upon binding of their ligand.

13.4 What Fc receptors exist, and how do they function?

Many different types of Fc receptors have been identified to date, and at least eight of them are known to participate in the effector phase of humoral immunity. Of these, the receptors that are capable of binding IgG antibodies (**Fcγ receptors**, or **FcγR**) serve to enhance the phagocytosis of opsonized particles. There exist three types of FcγR, called FcγRI, FcγRII, and FcγRIII. They differ greatly in the affinity of their interaction with IgG constant regions, and also in the pattern of their expression on leukocyte populations. All three types of the receptor bind preferentially IgG1 and IgG3 in humans, and IgG2a and IgG2b in mice.

The major high-affinity FcγR of phagocytes is the **FcγRI** (or **CD64**). This receptor is also expressed on eosinophils. The affinity of its interaction with constant regions of IgG antibodies is very high, the K_d being between 10^{-8} and 10^{-9} M. FcγRI consists of two subunits, the antibody constant region-binding alpha chain and the signal-transducing gamma chain. The former is a 40-kDa protein that has two immunoglobulin-like domains, and the latter is a 7-kDa polypeptide that is largely intracellular (with only a short stretch of amino acids outside) and homologous to TCR-associated ζ **chain** (see Chapter 7). The FcγRI gamma chain contains ITAMs in its cytoplasmic portion, and uses them to transduce the signal. The transduction involves activation of several PTK, and also of the enzyme oxidase, whose important role in the formation of ROI was discussed in Chapter 11. Transcription of the FcγRI gene is stimulated by **IFN-γ**, which is the reason why macrophages activated by this cytokine have higher numbers of this receptor per cell than resting monocytes.

The second FcγR, called **FcγRII** or **CD32**, possesses a much lower affinity for IgG constant regions (K_d 10^{-6} M), and is unable to bind monomeric IgG; for the binding to occur, IgG must be clustered in immune complexes or on a cell surface. Three separate genes, called A, B, and C code for FcγRII, and this molecule can exist in various isoforms resulting from alternative mRNA splicing. The most common isoforms are FcγRIIA (expressed on phagocytes, eosinophils, some endothelial cells, and platelets) and FcγRIIB (expressed on B lymphocytes). FcγRIIA can transmit activative signals, although it is generally thought that phagocyte activation after its triggering is rather weak. FcγRIIB is the B-cell-specific isoform of FcγRII that mediates the so-called **antibody feedback inhibition** (see Chapter 8).

The third FcγR, called **FcγRIII** or **CD16**, has already been mentioned when we discussed functions of NK cells (see Chapter 11). The actual receptor of NK cells is one of the two isoforms of FcγRIII (FcγRIIIA). An interesting feature of this receptor is that its signaling component can be represented by homodimers of the gamma subunit shared with FcγRI, or by heterodimers of the above gamma subunit and the ζ chain shared with the TCR receptor complex. CD16 does not participate in phagocyte activation but rather in **antibody-dependent cell-mediated cytotoxicity** (**ADCC**) (see later). The second isoform, FcγRIIIB, is expressed on neutrophils and transduces a weak activative signal.

The role of some other Fc receptors (e.g., FcαR, FcεR) has very special functions in the effector phase of humoral immunity. Their roles will be discussed later, in the appropriate sections.

13.5 What is ADCC, and what is its role in the effector phase of humoral immunity?

It is, essentially, the killing of a target cell by an effector cell, triggered by the binding of an Fc receptor on the effector to the C-region of an antibody bound to the target cell. Since this definition may sound too "Byzantine," it is perhaps easier to understand the concept of ADCC just by illustrating it with an example. NK cells are typical mediators of ADCC. They contain cytolytic granules and are able to release them when they are activated. If a microbial cell is precoated by IgG antibodies and the CD16 binds the clustered C-regions of these antibodies, the NK cell that expresses the CD16 becomes activated and releases the granules, killing the precoated cell. Thus, the killing is "*cell-mediated*" (because the killing granules originate from the NK *cells*), but also *antibody-dependent* (because it will take place only if the target is coated with *antibodies* of the appropriate isotype).

13.6 What is the role of complement in the effector phase of humoral immunity?

We have already introduced the concept of complement in Chapters 1 and 11, and mentioned that it is a complex system of zymogens that can be activated by the binding of its component either to specific antibody (classical pathway) or to bacterial cell wall (alternative pathway). The end result of both pathways of complement activation is the formation of membrane attack complex (MAC). This complex of self-assembling proteins serves as a "drill" that bores holes in bacterial cells coated with complement components, or any cells coated by antibodies to surface antigens. Another important function of complement components is opsonization, i.e., the enhancement of phagocytosis of the particles that bind them via special receptors (mostly CR1, see later). Yet another important function of complement, already mentioned in Chapter 8, is to promote B-cell antigen-induced activation. In addition, some products of complement activation serve as mediators of inflammation.

"**Complement**" is a French word that means, "adding a lacking or missing part," "additional," or "supplementing." This term was coined by J. Bordet, who discovered that while antibodies still bound the cells after the serum had been heated to 56°C, they lost their ability to lyse the bacteria. Bordet postulated that immune sera contain antibodies and an additional ("complement") substance that enabled the antibodies to lyse cellular targets. In older literature, one can find a parallel term "**alexin**," which means "protecting." This other term actually reflects the idea that the mere binding of antibodies to bacteria may not protect the organism from infection; to be protected, the host uses the "alexin" produced by its immune system and destroys bacteria with its help.

13.7 How are complement pathways actually triggered, and what is the sequence of events in complement activation?

Indeed, the pathways of complement activation differ in how they are triggered, and there are also substantial differences in the events that mark their course. Yet, both classical and alternative pathway (and also the lectin pathway mentioned in Chapter 11) have profound similarities and use some identical steps. All pathways can be subdivided into **early steps** that involve **proteolysis of the complement product called C3**, and **late steps** that lead to the **formation of a lytic complex**. The **central event** in all pathways is **proteolysis of C3** and the subse-

CHAPTER 13 Humoral Immunity

quent covalent **attachment** of a product of C3, called **C3b,** to microbial cell surfaces or to antibody bound to antigen (Fig. 13-1). During the early steps, an enzyme called **C3 convertase** is generated; this enzyme cleaves C3 and thus creates two products of its proteolysis, called C3a and C3b. The subsequent events are determined by the target of C3b attachment. If C3b binds to constant regions of **antibodies** bound to antigen, this steers the complement activation towards **classical pathway**. If C3b binds directly to the **microbial surface** (e.g., to LPS expressed on membranes of Gram-negative bacteria, or some other entities), this steers the complement activation towards the **alternative pathway**. However, the principal events that follow the attachment of C3b to whatever are again similar: they involve binding of other complement components to C3b and generation of another enzyme, called **C5 convertase**. The latter, as the name tells, cleaves the complement component C5 and initiates the above-mentioned late steps of complement activation, leading to the formation of a lipid-soluble macromolecular protein complex called MAC (from membrane-attack complex, see also Chapter 11).

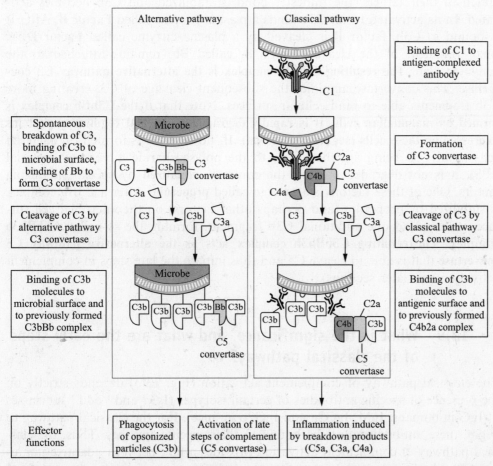

Fig. 13-1. Pathways of complement activation. An overview showing the alternative and classical pathways of complement activation.

13.8 What is the significance, and what is the exact sequence of events of the alternative pathway?

The name "alternative" stems from the fact that this pathway was discovered later than the antibody-dependent pathway; the latter was by that time already well dissected and therefore called "classical." The name "alternative," however, may be misleading because in fact, the antibody-independent pathway of complement activation appeared earlier in phylogeny and plays a crucial role in innate immunity.

The events that constitute the alternative complement pathway are shown in Fig. 13-2. Let us first concentrate on its early steps, because they are really peculiar and different from the early steps of the classical pathway. The unique feature of the alternative complement pathway during its early steps is the so-called **C3 tickover** – a continuous slow-rate cleavage of C3 in the fluid phase (blood plasma). Because of this cleavage, small amounts of C3b are always available in the circulation, and a small fraction of the circulating C3b binds microbial surfaces. The C3b molecule contains an internal **thioester bond** that is very unstable when C3b is circulating. However, when C3b molecules covalently attach to their targets, the thioester bond is stabilized and C3b becomes activated. In its activated form, C3b binds a plasma protein called **Factor B**. After it is bound to C3b, factor B is cleaved by a plasma enzyme called **Factor D**. A small fragment of the cleaved factor B, called **Bb**, remains attached to the activated C3b. **The resulting C3bBb complex is the alternative pathway C3 convertase**. This convertase amplifies the subsequent cleavage of C3, creating more C3b fragments able to bind cellular surfaces. Note that **if the C3bBb complex is formed on mammalian cells, it is rapidly degraded by special regulatory proteins expressed on these cells** (see later sections). If, however, it is formed on **bacterial** cells (with C3b being attached, e.g., to the polysaccharide moiety of bacterial LPS), it is not degraded but, on the contrary, **stabilized** by special stabilizing factors. One of these factors is a protein called **properdin**. After a C3Bb complex is stabilized by properdin (and perhaps other factors) on the microbial cell surface, it may bind one additional C3b fragment, forming the so-called **C3bBb3b** complex. **The resulting C3bBb3b complex acts as the alternative pathway C5 convertase** that is able to cleave C5 and thus initiate the late steps in complement activation (see later sections).

13.9 What is the significance, and what are the early steps of the classical pathway?

The classical pathway of complement activation (Fig. 13-3) depends strictly on the presence of specific antibodies of certain isotypes (IgM and "odd" subclasses of IgG in humans; IgM, IgG2a and IgG2b in mice). For the classical pathway to begin, these antibodies must be bound to their specific epitopes. Thus, the classical pathway of complement activation provides a link between adaptive immunity (antibodies bound to antigens) and innate immunity (complement components).

CHAPTER 13 Humoral Immunity

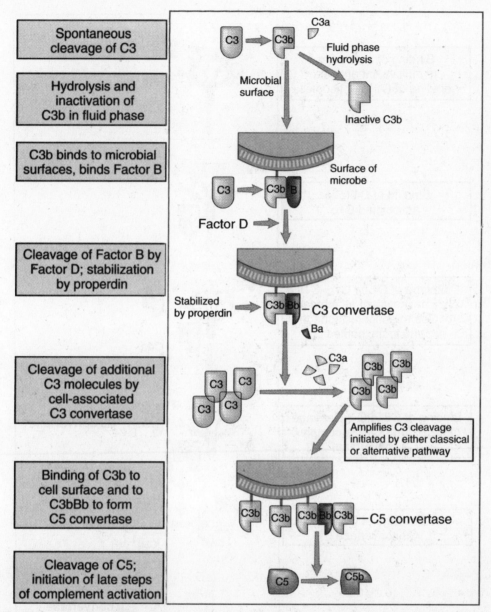

Fig. 13-2. The alternative complement pathway. This flow chart demonstrates the events that occur during the activation of the alternative complement pathway. See text for details. From Abbas, K., A.H. Lichtman, and J.S. Pober, *Cellular and Molecular Immunology*, fourth edition, W.B. Saunders, 2000, p. 318 (their Fig. 14-6).

The classical pathway (Fig. 13-3) is **initiated** when C_H2 domains (see Chapter 3) of antigen-bound IgG antibodies, or C_H3 domains of antigen-bound IgM antibodies are bound by the complement protein **C1**. The latter is a large multimeric protein complex that has the form of an umbrella. It consists of three components, called C1q, C1r, and C1s. C1q forms the radial part of the umbrella, which is an array of six chains ending with globular heads, and serves to bind antibody constant regions. C1r and C1s proteins are proteases.

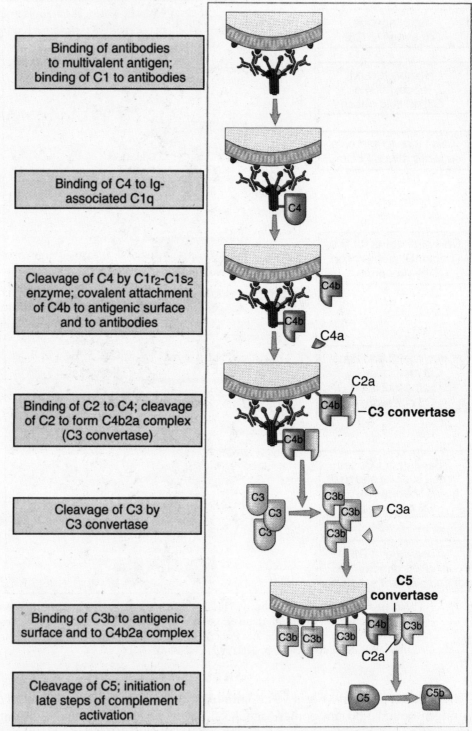

Fig. 13-3. The classical complement pathway. This flow chart demonstrates the events that occur during the activation of the classical complement pathway. See text for details. From Abbas, K., A.H. Lichtman, and J.S. Pober, *Cellular and Molecular Immunology*, fourth edition, W.B. Saunders, 2000, p. 321 (their Fig. 14-8).

CHAPTER 13 Humoral Immunity

One very important feature of the classical pathway initiation is the following: **in order to be activated, one C1q molecule must simultaneously bind to two antibody molecules.** In the case of IgG antibodies, C1q can bind only if at least two separate antibody molecules of this class are brought together. This happens when IgG antibodies bind a **multivalent** antigen (a "cartridge" of epitopes). In circulating IgM, which are pentameric (Chapter 3), the C1q binding sites are positioned inside the pentamer and inaccessible for the binding of C1q molecules. When IgM antibodies bind their epitopes, however, these sites become available for C1q binding. Because pentameric IgM can bind more C1q per antibody molecule than monomeric IgG, the former are much more efficient in binding (or "fixing") complement.

The early steps of the classical complement pathway are shown in Fig. 13-3. Binding of two or more of the globular heads of C1q to the constant regions of IgG or IgM leads to activation of C1r and then C1s. The activated C1s cleaves a complement protein called **C4**, leading to the appearance of two cleavage products, C4a and C4b. The larger of these two, **C4b**, attaches either to antigen–antibody–C1q complex or to the adjacent surface of the cell to which the antibody is bound. It is noteworthy that **C4 is chemically homologous to C3**, and C4b has an internal thioester bond that is stabilized when C4b attaches to target, just as in the case of C3b.

The bound C4b is then itself bound by a complement component called C2. When the C2–C4b complex is formed (Fig. 13-3), the nearby C1s cleaves C2, forming two fragments, C2a and C2b. The latter is small, soluble, and its function is unknown. On the contrary, C2a is larger; it remains attached to C4b, and these two bound molecules form **the C4b2a complex,** which acts as **the classical pathway C3 convertase**. The C4b component of the complex binds soluble C3, and the C2a component proteolytically cleaves it. The resulting C3b attaches to microbial cell surfaces as described in Section 13.8. It is important to remember that the complex of only one C4b and only one C2a can cleave many thousands of C3 molecules and thus be responsible for attachment ("deposition") of thousands of individual C3b fragments. Again, as in the alternative pathway (Section 13.8), some of the C3b fragments attach to the C3 convertase (i.e., the C4b2a complex) itself, thus forming the **C4b2a3b** complex, which **is the classical pathway C5 convertase**. This latter complex is responsible for cleavage of C5, which marks the beginning of the late steps in complement activation.

13.10 What is the lectin pathway of complement activation?

This pathway is very similar to the classical pathway of complement activation, although it develops in the absence of antibody. The essence of this pathway lies in the fact that a plasma lectin called "**mannose-binding lectin**" (**MBL**) is structurally and functionally homologous to C1q. When it binds to mannose residues of microbial surface polysaccharides, MBL triggers the sequence of events that is virtually indistinguishable from that in the classical complement pathway.

13.11 What is the sequence of events during late steps of complement activation?

As already mentioned, the late steps do not differ between the pathways and boil down, essentially, to the formation of MAC. C5 convertases **cleave C5** (which is a circulating complement component) into fragments called C5a and C5b. The former is soluble, and the latter attaches to cell surfaces and binds a complement component called C6. The C5b6 complex in turn binds C7, which is a hydrophobic protein. Thus, the complex becomes hydrophobic and inserts into plasma membrane. After this, it binds C8, which is also a hydrophobic protein, and then **C9**. The latter is a serum protein **homologous to perforin** (see Chapter 12) and capable of polymerization at the site of its insertion into cellular membrane. MAC is, essentially, the complex of C5b,6,7,8 and polymerized C9. It acquires a shape of a drill and bores holes in the membrane. These pores have the diameter of approximately 10 nm, and allow free movement of water and ions. The cell is killed either because of the swelling or because of the steep increase in calcium concentration.

13.12 What are complement receptors?

Several distinct types of receptors for complement components exist in higher vertebrates, and are expressed on various cell types. These receptors bind fragments of C3, C4, and C5 proteins. The best studied of these receptors are the following:

- **Type I complement receptor** (**CR1** or **CD35**) is a major complement receptor expressed on neutrophils, monocytes, red blood cells, eosinophils, lymphocytes (both T and B) and follicular dendritic cells. CR1 is a **high-affinity receptor for C3b and C4b**, and its principal functions are to promote phagocytosis of C3b- and C4b-coated particles and clearance of immune complexes from the circulation. We have already described the role of CR1 in phagocytosis in Chapter 11. The clearance of immune complexes is mediated mostly by red blood cells that trap these complexes on their surface via CR1 and deliver them mostly to the spleen and liver. There, the red blood cells together with the immune complexes are engulfed by resident macrophages. In addition, CR1 plays a role in the regulation of complement activation (see later sections).

- **Type 2 complement receptor** (**CR2** or **CD21**) is expressed on B lymphocytes, follicular dendritic cells, and cells of nasopharyngeal epithelium. Its principal functions are to **promote B-cell activation** by antigen and to **trap antigen–antibody complexes in germinal centers**. The first of these functions has been already described in Chapter 8, and the second in Chapters 2 and 4. The ligands of CR2 on B cells are the cleavage products of C3b called C3d, C3dg, and iC3b (see later sections), and on follicular dendritic cells – these are antigen–antibody complexes coated with iC3b and C3dg. In addition, CR2 serves in humans as the receptor for Epstein–Barr virus (EBV) – a virus that belongs to the family Herpesviridae and is capable of activating and malignantly transforming B lymphocytes.

CHAPTER 13 Humoral Immunity

- **Type 3 complement receptor** (**CR3** or **CD11bCD18**) is also known as **Mac-1** protein (because it was initially discovered on macrophages). Mac-1 is a member of the integrin family of cell adhesion molecules described in Chapters 7, 8, and 11. Its role in innate immunity was described in Chapter 11. In addition to macrophages, the cells that express Mac-1 include neutrophils, mast cells, and NK cells. The ligand for Mac-1 is iC3b.
- **Type 4 complement receptor** (**CR4** or **CD11cCD18**) is also known as p150,95. It is an integrin that uses the same beta chain (CD18) as Mac-1 protein, but a different alpha chain. Its function is very similar to that of Mac-1. One interesting feature of CR4 is that this molecule is abundantly expressed on **dendritic cells** and serves as a marker of this cell type.

13.13 How is activation of the complement system regulated?

It is indeed very important to keep complement pathways under strict regulation, for two main reasons. First, the above-described pathways must be triggered only when complement components bind **foreign** (i.e., bacterial) cells and **not host cells**. Second, the duration of complement activation must be limited even if complement components bind microbial agents. To achieve these two goals, many regulatory proteins are employed. Several of these proteins form a family of molecules called "**regulators of complement activity**" (**RCA**). All RCA members share structural homology and are encoded by adjacent and homologous genes (Fig. 13-4). These and some other proteins inhibit the formation of C3 convertases, break down and inactivate C3 and C5 convertases, and inhibit formation of the MAC. We will briefly describe several most important RCA molecules.

- **C1 inhibitor** (**C1 INH**) is an inhibitor of the classical complement pathway. It is a soluble plasma serine protease inhibitor ("serpin") that interferes with the activity of C1r and C1s (recall that these proteins are serine proteases, Section 13.9). C1 INH prevents the accumulation of enzymatically active C1r and C1s proteins, and limits the time for which these enzymes are available to activate subsequent steps in the complement cascade. C1 INH also inhibits other plasma serine proteases, including callicrein and coagulation factor XII. The genetic deficiency of C1 INH in humans causes the so-called hereditary angioneurotic edema (HANE), which is characterized by an increased breakdown of C4 and C2 and an accumulation of the so-called C2 kinin (a proteolytic fragment of C2) and bradykinin.
- **Factor H** is an inhibitor of the alternative complement pathway. It is a soluble plasma protein that inhibits formation of the alternative pathway C3 convertase by binding to C3b and displacing factor B (see Fig. 13-2), thus stopping the generation of the C3bBb complex.
- **Factor I** is an inhibitor of both classical and alternative pathways. It is a soluble plasma protein that binds to sialic acid residues and interferes with processing of C3b and C4b when these are "deposited" (fixed) on cell surfaces.

CHAPTER 13 Humoral Immunity

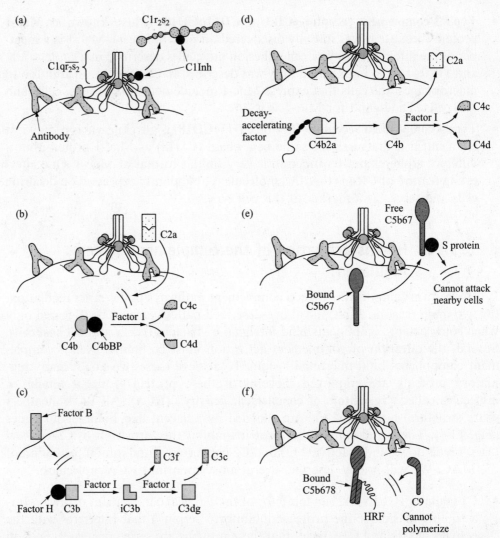

Fig. 13-4. Regulators of complement activity. This diagram shows functions of C1 inhibitor (C1Inh) (a), C4b-binding protein (C4bBP) (b), Factor H (c), decay-accelerating factor (DAF) (d), S protein (e), and homologous restriction factor (HRF) (f).

Mammalian cells have much more of sialic acid residues on their surface than microbial cells, which facilitates the attachment of the above RCA to mammalian surfaces and disfavors their attachment to microbes. Factor H acts as a cofactor to Factor I in its inhibition of C3b processing, and a number of other proteins (see below) act as its cofactors in its inhibition of C4b processing.

- **Complement receptor type 1 (CR1).** In addition to playing its important role in opsonization (see Section 13.12), this surface receptor acts as a cofactor to Factor 1, enabling the latter to interfere with processing of C3b and C4b.
- **C4b-binding protein (C4bBP)** is a plasma protein that acts as a cofactor to Factor I in its interference with C4b processing.
- **Membrane cofactor protein (MCP, CD46)** is a membrane protein expressed on leukocytes, epithelia, and endothelial cells. It acts as a cofactor to Factor I.

CHAPTER 13   Humoral Immunity

- **Decay-accelerating factor** (**DAF**) is a membrane protein expressed on leukocytes, red blood cells, epithelia, and endothelial cells. It acts as a cofactor to Factor I, and is believed to be of a greater importance than MCP; more DAF than MCP is expressed on membranes. The genetic deficiency of DAF in humans results in a disease called paroxysmal nocturnal hemoglobinemia. This disease manifests in attacks of hemolysis that are believed to be caused at least in part by deregulated complement activation on the surface of red blood cells.
- **"Homologous restriction factor"** (**HRF**) or **CD59** is a membrane protein with the pattern of expression similar to that of DAF, but present in higher numbers. Its function is to interfere with formation of the MAC. CD59 incorporates into growing MAC complexes and thus abrogates the polymerization of C9.
- **S-protein** is a plasma protein whose function is to interfere with formation of the MAC. Unlike CD59, S-protein does not insert in the cell membrane itself, but rather prevents the insertion of the hydrophobic C7 molecule.

13.14 What components of complement (or their cleavage products) are instrumental in particular functions of the complement system: cell lysis, opsonization, promotion of antigen-induced B-cell activation, or mediation of inflammatory responses?

As mentioned previously, complement-dependent cell **lysis** is mediated by the **MAC**, which assembles from complement components C5b (a cleavage product of C5), C6, C7, C9, and polymerized C9. **Opsonization** is mediated by **C3b**, **iC3b**, and **C4b**, which coat microbial particles and bind to complement receptors on macrophages and neutrophils (C3b and C4b to CR1, and iC3b to Mac-1). Such opsonization is especially pronounced if FcγRI expressed on the same macrophage is occupied by the antigen-bound IgG antibody's constant region. Interestingly, knockout mice lacking C3 and C4 show no defects in IgG antibody-mediated phagocytosis, while knockout mice that lack FcγRI do show such defects. These data indicate that in phagocytosis mediated by antibody constant regions, Fc receptor-mediated reaction may be more important than complement-mediated reactions. The proteolytic complement fragments **C5a**, **C4a**, and **C3a** induce acute **inflammation** by binding to mast cells and neutrophils. The binding of these fragments to mast cells causes their degranulation (see later). In neutrophils, C5a stimulates responses to chemokines, adhesion and (in high doses) respiratory burst. C5a also stimulates the expression of P-selectin by endothelial cells, thus strengthening neutrophil adhesion. Specific receptors of C5a and C3a have been identified. The C5a receptor is a member of the seven alpha-helical transmembrane receptors (serpentines) that signals through GTP-binding proteins. It is expressed on many cell types, including many types of blood cells and also muscle cells and astrocytes (elements of neuroglia). The C3a receptor is somewhat less well studied, but it is thought that it also signals through GTP-binding proteins and promotes inflammatory responses upon ligand binding.

13.15 If effector functions of cell-mediated immunity (e.g., DTH) can be injurious to host tissues, is this also true for the complement-mediated effector reactions?

Yes. Complement activation may result in the lysis of host cells and acute or chronic inflammation. This clearly happens in patients with deficient CA (see above), but can also happen even when RCA are qualitatively and quantitatively sufficient, e.g., during autoimmune diseases (see later chapters).

13.16 What are the effector functions of antibody classes other than IgM and IgG?

Obviously, IgD has no effector function because it is never secreted (Chapter 3). IgA and IgE antibodies play important roles in the effector phase of humoral immunity. IgE antibodies mediate a particular type of immune reaction called immediate hypersensitivity, which will be discussed at the end of this chapter. IgA is important for the development of immune responses in certain anatomic locations.

13.17 What are the locations where IgA plays its role, and what is this role?

IgA is produced in larger amounts than any other isotype. An adult human produces about 2 g of IgA per day, which accounts for 70% of all antibodies. Yet, the concentration of IgA in the serum is modest, and it is a rather minor component of systemic immune responses. The explanation for this apparent discrepancy lies in the fact that IgA is instrumental in **mucosal immunity**. It is **the major class of antibody that is produced in the mucosal compartment of the immune system**. B lymphocytes and plasma cells that produce this class of antibody are located predominantly in lymphoid agglomerates that lie underneath the epithelia of the gastrointestinal and respiratory tracts. Some of these agglomerates, e.g., tonsils and Peyer's patches, have a structure that somewhat resembles lymph nodes with follicles (see Chapter 2). The IgA secreted in these agglomerates is efficiently transported across the epithelium of the gastrointestinal and respiratory tracts with the help of the so-called **poly-Ig receptor**. Therefore, the vast majority of the produced IgA molecules locate ultimately in mucosal secretions. There, the IgA antibodies are used to neutralize various microbes that enter the body through gastrointestinal and respiratory portals of entry.

B lymphocytes located in the mucosal compartment of the immune system tend to switch to IgA upon activation. In part, this is so because mucosal T cells, more than T cells located elsewhere, produce two cytokines that stimulate IgA class switch. These cytokines are TGF-β and IL-5 (see Chapter 10). In addition, it may be that IgA-producing B cells have a special propensity to home into the mucosal compartment.

CHAPTER 13 Humoral Immunity

13.18 How exactly is IgA transported into the lumen of the gastrointestinal and respiratory tracts?

The above-mentioned active transport of IgA across the epithelia towards the lumen of the gastrointestinal and respiratory tracts occurs as follows. The secreted IgA at first accumulates in the lamina propria (the connective tissue lying immediately underneath the epithelia) in the form of a dimer that is held together by the J chain (see Chapter 3). From here, the IgA molecules are "picked up" by the poly-Ig receptor, which is produced by the epithelial cells and expressed on their basal and lateral surfaces. The IgA binds with the poly-Ig receptor, and their complex is endocytosed and actively transported in vesicles to the luminal surface. When the bound IgA reaches the luminal surface, the poly-Ig receptor is proteolytically cleaved and its extracellular domain leaves the cell, together with the bound IgA dimer. The fragment of the poly-Ig receptor that remains associated with IgA is called the **secretory component**. The above-described process is called **transcytosis**. The poly-Ig receptor facilitates the **transcytosis** of IgA into the mucosal secretions of the gastrointestinal and respiratory tracts, including saliva, bile, and sputum. The same mechanism is accountable for the appearance of IgA in milk. In addition to the transcytosis of IgA, the poly-Ig receptor is also responsible for transcytosis of some IgM.

13.19 Can IgG also be actively transported through epithelial layers?

Yes. It happens during the late fetal and neonatal stages of mammalian development. Fetuses and neonates do not mount their own antibody responses, but rather depend on maternal IgG transported across the placenta and (after birth) on maternal IgG and IgA contained in breast milk. Transport of IgG across the placenta and across the neonatal intestinal epithelium is mediated by a special receptor called the **neonatal Fc receptor (FcRn)**. This receptor resembles a Class I MHC antigen (see Chapter 5) rather than other Fc receptors, but it does not utilize its peptide-binding domain for the interaction with the constant region of IgG antibodies. Interestingly, adults retain the FcRn, which in an adult organism serves to protect IgG antibodies from catabolic degradation. Circulating IgG molecules can diffuse from blood vessels into epithelial layers and bind FcRn expressed there. When it happens, the IgG molecules are endocytosed, "stored" inside epithelial cells in a form that protects them from degradation, and then returned into circulation.

13.20 What is the role of IgE in the effector phase of humoral immunity?

IgE is a class of antibodies whose role is unique. This class of antibodies is the principal mediator of the reaction called **immediate hypersensitivity**. The immediate hypersensitivity reaction is the immunological correlate of a clinical phenomenon known as **allergy**. IgE mediates this reaction because the high-affinity receptor for the constant region of this antibody (called FcεRI) is expressed on mast cells, basophils, and eosinophils. The IgE trapped on the surface of mast cells and basophils via the FcεRI can be cross-linked by the antigen to which the IgE is specific. When this happens, mast cells and basophils release their granules, which, in turn, causes such well-known phenomena attributable to allergies as soft swelling, local redness and hyperthermia, bronchoconstriction, paralysis of peristalsis, and others.

As was discussed in Chapters 2, 9, and 10, the class switch to IgE is mediated by **IL-4**, which is a "signature cytokine" of **T_h2** lymphocytes. Another T_h2-derived cytokine, IL-5, serves as a potent activator of eosinophils, which also play a distinct role in the development of the immediate hypersensitivity. Because of that, the immediate hypersensitivity and allergic reactions can be regarded as manifestations of the effector phase of T-cell-mediated immunity that has been skewed towards the T_h2 pathway of T-cell differentiation. Yet, it also makes sense to discuss them in the context of the effector phase of humoral immunity because of the central role that IgE antibodies play in these reactions.

13.21 How was it discovered that IgE mediates allergies?

The first demonstration of the role of antibodies in the development of allergies was provided in the 1920s by Praustnitz and Kustner, who showed that a non-allergic individual injected with the serum taken from an allergic individual developed an allergic reaction to the same substance that was allergic to the donor of the serum. The reaction in the recipient developed at the site of the intradermal injection of the serum. Later, after the introduction of a technique called **adoptive transfer** (see Chapter 1), it was shown that allergic reactions in laboratory rodents can be adoptively transferred only by one special class of antibodies. Initially, these antibodies were called **reagins**, although later this term was almost completely replaced by the term IgE. Another evidence supporting the role of IgE in allergy comes from experiments with anti-IgE antibodies. It was shown that the injection of such antibodies can mimic the antigen that causes the immediate hypersensitivity, similarly to anti-IgM antibodies mimicking antigen in inducing polyclonal B-lymphocyte activation (Chapter 8).

13.22 Why are the reactions mediated by IgE called "immediate" hypersensitivity reactions?

Like DTH, immediate hypersensitivity is an antigen-specific and clinically observable reaction that points to the fact of a previous encounter with the antigen ("sensitization"). The reason for using the adjective "immediate" is to stress the fact that the symptoms of immediate hypersensitivity indeed develop much faster than the symptoms of DTH. In sensitized individuals, a second encounter with the antigen (or "allergen") can cause the above-mentioned symptoms of hypersensitivity in minutes or even seconds after the introduction of the allergen. (Recall that in DTH, the principal symptoms develop over 24–48 hours after a second exposure to the antigen.)

CHAPTER 13 Humoral Immunity

13.23 How are individuals sensitized to allergens?

Obviously, not all individuals are sensitized to allergens. A variety of antigens can be allergenic and sensitize individuals who are called "**atopic**." Such individuals are predisposed to the skewing of their T-cell responses to particular antigens towards the T_h2 pathway of differentiation. Because of that, the production of IL-4 (and also IL-5) in these individuals is enhanced compared to those who are nonatopic, and IgE antibodies are secreted in these individuals in quantities that are larger than those in nonatopic individuals. In addition, atopic individuals may have an enhanced expression of the FcεRI on their mast cells and basophils. Atopy is known to "run in families," and is thought to be associated with a number of autosomal genes, e.g., a locus on chromosome 5q near a cluster of cytokine genes or a locus on chromosome 11q13 near the gene that codes an FcεRI beta chain. The exact pattern of inheritance of the atopic phenotype is not known however, and neither are all genes that might contribute to it.

13.24 What are the most common allergens?

Since the immediate hypersensitivity depends on T_h cells, the antigens that induce it (allergens) must be T-cell-dependent protein antigens. The exact structural features that determine the allergenic nature of a protein are not known, although there are some noticeable similarities between many known allergens. For example, most known allergens are highly **glycosylated** proteins of a relatively **small molecular weight**. In addition, many common allergens (for example, major allergens of house dust mites and bee venom) are **enzymes**. How exactly these features contribute into the development of allergy is not known. Some nonprotein substances, e.g. penicillin, are highly allergenic to some people. It is thought that these substances serve as **haptens**, becoming conjugated to self protein carriers; again, the exact mechanism of their action has not been elucidated. Apparently, repeated encounter with allergens, regardless of their exact structure, plays a major role in sensitization against them.

13.25 What are the properties of the FcεRI, and how does its cross-linking cause mast cell and basophil degranulation?

FcεRI is constitutively expressed on mast cells and basophils. **The affinity of its binding to the constant region of IgE is extremely high**; the K_d of their interaction is 10^{-10} M. Because of such a high affinity, the serum IgE can be efficiently trapped on FcεRI, even given that the concentration of secreted IgE the antibody in the serum is low (less than 1 μg/ml in nonatopic individuals, and several micrograms per milliliter in atopic individuals). In addition to mast cells and basophils, FcεRI is expressed on some Langerhans cells and dermal macrophages (where its function is unknown), and also on some blood monocytes and activated eosinophils, where it mediates ADCC (see later sections).

FcεRI consists of three polypeptide chains. Its 25-kDa IgE-binding α subunit has an extracellular domain that folds into two tandem Ig-like domains, and is associated with the 27-kDa β chain, which spans the plasma membrane four times, and two identical disulfide-linked 7-kDa γ chains. The latter are shared with other Fc receptors and homologous to the T-cell-receptor-associated ζ chain (see Section 3.4). There is one ITAM on the β chain, and one on each of the two γ chains. The binding of IgE to its binding site on the FcεRI alpha subunit up-regulates the expression of the receptor.

The **cross-linking** of FcεRI can occur when two neighboring molecules of this receptor on the same cell bind IgE molecules of the same antigenic specificity, and these two IgE molecules bind their antigen. This situation is common in atopic patients who accumulated large amounts of bound IgE specific to a particular allergen, to which these individuals have been repeatedly exposed. The cross-linking of FcεRI initiates in mast cells and basophils a signaling cascade that involves PTK and is similar to the cascade triggered by antigen (see Chapters 7 and 8). Upon cross-linking, the ITAMs of the receptor are phosphorylated by the lyn kinase, which is constitutively associated with the receptor's β chain. This facilitates the recruitment of syk kinase, which triggers the downstream events described in Chapters 7 and 8, in particular, activation of PLC-γ, breakdown of membrane phospholipids, elevation of $[Ca^{2+}]_i$, and activation of PKC. The activated PKC in the presence of high calcium concentrations phosphorylates cytoplasmic myosin chains and thus dissociates the actin–myosin complexes present underneath the plasma membrane of mast cells and basophils. The latter event allows the special granules that have been produced in advance to come into proximity with the plasma membrane, to fuse with it, and to be expunged from the cells (exocytosed). This mast cell and basophil degranulation is facilitated by GTP-binding proteins, and can be inhibited by cAMP-dependent proteinase A. In addition, the FcεRI cross-linking leads to the secretion of lipid mediators (see later sections) and (in mast cells) cytokines, notably IL-4, IL-5, IL-6, TNF, IL-3, GM-CSF, and some chemokines (but, in contrast with T cells, not IL-2).

13.26 What substances do the exocytosed granules contain, and what is the biological effect of their release?

The biologically active substances that are stored in the granules of mast cells and basophils can be classified in three categories: biogenic amines; enzymes; and proteoglycans. **Biogenic amines**, sometimes called vasoactive amines, are nonlipid, low-molecular weight compounds that share an amine group. Human mast cells contain mostly one representative of this group, called **histamine**, while rodent mast cells contain approximately equal amounts of histamine and **serotonin**. Histamine exerts its effects upon binding to its receptors expressed on various cells and designated H1, H2, H3, etc., according to their pharmacological inhibitors. Binding of histamine to its receptors on endothelial cells causes these cells to produce smooth muscle relaxants, which results in **vasodilation**; it also causes the contraction of endothelial cells and **leakage of plasma** into the tissues. Thus, histamine causes what is usually called the **"wheal and flare" reaction**: a rapid swelling of the tissue

CHAPTER 13 Humoral Immunity

without induration (a difference with DTH), and an increased blood influx through dilated blood vessels that manifests as redness and local hyperthermia. Histamine also causes constriction of intestinal and bronchial **smooth muscle**. This may result in the **increased peristalsis** (especially when allergens are ingested) and **bronchospasm** (when they are inhaled, e.g., in asthmatics).

The **enzymes** stored in the granules of mast cells include tryptase, chymase, carboxypeptidase A, and cathepsin G. Some of these enzymes, e.g., tryptase, are not known to be produced by any other cell type. Tryptase is known to activate collagenase and thus to contribute into connective tissue lesions, while chymase plays a role in vascular reactions by converting angiotensin I to angiotensin II. Basophil granules also contain several enzymes, including neutral proteases (similar to the enzymes in mast cells) and major basic protein and lisophospholipase (absent from mast cells, but present in eosinophils).

The two major **proteoglycans** stored in the granules of mast cells and basophils are **heparin** and **chondroitin sulfate**. Because these molecules have a strong net negative charge, they serve as storage matrices for positively charged biogenic amines, proteases, and some other molecules. When released from the granules, proteoglycans "let go" of the molecules bound to them, albeit with varying ratio: for example, biogenic amines dissociate more rapidly than proteases. Thus, proteoglycans serve as **regulators of the kinetics** of immediate hypersensitivity reactions.

13.27 What lipid mediators do mast cells and basophils secrete, and what are their biological effects?

The most important of these de-novo-synthesized lipid mediators are the **metabolites of arachidonic acid**. This compound undergoes chemical transformations catalyzed either by **cyclooxygenase** or by **lipoxygenase**. The major mediator produced by the cyclooxygenase pathway in mast cells (and to a much lesser extent, in basophils) is **prostaglandin D_2 (PGD_2)**. Like histamine, PGD_2 acts as a vasodilator and bronchoconstrictor. In addition, it promotes neutrophil chemotaxis and accumulation in inflammatory sites. The major mediators produced by the lipoxygenase pathway in mast cells and basophils are **leukotrienes**, especially **leukotriene C_4 (LTC_4)**. These compounds bind to their specific receptors in smooth muscle cells and cause severe and prolonged bronchoconstriction. LTC_4 and its degradation products LTD_4 and LTE_4 together used to be called (in older literature) the "**slow-reacting substance of anaphylaxis**" (**SRS-A**). Leukotrienes are very dangerous for atopic patients, especially asthmatics, because of their effect on airways. Hence, physicians should be very careful not to prescribe to these patients drugs that might exacerbate the synthesis of leukotrienes. One such drug is aspirin, which inhibits cyclooxygenase and thus shunts the metabolism of arachidonic acid towards the lipoxygenase pathway (Fig. 13-5). In addition to the metabolites of arachidonic acid, mast cells and basophils secrete the so-called **platelet-activating factor** (**PAF**) − a hydrophobic phospholipid that induces platelet aggregation, bronchoconstriction, and accumulation of inflammatory leukocytes. Endothelial cells under the influence of histamine and leukotienes can also secrete PAF.

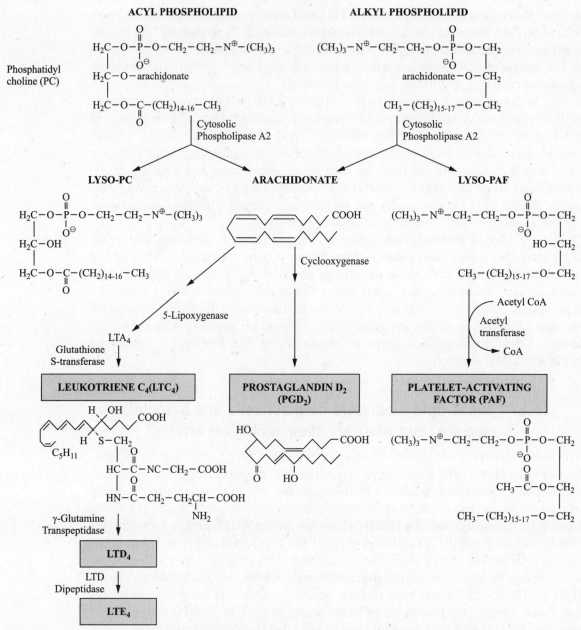

Fig. 13-5. Biosynthesis of lipid mediators. Breakdown of membrane phospholipids by cytosolic phospholipase A2 leads to generation of leukotriene C_4, prostaglandin D_2, and platelet-activating factor.

13.28 What is the role of cytokines produced by activated mast cells and basophils?

The role of these cytokines, compared to the role of the same cytokines produced during the effector phase of CMI by activated T_h2 lymphocytes, is not clear. It seems, however, that mast cell- and basophil-derived cytokines are important for the development of the so-called late phase reaction, which is mediated predomi-

CHAPTER 13 Humoral Immunity

nantly by TNF and chemokines and involves massive recruitment of inflammatory leukocytes. Of these leukocytes, **eosinophils** are of the greatest importance for the development of allergic reactions.

13.29 What is the "late phase reaction," and what is the role of eosinophils in it?

The "late phase reaction" is the sequence of events that follow the immediate effects of mast cell degranulation such as vasodilatation, bronchoconstriction, "wheal and flare" reaction, etc. While these effects are seen within minutes after the challenge with allergen, the "late phase" events unwind more slowly and are fully developed within several hours after the challenge. Principally, the "late phase reaction" is the massive local infiltration of the tissue with leukocytes and especially with **eosinophils**.

A description of eosinophils has already been given in Chapter 2. Here, we will emphasize that eosinophils are similar to mast cells and basophils in that they contain granules that can be released and injure local tissues, and FcεRI. In addition, unlike mast cells and basophils, eosinophils express Fcγ and Fcα receptors. One important difference between mast cells and basophils on one hand and eosinophils on the other is that eosinophils are not as easily triggered to release their granules by FcεRI cross-linking. Rather, these cells bind to their targets and kill these targets with the help of their granules, in a way acting similarly with natural killer cells. Their major function is, actually, ADCC (see Section 13.5) mediated by FcεRI and perhaps FcαR. This function is thought to represent the major protective component of immediate hypersensitivity reactions because **the ADCC performed by eosinophils is efficient against such common parasites as helminthes and arthropods (ticks and mites)**. These parasites are very resistant to virtually all other effector immune reactions. In addition to performing ADCC, eosinophils also secrete lipid pro-inflammatory mediators similar to those described in Section 13.27, and some cytokines (notably IL-4, IL-5, IL-10, IL-1, IL-3, IL-67, TNF, GM-CSF, and chemokines). The biologic significance of eosinophil cytokine production is still unknown.

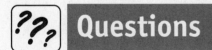
Questions

REVIEW QUESTIONS

1. Why is it important for an antibody to have a high affinity to its epitope in order to serve in antigen neutralization efficiently?

2. The so-called "hyper-IgM syndrome" is an X-linked recessive disorder when patients do not express functional CD40 ligand. Which of the following effector functions of antibodies will be severely impaired in a boy suffering from this disease: (a) neutralization of bacterial toxins; (b) complement-dependent lysis of microbial particles; (c) phagocytosis; (d) ADCC; and (e) immediate hypersensitivity?

3. Explain, as briefly as possible, why Fcγ receptors associate with antibodies that are bound to their epitopes better than with antibodies that are unbound.

4. Which effector functions of antibodies do you expect to be lacking in IFN-γ knockout mice?

5. During the alternative pathway of complement activation, which event turns C3 tick-over into actual activation of complement cascades?

6. What is the role of internal thioester binds in complement activation?

7. What exactly does C1 INH inhibit, how does it do it, and what consequences does its action have?

8. In addition to the dysfunction of the effector phase of immunity, what other dysfunctions of the immune system do you expect in individuals genetically deficient for CR2?

9. Why do we say that complement is also a mediator of inflammation?

10. Explain why IgA is not the most prevalent antibody isotype in the serum, although more of it is produced than of any other isotype.

11. How do infants and adults benefit from having FcRn?

12. What did the Praustnitz–Kustner experiment prove?

13. If you knew the way to completely shut down the T_h2 pathway of T-cell differentiation, would you recommend doing it in order to treat patients with allergies?

14. What is the role of myosin chains in immediate hypersensitivity reactions?

15. As detailed in the text, aspirin exacerbates the symptoms of asthma because it blocks cyclooxygenase and thus shunts the metabolism of arachidonic acid towards the lipoxygenase pathway. If patients receiving aspirin were to be given some strong stimulators of cyclooxygenase pathway, would they still benefit?

16. Summarize the harmful and the protective sides of immediate hypersensitivity.

MATCHING

Direction: Match each item in Column A with the one in Column B to which it is most closely associated. Each item in Column B can be used only once.

Column A

1. Allosteric effect
2. CD32
3. ADCC
4. C3 convertase
5. Factor B
6. Properdin
7. MBL
8. S-protein
9. Dimer
10. Atopy

Column B

A. Early step of complement pathways
B. Severe bronchoconstriction
C. IgA in mucosal secretions
D. Mast cell granules
E. Stabilizes C3bBb
F. FcγRIIB
G. Antibody binds outside the receptor
H. FcγRIII
I. Cleaved by Factor D
J. Lectin pathway of complement

CHAPTER 13 Humoral Immunity

11. Serotonin
12. LTC4

K. Prevents membrane insertion of C7
L. Increased IgE production

Answers to Questions

REVIEW QUESTIONS

1. Because if a microbial receptor or toxin has a high affinity to its ligand and the antibody has a low affinity, it will not be able to compete with the ligand for the microbe or its toxin on the host's cells.
2. (c), (d), and (e) will be completely missing because in the absence of CD40L-mediated cognate T–B-cell interaction these is no class switch to downstream isotypes, and therefore no IgG, IgA, and IgE. (a) will be detectable but weak, because IgM antibodies on average have lower affinity than IgG antibodies. (b) will be intact because IgM antibodies are the best complement binders.
3. The affinity of FcγRI is high enough to bind soluble monomeric IgG antibodies; yet, when several antibody molecules are bound to closely positioned epitopes, it creates the multivalent array that can interact with the Fc receptor with high avidity. The affinity of FcγRII and -III is so low that they cannot bind soluble monomeric IgG antibodies.
4. Phagocytosis and NK-mediated ADCC.
5. Binding of C3b to microbial surface (because this event begins the creation of the alternative pathway C3 convertase).
6. They stabilize C3b when it binds to cellular proteins of polysaccharides, and thus prevent it from hydrolysis.
7. Serine proteases, in particular C1r and C1s. It does so because it is a serine protease inhibitor ("serpin") by nature. The consequence is inhibition of the classical pathway of complement activation.
8. Lack of affinity maturation (because CR2 is expressed on follicular dendritic cells in germinal centers and participates in the selection of high-affinity mutants). Lack of EBV-induced B-cell activation *in vitro* (because CR2 is also the receptor for EBV).
9. Because products of its cleavage (C3a, C4a, C5a) serve as recruiters and activators of inflammatory leukocytes.
10. Because it is produced in the mucosal compartment of the immune system, from where it is actively transcytosed into mucosal secretions (saliva, mucus, sputum, etc.).
11. Infants use it to transport maternal IgG across their intestinal epithelium; adults use it to protect their IgG against catabolic degradation.
12. That serum antibodies are responsible for allergic reactions.
13. No, because at least some T_h2 cells are needed to prevent the hyperactivation of T_h1 cells and its consequences (e.g., granulomatous inflammation).

14. When the signal from cross-linked FcεRI is transduced, they dissociate from actin–myosin complexes and promote the exocytosis of mast cell and basophil granules.
15. No, because that would defeat the purpose of prescribing the drug. Aspirin inhibits fever and inflammation exactly because it inhibits the formation of PGD_2 through the cyclooxygenase pathway.
16. Harmful: tissue swelling, bronchoconstriction, enhancement of peristalsis because of the release of granules and secretion of lipid mediators (and probably cytokines). Beneficial: "late phase reaction" (eosinophilic infiltrate) and eosinophil-mediated ADCC.

MATCHING

1, G; 2, F; 3, H; 4, A; 5, I; 6, E; 7, J; 8, K; 9, C; 10, L; 11, D; 12, B

CHAPTER 14

Immunity to Microbes

Introduction

Historically, immunity has been viewed as protection against infectious diseases caused by microbes. Although our current definition of immunity is broader (see Chapter 1), we still appreciate that reactions aimed at the elimination of potentially or actually dangerous microbial agents are perhaps the most important physiologic function of the immune system. Innate and adaptive immunity normally mounts efficient responses against such microbes as bacteria, viruses, fungi, and parasites. Yet, microbes evolve to evade immune responses, using a variety of adaptations that often allow them to thrive in hosts with perfectly normal immune system. In this chapter, we will discuss the ways in which the immune system deals with microbes, and the ways in which the microbes escape the immune attack. In addition, we will highlight the point that the immune response to microbes can itself be injurious to host tissues. Moreover, there are cases when the aggravated response of the host's immune system to a microbe causes much more damage to the host than the microbe as such. This point will be illustrated by a number of examples.

Another point that we will highlight in the discussion is that some phenomena discovered initially as pertaining to microbial immunity are in fact very significant for immunity in general. We will illustrate this point, discussing such molecular mechanisms as the action of superantigens, antigen variation, and gene conversion.

In the final sections of the chapter, we will discuss the concept of vaccines and vaccination, analyze different kinds of vaccines, and briefly highlight some modern trends in vaccine development.

Discussion

14.1 Is there a way to classify microbes according to their relationship with the immune system?

Yes. The variety of microbes is tremendous, and their exact taxonomy is highly complicated and goes beyond the scope of an immunology course. However, it is useful to organize the discussion of antimicrobial immunity, separating the following groups of microbes: (a) extracellular bacteria; (b) intracellular bacteria; (c) fungi; (d) viruses; and (e) parasites. Each of these groups of microorganisms is characteristic in its own way to trigger innate and adaptive immune responses, and also to evade them.

14.2 What are the principal characteristics of the innate immunity against extracellular bacteria?

Extracellular bacteria replicate outside the host's cells and can harm the host by triggering inflammation or producing toxins. The **innate** immune system of the host often responds to their invasion strongly enough to contain the infection. Gram-positive bacteria contain a surface peptidoglycan, and Gram-negative bacteria a lipopolysaccharide (LPS). These molecules are known to activate the alternative complement pathway (see Chapter 13). Some bacteria express molecules with mannose residues, and are thus able to activate the lectin complement pathway through MBL (Chapter 13). The consequences of complement activation are **opsonization** (enhanced phagocytosis of bacteria coated by complement protein fragments), enhancement of local **inflammation**, and direct **lysis** of the bacteria by the MAC (see Chapter 13). In the absence of opsonins, macrophages still exert some protection against the extracellular bacteria because they can recognize and bind these microorganisms through their scavenger receptors (see Chapter 11). In addition, activated macrophages produce cytokines, in particular, TNF, IL-1, and chemokines, thus promoting recruitment of inflammatory leukocytes (Chapter 10). As was detailed in Chapters 10, 11, and 12, all of the above-mentioned processes can injure local tissues and also cause systemic effects like fever and the synthesis of acute-phase proteins.

14.3 What are the main features of the adaptive immunity against extracellular bacteria?

Speaking of the **adaptive** immunity against extracellular bacteria, one should remember that most of these bacteria are encapsulated, and that their capsules consist mostly of polysaccharides. For that reason, **humoral immunity** – and especially **T-cell-independent** B cell responses (Chapter 8) – play the most important role in the protection against these microorganisms. The T-cell-independent antibody responses to bacterial capsular polysaccharides usually involve predominantly IgM antibodies of relatively low affinity, and show little or no immune

CHAPTER 14 Immunity to Microbes

memory. However, responses to some extracellular encapsulated bacteria, e.g., pneumococci and *H. influenzae* type b, involve antibodies that have downstream isotypes, mostly IgG2 and also IgA. It has recently been shown that some of these antibodies show moderate somatic mutation, especially in their heavy chain V-region, and may be subject to some selection in peripheral lymphoid organs (possibly in the spleen). Antibody responses can also be triggered against bacterial toxins. Since both endo- and exotoxins are proteins, these latter antibody responses are typical T-cell-dependent responses that show extensive class switch and affinity maturation (see Chapters 4 and 8).

Antibodies against bacterial toxins, especially high-affinity IgG (and also IgA), can efficiently **neutralize** these toxins. IgM, IgG1, and IgG3 antibodies can **bind complement** and thus promote the **lysis** and the complement-mediated phagocytosis of the targets.

Protein antigens of extracellular bacteria can be processed and presented to **T lymphocytes**, mediating typical **T_h responses**. The responding cells are predominantly CD4$^+$ T_h cells that differentiate into cytokine-producing T_h1 and/or T_h2 cells. In the case of Gram-negative bacteria, the response is usually T_h1, probably because the bacterial LPS stimulates macrophages to produce large quantities of T_h1-stimulating cytokine IL-12 (see Chapter 10). In severe cases, hyperproduction of TNF, IL-1, and LT by the macrophages activated by T_h1 cells can result in septic shock (see Chapters 10 and 11). Also, a unique feature of T-lymphocyte responses to extracellular bacteria is the phenomenon of **superantigens**.

14.4 What are superantigens, and how does the immune system respond to them?

Superantigens are proteins that have the ability to stimulate lymphocytes by binding a molecular moiety that is still within the V-region of the lymphocyte antigen receptor, but just outside of the site of interaction with conventional antigens. Recall from Chapters 4 and 7 that antigen receptors interact with antigens through CDRs, which are unique for the given lymphocyte clone. Because of that, a clone of T or B lymphocytes has its unique antigen specificity. Superantigens do not react with CDRs, but rather bind to consensus sequences in the FRs. For example, a staphylococcal protein called enterotoxin B (SEB, from staphylococcal enterotoxin B) binds a conserved amino acid sequence shared by many V-regions of the TCR β chain (Vβ3, 12, 14, 15, 17 and 20 in humans, and Vβ 7, 8.1, 8.2, 8.3, and 17 in mice). This sequence of amino acids is coded by a TCR Vβ framework region. Because of such broad binding pattern, SEB and other staphylococcal enterotoxins are extremely powerful T-cell mitogens (proliferation inducers). They can activate T lymphocytes at concentrations as small as 10^{-9} M. It has been shown that as many as one out of five murine splenic or human peripheral blood T lymphocytes may respond to a particular endotoxin. Such a tremendously strong T-cell response may result in hyperproduction of many different cytokines by strongly activated macrophages, and sometimes lead to septic shock and other complications.

One interesting feature of SEB and other bacterial T-cell superantigens is the peculiar character of their interaction with MHC molecules. Superantigens are not processed by antigen-presenting cells, but rather **bind directly to MHC Class II molecules** expressed on the APC surface. Moreover, each molecule of staphylococcal superantigen has two binding sites for MHC Class II molecules, which allows the superantigens to **cross-link** MHC Class II molecules on the APC surface. The cross-linked MHC molecules, in turn, engage multiple TCRs so that two neighboring TCRs on the same T cell can be bound simultaneously by each superantigen–MHC dimer. This results in strong T-cell activation, even in the absence of co-stimulators.

14.5 How do extracellular bacteria evade immune responses?

These microorganisms can evade innate immunity by **resisting** or **inhibiting phagocytosis**, and by **inhibiting complement activation**. Both of these goals are achieved by encapsulation. Bacteria coated with thick polysaccharide capsules are difficult for phagocytes to digest. In addition, sialic acid residues contained in these capsules inhibit complement activation by the alternative pathway.

One major adaptation used by extracellular bacteria (as well as other microbes) to evade adaptive immunity is **antigenic variation**. This means that some surface antigens of many bacteria are subject to continuous structural change, because the genes that code for these antigens undergo gene conversion. The latter is a process of modifying structural genes by adding to them pieces of varying length that are borrowed from the so-called donor sequences (see Section 5.13). One good example of antigenic variation is a bacterial surface protein called **pilin**. This protein is contained in the pili, i.e., the small appendages that bacterial cells use to attach to eukaryotic cells. In gonococci, the pilin gene undergoes such an extensive conversion that the progeny of one bacterial cell can express as many as 10^6 structurally and antigenically distinct pilin molecules. Encapsulated bacteria (e.g., *Haemophilus influenzae*) can also modify their external polysaccharides through gene conversion in the sequences that code for glycosidases.

14.6 What are the main features of innate immunity against intracellular bacteria?

The innate immune response to intracellular bacteria (which are adapted to survival inside cells, including phagocytes) consists predominantly of **phagocytosis** and **NK cell activation**. As follows from the biology of these microorganisms, their phagocytosis is often incomplete. However, when macrophages engulf these bacteria, they can become activated and produce IL-12, which strongly activates NK (see Chapter 10). NK bind and lyse these bacteria, and also produce IFN-γ, which activates macrophages and promotes the digestion and killing of the bacteria by their ROI (Chapter 11). Of note is the fact that mice with severe combined immunodeficiency, and which have no functional lymphocytes, can contain *L. monocytogenes* infection by the

CHAPTER 14 Immunity to Microbes

combined action of macrophages and NK cells. Yet, innate immunity cannot eradicate intracellular bacteria, it can only stop the spreading of infection for a short time.

14.7 How do the adaptive immune responses against intracellular bacteria work?

Since intracellular bacteria are out of reach for antibodies, the major protective adaptive immune responses against them are cellular immune responses. Both **T_h** and **CTL** participate in these responses. Microbial peptides generated in the phagolysosome are presented in the context of MHC Class II molecules and activate predominantly T_h cells. The major protective T_h cell response against intracellular bacteria involves **T_h1 differentiation** and T_h1-mediated **macrophage activation**. As was discussed in Chapter 12, T_h1 cells activate macrophages by producing cytokines, especially IFN-γ, and by contact with macrophages through CD40L and CD40. The activated macrophages kill intracellular bacteria by up-regulating the production of their lysosomal enzymes and ROI. The IFN-γ that they produce also activates class-switch to antibody isotypes that bind complement and opsonize macrophages (e.g., IgG2a in mice). The T_h1 pathway itself is up-regulated by **IL-12**, which is produced by macrophages that are activated after their initial contact with the bacteria. The T_h2 pathway can also be involved (see later). CTL responses against intracellular bacteria, mediated predominantly by CD8$^+$ T lymphocytes, are stimulated when these microorganisms or their protein components escape from phagolysosomes, and the microbial peptides are generated in the cytosol.

14.8 How do immune responses against intracellular bacteria injure the host's tissues?

We have already discussed some injurious consequences of innate and adaptive immune responses against intracellular bacteria in Chapters 11 and 12. Recall that hyperactivation of phagocytes (especially neutrophils) can result in the formation of abscesses, and chronic activation of macrophages – in **granulomatous inflammation**. The latter features very prominently during the infection with *M. tuberculosis*. Although the bacterium, formally speaking, is the cause of the disease, the actual tissue injury in the affected organs (especially lung) is mediated by activated macrophages that form granulomas. Another example of tissue injury mediated by the host's immune system is infection with *M. leprae*. This microorganism can elicit a strong response of T_h2 cells that produce IL-4 and stimulate humoral immunity (Chapter 10). In this case, the so-called lepromatous form of lepra develops, characterized by extensive skin lesions. The lesions develop because activated T_h2 cells inhibit the activation of macrophages, which leads to the spread of infection. In the patients with the so-called tuberculoid form of lepra, the T_h1 pathway is more pronounced, and the bacteria are contained in macrophages.

14.9 How do intracellular bacteria evade immune mechanisms?

The major adaptation that allows intracellular bacteria to survive and often successfully escape immune responses is their **intracellular reproduction**. In addition, various intracellular bacteria actively inhibit the function of phagocytes. For example, *M. tuberculosis* is able to inhibit the formation of phagolysosomes, and *L. monocytogenes* can disrupt the phagolysosome membrane with the help of its specialized protein called hemolysin. *M. leprae* produces a phenolic glycolipid that disrupts the formation of ROI. Because of these mechanisms, intracellular bacteria often cause chronic infections that might be asymptomatic, but are reactivated without any apparent cause.

14.10 What are the main features of immunity to fungi?

Fungal infections began to attract much attention since it was discovered that they are a major complication of AIDS and other immunodeficiencies. May fungal infections are opportunistic, i.e., they show no symptoms in individuals with a healthy immune system, but are rampant in immunodeficient individuals. This partly explains, why we know so little about anti-fungal immunity: *individuals who show signs of fungal infections often are unable to mount immune responses*.

In healthy individuals, both innate and adaptive immunity are important to protect the organism against the spread of fungi. The major participants in the **innate** immunity are neutrophils and macrophages. Fungi are killed after phagocytosis by lysosomal enzymes, or by ROI. Neutrophils also release lysosomal enzymes, and ROI, thus killing fungi that are located in extracellular space. **Cell-mediated immunity (CMI)** is the major mechanism of **adaptive** immunity against fungi. Both $CD4^+$ and $CD8^+$ T lymphocytes participate in anti-fungal responses; $CD4^+$ helper T cells may differentiate into T_h1 or T_h2 cells. In some cases (e.g., infections with *Candida albicans*), T_h1 responses are protective and T_h2 responses are detrimental to the host, perhaps because the latter inhibit macrophage activation. Fungi are known to elicit antibody responses, but the protective role of the latter has not been demonstrated.

14.11 What are the main features of innate immunity against viruses?

As has already been mentioned in Chapters 10 and 11, the principal mechanisms of **innate** immunity against viruses are inhibition of viral infection by **type I interferons** (IFN-α and IFN-β) and **NK cell-mediated killing** of virally infected cells. As we discussed, type I interferons induce the so-called antiviral state by inhibiting some enzymes crucial for viral replication, and up-regulate MHC Class I molecules, thus making infected cells more suitable targets for CTL. NK cells lyse virally infected cells that down-regulate their MHC Class I molecules, because that releases their status of inhibition, normally conferred by the interaction of their inhibitory receptors with the above molecules (Chapter 11).

CHAPTER 14 Immunity to Microbes

14.12 What are the main characteristics of adaptive immunity against viruses?

The main mechanisms of adaptive immunity to viruses are **antibody responses** and **CTL responses**. Antibodies are effective against viruses only during the extracellular stages of life of these microorganisms. They can block the binding of a virus to a host cell or the entrance of a virus into a host cell. The most efficient and protective of these **neutralizing antibodies** are **high-affinity IgG and IgA** antibodies. The latter are produced mostly in the mucosal compartment of the immune system (Chapter 13) Very efficient antibodies elicited by some common protective antiviral vaccines administered by mouth, e.g., polio vaccine, belong to this category. In addition, antiviral antibodies can **opsonize** viral particles and **activate complement**. Direct lysis of some viral particles by the MAC (especially those with lipids in their envelope) was shown to be protective against viral infections. Of course, antibodies cannot reach viruses that replicate inside host cells; therefore, humoral immunity cannot eradicate established viral infections.

T lymphocytes, and especially **CTL**, are an important tool that allows the host to kill virally infected cells and thus eradicate viral infections. CTL recognize viral peptides and are activated mostly due to the phenomenon of **cross-priming** described in Chapter 12. (Recall, this means that if a virus infects a nonprofessional APC that has no co-stimulators, the infected cell is phagocytosed by a professional APC that presents viral peptides and expresses co-stimulators, making T-cell activation possible.) As was outlined in Chapter 12, activated $CD8^+$ CTL undergo massive clonal expansion during viral infections, so that a substantial fraction of all circulating $CD8^+$ T lymphocytes is specific to viral antigens. CTL lyse virally infected cells with the help of their granules. In addition, as was mentioned in Chapter 12, granzyme B that is contained in cytotoxic granules activates cellular nucleases (through the activation of effector caspases), and thus destroys the infected cells' genomes.

14.13 Can antiviral immune responses injure host tissues?

Yes. This is characteristic mostly for noncytopathic viruses, i.e. viruses that do not lyse the infected cells. One classical example of tissue injury by antiviral immune response is the injury made by **lymphocytic choriomeningitis virus (LCMV)**. In mice, CTL specific to LCMV peptides can lyse meningeal cells and cause meningitis. Paradoxically, T-cell-deficient mice infected with LCMV are clinically healthy carriers, while normal LCMV-infected mice are clinically ill due to the T-cell antiviral response. In humans, hepatitis B virus causes latent infection in immunodeficient individuals, while in healthy individuals it triggers vigorous CTL response that damages the liver. Hepatitis B virus can also trigger antibody responses that lead to the formation of immune complexes, and may cause systemic **vasculitis** (injury of small blood vessels) because of their insolubility and deposition in these vessels.

14.14 How do viruses evade immune responses?

Viruses are very ancient organisms and, as such, have evolved to be very "crafty" in adapting to their host and evading their host's responses. The various ways viruses employ to escape immune reactions can be grouped as follows:

- **Antigenic variation**. Viruses constantly change their antigens by mutating or reassorting their genes. This phenomenon accounts for the tremendous number of viral strains, and makes protective immunization against viruses so difficult.

- **Inhibition of antigen processing and presentation** (Fig. 14-1). For example, adenoviruses code for a protein called **E1A**, which is able directly to inhibit the transcription of MHC Class I genes. Herpes simplex virus codes for a protein called **ICP-47** that binds to TAP (see Chapter 6) and interferes with the active transport of cytosolic peptides into the endoplasmic reticulum. Some adenoviruses code for a protein called **E3** that binds to MHC Class I molecules in the endoplasmic reticulum and does not let them exit the reticulum when they are loaded with peptides. Proteins coded in the human and murine cytomegaloviruses (CMV) have similar properties – they interfere with the expression of loaded MHC Class I molecules, and **Nef** protein coded by the human immunodeficiency virus (HIV) forces MHC Class I molecules to be internalized.

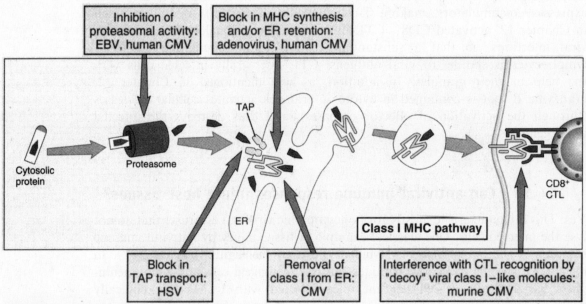

Inhibition of antigen presentation by viruses.
The pathway of class I major histocompatibility complex (MHC)–associated antigen presentation is shown, with examples of viruses that block different steps in this pathway. CMV, cytomegalovirus; CTL, cytolytic T lymphocyte; EBV, Epstein-Barr virus; ER, endoplasmic reticulum; HSV, herpes simplex virus; TAP, transporter associated with antigen processing.

Fig. 14-1. Inhibition of antigen presentation by viruses. This figure shows some examples of interference of viruses with the Class I MHC pathway of antigen presentation. See text for details. From Abbas, K., A.H. Lichtman, and J.S. Pober, *Cellular and Molecular Immunology*, fourth edition, W.B. Saunders, 2000, p. 355 (their figure in Box 15-3).

CHAPTER 14 Immunity to Microbes

- **Production of molecules that inhibit immunity**. Poxviruses produce molecules that are structurally similar to the receptors for several cytokines (including IFN-γ, TNF, and IL-1), but are secreted and thus function as soluble competitive inhibitors of cytokine–cytokine receptor interaction. Similarly, CMV produce a molecule that resembles MHC Class I protein and may act as a competitor to the "real" MHC molecule for peptide binding. EBV produces a protein that is homologous to IL-10 (recall from Chapter 10 that this cytokine is a potent suppressor of macrophage activation and cell-mediated immunity).
- **Destruction of immunocompetent cells**. The obvious example is HIV (see Chapter 18).

14.15 What are the main features of immunity to parasites?

It should be noted here that parasitic infestations are a severe problem for humankind, especially in developing countries. As much as 30% of the world's population is believed to be infested by parasites; approximately 100 million people are suffering from malaria, and about 1 million die of this disease every year. These facts explain why immunologists are so interested in body defense against parasites, and are searching for ways to boost antiparasitic immunity. Parasites usually go through complex life cycles and change their hosts, as well as their structural and antigenic properties. Because of that, the immune response against parasites is often **inefficient**. In addition, many parasites are resistant to such effector mechanisms of immunity as phagocytosis and complement-dependent, NK-mediated and T-cell-mediated lysis.

14.16 How does the innate immunity to parasites work?

Both protozoan and metazoan (helminthic) parasites trigger such innate immune mechanisms as **inflammation, phagocytosis**, and **complement activation** through the alternative pathway. However, these responses are seldom efficient because parasites are often coated with thick teguments and cannot be killed by phagocytes. Many protozoan parasites in fact replicate inside macrophages. In addition, some parasites can inactivate complement cascades and are resistant to complement-mediated lysis.

14.17 What are the main features of adaptive immunity to parasites?

The principal mechanism of adaptive immunity against protozoan parasites is cell-mediated immunity, especially macrophage activation by T_h1 cell-derived cytokines. The importance of T_h1 pathway of T-lymphocyte differentiation for resistance to protozoan parasites became clear from experiments where inbred mice were infected by an intracellular protozoan parasite, *Leishmania major*. Some inbred strains, e.g., BALB/c, are very susceptible to the *L. major* infection and develop characteristic skin lesions. Many other strains are resistant to the infection. It has been shown that in the susceptible strains, T-lymphocytes specific to *L. major* antigens differentiate into T_h2, while in the resistant strains they differentiate into T_h1. The infection is exacerbated by IL-4 and inhibited by IFN-γ. However, not all protozoans are efficiently dealt with by T_h1 cells. Some unicellular protozoan parasites are eliminated more efficiently by $CD8^+$ CTL. A practical, and very important, example of this is the immunity against malaria.

14.18 What is important to know about immunity to malaria?

Malaria is a severe disease caused by a protozoan called *Plasmodium falciparum*, and transmitted by the mosquito, *Anopheles maculipennis*. Infection of humans is initiated when they are bitten by the mosquito, and numerous **sporozoites** (a form of individual cells of *P. falciparum*) enter the bloodstream. The sporozoites are quickly eliminated from the blood and accumulate in liver, where they enter hepatocytes and turn into **merozoites**. These lyse hepatocytes and invade red blood cells. In some blood cells, merozoites turn into sexual stage **gametocytes**; mosquitoes then pick these up during their blood meal, completing the cycle. The symptoms of malaria include fever spikes, anemia, and splenomegaly. Of note is the fact that many pathologic manifestations of malaria may be due not to the action of sporozoites or merozoites on human cells, but rather to the activation of T cells and macrophages and consequent production of cytokines.

The immune response to malaria is rather complex and **stage-specific**. This means that immunization against sporozoites does not protect against merozoites or gametocytes, etc. A protective antibody response may be triggered by vaccination with irradiated sporozoites; in this case, the elicited antibodies include those specific to the so-called **circumsporozoite (CS) protein**. The CS protein is used by sporozoites to bind to hepatic cells. Thus, the CS-specific antibody is an example of neutralizing antibody (see Chapter 13). The efficiency of this antibody response depends, however, on T cell help. The epitope bound by neutralizing antibody is a tandem repeat asparagine-alanine-asparagine-proline (or, according to one-letter amino acid nomenclature, **NANP**), repeated approximately 40 times. The immunodominant CS epitopes recognized by T cells, however, lie outside of the NANP region and, therefore, the T cell help is not very efficient. A protein expressed on merozoites and called **MSP-1** (from **m**erozoite **s**urface **p**rotein-**1**) has recently attracted much attention as eliciting strong and possibly protective antibody responses.

$CD8^+$ CTL responses became much more appreciated within recent years as the responses that confer protection against malaria and stop its transmission. The protective CTL responses involve direct lysis of sporozoite-infected hepatocytes and secretion of IFN-γ (see Chapter 10). The latter can induce the infected hepatocytes to produce nitric oxide and other substances that can kill *Plasmodium* cells. Efforts of immunologists are currently concentrated on creation of new vaccines that would trigger strong CTL responses against *Plasmodium falciparum* antigens.

CHAPTER 14 Immunity to Microbes

14.19 How do parasites evade immune responses?

Parasites indeed have a remarkable capacity to evade immunity by employing a variety of adaptations. First of all, many protozoan parasites show **antigenic variation** that we have already described in Sections 14.5 and 14.14. This antigenic variation may be **stage-specific**, as in *Plasmodium falciparum*, where antigens of one of the stages in the life of parasites (e.g., sporozoites) are distinct from antigens of another stage (e.g., merozoites). Alternatively, the antigenic variation may be **continuous**, as in the case of African trypanosomes. In these organisms, in particular *Trypanosoma brucei* and *Trypanosoma rhodesiense*, the genes that encode the major surface protein (called **VSG**, from **v**ariable **s**urface **g**lycoprotein) undergo peculiar "waves" of expression. These genes are located in tandem, and the expression of each VSG is associated with a duplication of a VSG gene and its translocation to a place in the DNA where this gene becomes transcriptionally active. The number of individual VSG genes is huge (perhaps more than 1,000), and so is the extent of antigen variants produced in result of expression of each individual VSG gene.

Another mechanism of evasion employed by parasites is biochemical **modification of their surfaces** during their interaction with the host. For example, schistosome larvae develop thick teguments that make them resistant to complement and phagocytosis when they travel to the lungs of infected animals. The exact mechanism of this modification is unknown. An interesting example of surface modification is antigen shedding seen in many parasites; it can be spontaneous or induced by specific antibodies. Finally, **parasites may actively inhibit innate and adaptive immunity**. More or less generalized immunosuppression has been observed in malaria, infections with schistosomes, trypanosomes, filariasis, etc. Although the exact mechanisms of parasite-induced immunosuppression are not known, it is thought that this phenomenon can be at least partially explained by production of inhibitory cytokines (e.g., IL-10, TGF-β) by activated macrophages and T lymphocytes.

14.20 What is a "vaccine," and what kinds of vaccines exist?

The terms "vaccine" and "vaccination" appeared after Edward Jenner's success in conferring antismallpox immunity by injecting a material from cowpox-induced lesions. By the term **vaccine**, we now mean a preparation of microbial antigen, often mixed with adjuvants (see Chapters 9 and 11), that is administered to individuals to confer protective immunity against infections. The following kinds of vaccines are currently used in preventive medicine or in experimental or clinical trials: (a) live but avirulent (attenuated) microorganisms; (b) killed microorganisms; (c) purified macromolecular components of microorganisms; (d) synthetic vaccines; (e) live viral vectors; and (f) plasmids containing cDNA encoding for microbial antigens.

14.21 What are live vaccines, and how do they work?

Live, attenuated vaccines were introduced by Louis Pasteur in the 1880s. Pasteur experimented with inducing cholera in chickens by injecting them with cultured

Vibrio cholerae. Some of the bacterial cultures which accidentally became old and overgrown (they were forgotten while Pasteur was on vacation) failed to induce cholera, and were called **attenuated** (i.e., such that they lost their aggressiveness, or were "tamed"). Remarkably, if chickens were injected with fresh, virulent cultures after they had been injected with attenuated cultures, they remained disease-free (immune). Later, Pasteur and others obtained and used a number of attenuated vaccines that served as life-savers for innumerable millions of people. Among them were vaccines against human cholera, tuberculosis, anthrax, and rabies, to mention just a few. The great advantage of attenuated vaccines is that they cause exactly the same immune response that the microbe itself causes. However, there are some serious limitations for their use. First of all, many live attenuated bacterial vaccines trigger rather weak and short-lasting immunity. Live, attenuated viral vaccines are somewhat more effective; Pasteur's rabies vaccine and Jonas Salk's polio vaccine are classical examples. Recently, deletion and temperature-sensitive mutant viruses began to be used as vaccines with very promising results. The other serious limitation in the use of live, attenuated vaccines is the concern about their safety because of incomplete attenuation and possible return of pathogenicity.

14.22 What are killed vaccines, and how do they work?

Killed bacterial vaccines work essentially in the same way as live, attenuated vaccines. Bacterial cultures that are used for the preparation of this type of vaccines are usually treated with formalin or other toxic agents.

14.23 What vaccines are based on purified macromolecular components of microorganisms, and how do they work?

The vaccines that are prepared from purified microbial macromolecular components (proteins and/or polysaccharides) are called **subunit vaccines**. They can be prepared from the components of microbes themselves, or from inactivated microbial toxins. The inactivated microbial toxins used for vaccine preparation are called **toxoids**. Two classical examples of extremely useful bacterial toxoids are diphtheria toxoid and tetanus toxoid, which helped to control these two dangerous infectious diseases. Examples of subunit vaccines prepared from bacterial cell walls include polysaccharide vaccines against *H. influenzae* type b and pneumococci. As typical T-cell-independent antigens, these bacterial polysaccharides elicit mostly low-affinity antibodies and are rather weak in inducing immune memory. In addition, infants do not mount strong responses to *H. influenzae* polysaccharide until the age of approximately 4 years. Nevertheless, polysaccharide vaccines confer long-lasting protection, probably because polysaccharides are not degraded easily and persist for years in secondary lymphoid organs, especially in the marginal zone of the spleen (Chapter 2). High-affinity antibodies to *H. influenzae* and pneumococcal polysaccharides can be obtained if the latter is chemically conjugated with a carrier protein, e.g., mutant avirulent diphtheria toxoid or purified

CHAPTER 14 Immunity to Microbes

meningococcal protein. These conjugate vaccines are now widely used for immunization with excellent results. They proved to be very efficient against *H. influenzae* even in very small infants. It is noteworthy that purified protein vaccines elicit strong antibody and T_h responses but, as exogenous proteins, they are limited in their ability to trigger CTL responses (see Chapter 6).

14.24 What are synthetic vaccines, and how do they work?

Synthetic vaccines are usually oligo- or polypeptides synthesized in laboratories, and are based on knowledge of the amino acid sequence of microbial antigens. In the past, such knowledge was derived from direct partial amino acid sequencing of purified proteins, and the vaccines created were usually linear or branched polymers of three to ten amino acids. Currently, the complete amino acid sequence of many microbial proteins can be deduced from the DNA sequence of their genes. This opens new avenues in the usage of synthetic vaccines. The other major avenue is testing overlapping peptides containing certain amino acid substitutions induced by mutations. This can lead to establishing the exact epitopes recognized by antibodies and TCR, and the epitopes bound to various allelic forms of MHC molecules.

14.25 What are live viral vectors, and how are they used?

The use of live viral vectors as vaccines is based on the observation that genes that code for mocrobial antigens can be ligated into vaccinia viruses and inoculated into experimental animals, where they code for the antigen and trigger protective immune responses. A variety of live viral vectors are currently under trial. A limitation of these potentially very useful vaccines is that vaccinia viruses infect host cells, and their antigens elicit undesirable immune responses in vaccinees (especially CTL responses).

14.26 What are the main properties of plasmids with incorporated cDNA as vaccines?

Plasmids that harbor cDNA coding for microbial antigens are the newest generation of vaccines, and have attracted much attention in recent years. Extensive evidence indicates that inoculation of such plasmids in a host leads to strong and long-lasting protective immunity to the antigens that the cDNA codes for. A likely explanation of the unusual efficacy of such vaccines is that the plasmids transfect the host's professional antigen-presenting cells. The microbial antigens are synthesized from the transfected DNA at high rate, and peptides are generated from the resulting protein in the APC's cytosol, leading to strong CTL responses. A variety of approaches are currently under trial, aimed at further enhancing the efficacy of these DNA vaccines. Among these approaches, it is worth mentioning attempts to co-transfect APC with antigens of interest together with various cytokines and co-stimulators. Importantly, DNA vaccines attract the close attention of immunologists because, in the future, they may be used as prophylactics for not only infectious diseases but also cancers. This can be made possible if tumor-specific and tumor-associated antigens become better characterized (see Chapter 16).

Questions

REVIEW QUESTIONS

1. How, in your opinion, will immunity to extracellular bacteria be changed in mice that are knockout for the macrophage scavenger receptor?
2. In Chapters 1 and 3, it was mentioned that von Boehring and Kitasato discovered antibodies when they mixed diphtheria toxin with sera of immune guinea pigs. How would you explain their experiment in precise molecular terms?
3. If an antigen binds TCR without co-stimulation, the T cell is anergized. Does the same happen when a T cell is bound by a staphylococcal superantigen? Why, or why not?
4. How is antigenic variation in encapsulated bacteria different from antigenic variation in bacteria that lack an external capsule?
5. What is the role of phagolysosome in immunity against intracellular bacteria? By using hemolysin as an example, explain how an intracellular bacterium can evade immunity.
6. What is the main reason for our poor knowledge about immunity to fungi?
7. Rhinovirus (the virus that causes common cold) infects cells of the mucosal epithelium in the upper respiratory tract. These cells are not professional APC and, as such, express little or no co-stimulators. Yet, it is known that T lymphocytes respond to rhinovirus peptides very strongly. Does this not contradict the concept of tolerance induced by the absence of signal 2 (Chapter 10)?
8. What property of viruses simultaneously helps them to evade immunity and to fall victim of immunity?
9. How do molecules coded by poxviruses allow these viruses to evade the immune attack?
10. Why exactly is the T_h2 pathway of T-lymphocyte differentiation harmful for patients with *L. major* infection?
11. What is the most apparent obstacle for the creation of an efficacious antimalaria vaccine?
12. How does the genetic coding of the VSG protein help African trypanosomes to evade immunity?
13. What is the main reason for general immunosuppression in patients infested with helminthes?
14. What is the main advantage of live, attenuated vaccines over subunit vaccines?
15. What is the mechanism of action of DNA vaccines?

CHAPTER 14 Immunity to Microbes

MATCHING

Direction: Match each item in Column A with the one in Column B to which it is most closely associated. Each item in Column B can be used only once.

Column A

1. Endotoxins
2. SEB
3. Genes undergo conversion
4. Skin lesions in lepra
5. Hemolysin
6. Dangerous for immunodeficient patients
7. Can eradicate viruses
8. ICP-47
9. Circumsporozoites
10. Toxoid
11. Can transfect APC

Column B

A. Due to lack of IFN-γ
B. Fungi
C. DNA vaccines
D. Express NANP
E. Inactivated toxin
F. Not secreted
G. Disrupts phagolysosome
H. CTL responses
I. Binds TAP
J. Pilin
K. Binds Vβ

 Answers to Questions

REVIEW QUESTIONS

1. Macrophages in these mice will not be able to bind to extracellular bacteria through the missing receptor. That might weaken their activation and production of IL-12, thus making the T_h1 pathway of T-cell differentiation less pronounced and the T_h2 pathway more pronounced. However, other nonspecific macrophage receptors might at least partially compensate for the lack of scavenger receptors.

2. Perhaps guinea pigs produced neutralizing antibodies that bound the moiety of the diphtheria toxin used to bind to mammalian cell receptors.

3. No, because SEB cross-link MHC Class II molecules and thus consequently cross-link TCRs on the same cell, mimicking the binding of TCR and co-stimulator.

4. In the former, the conversion affects enzymes that modify carbohydrates (e.g., glycosidases in *H. influenzae*): in the latter, protein antigens are modified.

5. Phagolysosome contains proteolytic enzymes necessary to generate peptides that will be recognized by T cells. Hemolysin helps *L. monocytogenes* to evade T-cell-mediated immunity because it disrupts the membrane of phagolysosomes.

6. Fungal infections are pronounced in individuals who are unable to mount immune responses.

7. No, because of cross-priming. Professional APC that do express co-stimulators phagocytose rhinovirus-infected cells and present viral peptides to T cells, together with co-stimulators.

8. Down-regulation of MHC Class I molecules. It allows the viruses to evade CTL responses, but makes them more susceptible to NK responses.
9. They produce molecules that are structurally similar to cytokine receptors, but capable of being secreted. These soluble homologues are competitive inhibitors of cytokine receptors.
10. Because T_h2 cells produce IL-4, which inhibits the expansion of T_h1 cells and subsequent activation of macrophages. The infected individual suffers because macrophages are the major effector cells in anti-*L. major* responses.
11. The fact that the immune response to *P. falciparum* is stage-specific, and infected individuals usually contain the parasites that are at several different stages of their life cycle.
12. These parasites have more than 1,000 tandemly arranged VSG genes, each of which can be translocated and expressed, coding a slightly modified form of VSG. Antibodies and T cells specific to one of these isoforms will not be helpful if the progeny of the parasite expresses another isoform.
13. Production of inhibitory cytokines like IL-10 and TGF-β.
14. The former induce exactly the same immune response as the microbe itself induces; the latter may induce responses to parts of microbes that are not immunodominant, and therefore be less efficient.
15. They perhaps transfect professional antigen-presenting cells and thus strongly prime T cells when the coded microbial antigen becomes expressed.

MATCHING

1, F; 2, K; 3, J; 4, A; 5, G; 6, B; 7, H; 8, I; 9, D; 10, E; 11, C

Transplantation Immunology

Introduction

Transplantation can be defined as the process of moving cells, tissues, or organs from one individual to another. The object that is being transplanted is called a **graft**; the individual that provides the graft is usually called the **donor**, and the individual that receives the graft is the **recipient** or host. In clinical practice, transplantation is used to correct a functional or morphological deficiency in the patient. A special case of transplantation is **transfusion**, i.e., transplantation of circulating blood or its components. Attempts to transplant organs or tissues have been made since times immemorial, and almost always failed because the donor inevitably rejected the graft. Only in the middle of the 20th century did scientists come to a clear understanding that graft rejection is a result of immune reaction of the host. In this chapter, we will discuss some classical experiments that helped firmly to establish this notion.

A major concept that helped to advance both transplantation and basic immunology is that **MHC controls the reactions of graft rejection**. Transplantation is a completely artificial, man-made phenomenon and, therefore, one cannot say that the control of graft rejection is the principal biological role of MHC. Rather, this function of MHC is "accidental" and mimics its true biological role in T-cell immunity against microbial or cancerous antigens. However, dissection of the role of MHC in graft rejection led to many practical achievements (e.g., matching of donors and recipients for transplantation), as well as to some important discoveries in basic molecular immunology. The role of MHC in transplantation immunology will be discussed in this chapter from these standpoints.

Rejection of the graft may take many forms. It is important to know the general mechanism of graft recognition and of host response to the graft, but it is perhaps equally important to know the morphologic and clinical details of different forms of graft rejection in order to be able to diagnose them promptly. In this chapter, we will provide characteristics of hyperacute, acute, and chronic graft rejection, and will show how effector mechanisms of innate and adaptive immunity act during the rejection that takes either of these forms. We will also discuss some basic approaches in graft rejection diagnostics and therapy.

At the end of the chapter, we will discuss a special case of transplantation – bone marrow transplantation. We will analyze the efficiency of this promising new approach to treatment of various forms of cancer and immunodeficiency, and the limitations of this method. We will also define and characterize a major complication of bone marrow transplantation, which is the so-called graft-versus-host disease. Finally, we will discuss the pathogenesis of immunodeficiency in long-term survivors of bone marrow transplantation and highlight some new and interesting experimental findings that shed light on the origins and possible prevention of this phenomenon.

Discussion

15.1 The literature contains special terms referring to combinations of donor and recipient during transplantation. What are these terms?

These terms are all based on the following Greek words or word parts: *autos*, the same; *syn-*, a prefix that points to two things being identical; *allos*, other; and *xenos*, foreign or alien. The terms derived from these Greek words and used in transplantology are as follows. A graft transplanted from one individual to the same individual is called an **autologous graft** or **autograft**. (Autografting is used in experiments, and also sometimes in clinical practice, e.g., during bone marrow transplantation; see later sections.) A graft transplanted between two genetically identical individuals (e.g., from one animal from an inbred strain to another animal of the same strain, or from one identical twin to another) is called a **syngeneic graft**. A graft transplanted between two individuals that are genetically different, but belong to the same species (e.g., from an animal that belongs to one inbred strain to an animal that belongs to another inbred strain, or from one randombred animal or human to another) is called an **allogeneic graft** or **allograft**. A graft transplanted between two individuals that belong to two different species is called a **xenogeneic graft** or **xenograft**. Allografting is the most widely used in clinical practice, although, as we will detail later, xenografting attracts much attention for its surprisingly high potential in certain clinical situations.

15.2 What experiments, performed in the mid-20th century, shed light on the immunological nature of graft rejection?

Although the phenomenon of almost inevitable rejection of transplants was known for centuries, the nature of this rejection was not appreciated until the

CHAPTER 15 Transplantation Immunology

1950s. The decisive experiments that proved graft rejection to be a form of immune reaction were performed in Great Britain, between 1940 and 1955, by a group of scientists led by Peter Medawar. Medawar and his co-workers grafted skin from mice that belonged to certain inbred strain, to other mice. Recall from Chapter 5 that approximately at the same time, another group (led by George Snell, in the USA) was carrying out similar experiments. However, while Snell *et al.* were interested in the genetic basis of transplant rejection or acceptance, Medawar and his group sought to establish the actual mechanism of transplant rejection.

The first basic observation that Medawar and his co-workers made was that a second graft transplanted to an allogeneic recipient was rejected **faster** than the first. In addition, this **second-set rejection** was qualitatively **stronger** than the first: the graft underwent necrosis before it fell off. Second-set rejection was **specific**: if the recipient received a graft from one allogeneic strain and then a graft from another strain, both grafts were rejected in a "first set" fashion. This observation made Medawar and his colleagues think that **rejection of the graft resembles an immune reaction** in that it shows specificity and is faster and stronger when the host encounters the graft, like an antigen, repeatedly. The second basic observation was based on the technique of adoptive transfer. The experimenters showed that adoptively transferred lymphocytes from a graft recipient can "sensitize" naïve syngeneic mice. In other words, if a mouse that belongs to strain A bears a graft from strain B, and the graft-bearing A-mouse lymphocytes are adoptively transferred into an A-mouse without graft, the mouse that adoptively transfers the lymphocytes will reject the B graft in a "second-set fashion," although it had never been transplanted to it before. Taken together, these two observations proved that **graft rejection is a form of immune reaction, in which lymphocytes specifically recognize graft as foreign**.

15.3 What exactly is the role of MHC molecules in the immune system's reaction to the graft?

This role can be briefly summarized as follows: **MHC molecules, or peptides generated from these molecules, serve as the main entity recognized in the graft by the host**. In addition, **MHC molecules serve as the main target of the immune attack on the graft developed by the host**. There are several lines of evidence that support this notion. First, in case of allogeneic transplantation, the discrepancy in MHC alleles always results in graft rejection, and the greater the discrepancy, the stronger the rejection. Second, physical removal of antigen-presenting cells ("passenger leukocytes," see later) from the graft substantially diminishes the strength of the rejection, and delays its time. Finally, the recognition of the graft by the host and the immune attack of the host on the graft can be modeled *in vitro* (see later sections). In these *in-vitro* models, both the recognition and the immune attack can be completely blocked by anti-MHC antibodies.

15.4 What are the main features of the immune recognition of allogeneic MHC molecules?

Although antibodies do have a role in transplant rejection (see later), **T-cell recognition of allogeneic MHC molecules is of the greatest significance in the fate of the allograft**. Principally, allogeneic MHC molecules can be recognized by the host's T lymphocytes following either direct or indirect presentation. The former means that T cells of the host recognize intact MHC molecules expressed on the allograft's antigen-presenting cells. The latter means that antigen-presenting cells of the host internalize and process MHC molecules of the graft and present their peptides to self T lymphocytes in association with self MHC molecules. In both cases, **the reaction of T lymphocytes against the allograft is extremely vigorous**, and **is characterized by unusually high frequency of T-cell clones of the host responding to each particular allelic form of the MHC molecules of the graft**.

15.5 What are the reasons for the frequency of alloreactive T-cell clones being unusually high?

It has been estimated that **approximately 2% of all T lymphocytes of an individual who responds to a directly presented allogeneic MHC molecule are recognizing and responding to this molecule**. In contrast, only 0.001% or less of all of the individual's T lymphocytes respond to any given molecule of a foreign antigen other than MHC. The unusually high fraction of T cells involved in the response to all-MHC resembles a response to superantigens (see Chapter 14) rather than responses to antigens. It definitely implies that many distinct T-cell clones recognize and respond to the allogeneic moiety of the MHC molecule in a cross-reactive fashion.

The molecular mechanism that accounts for such a wide-ranged **cross-reactivity** of the host's T lymphocytes is rather complex. It might be better understood if we recall that the thymus, where T-lymphocyte precursors undergo their "education" and are selected, has virtually no foreign MHC molecules (because foreign antigens are trafficked from their portals of entry directly into peripheral lymphoid organs). Therefore, no mature T cells are selected to recognize foreign MHC molecules as such, or peptides generated from these molecules as such. **The ability of a T cell to bind a foreign MHC molecule is due exclusively to cross-reactivity**. The T cells of the host that originally are selected to recognize certain foreign peptides in association with self MHC cross-react with: (a) portions of foreign MHC molecules that are structurally identical to the foreign-peptide–self MHC combination that the T cells normally recognize; and (b) portions of foreign MHC molecules associated with wide variety of self peptides that, again, become, due to this unusual combination, identical in their amino acid sequence to various foreign peptide-self MHC combinations. In addition, many T-cell clones that respond to foreign MHC molecules are cross-reactive memory T cells that originally recognized microbial peptides in association with self MHC.

CHAPTER 15   Transplantation Immunology

15.6 Is the frequency of T cells reacting to indirect allo-MHC presentation equally high?

Perhaps. Indirect presentation of foreign MHC means that they are recognized as "regular" antigens, i.e., processed in professional APCs and presented to T cells of the host as peptides associated with the host's own MHC molecules. In this case, $CD4^+$ MHC Class II restricted cells of the host are predominantly involved, because the cells that express foreign MHC molecules are phagocytosed by the host's cells, and the foreign MHC molecules themselves are processed in the host's cells' endosomal-lysosomal compartment (see Chapter 6). This is a case of "cross-priming" discussed in Chapter 14. Since MHC molecules are extremely polymorphic, and each allogeneic MHC molecule can give rise to multiple foreign peptides, again, very many different T-cell clones can recognize and respond to a single allo-MHC molecule.

15.7 Are MHC molecules the exclusive target of the host's T cells?

No. There are some other polymorphic gene systems in the donor of the graft, coding for proteins that might be allelically "foreign" to the host. These antigens usually induce, if any, slower and weaker responses of the host than MHC molecules, and are therefore called **minor histocompatibility antigens**. These antigens are especially important for the development of some complications in bone marrow transplantation, which will be discussed later in this chapter.

15.8 How can the recognition of MHC molecules and the immune attack on them be modeled *in vitro*?

As was discussed in Chapter 5, the two most versatile and popular tests used for this purpose are the **mixed lymphocyte reaction (MLR)** and **cytolytic (CTL) assays**. Recall that MLR involves mixing donor and host's mononuclear cells in a culture dish and assaying proliferation by incorporation of ^{3}H-thymidine. If one of the two cell populations is rendered incapable of proliferation (usually by gamma-irradiation), the MLR becomes "one-way," such that the nonproliferating population serves exclusively as **stimulators** (i.e., presents the antigen), and the other serves as **responders** (develops the response to recognized antigen). Application of antibodies that block MHC molecules on the stimulator cells, especially the N-terminal peptide-binding regions of these molecules where allelic differences actually concentrate (see Chapter 5), will strongly inhibit or abolish the proliferation of responders. In the CTL assay, radioactive chromium escapes stimulator cells only if they are damaged by activated and differentiated responder CTL that recognize allelic differences in the stimulator's MHC molecules. Again, CTL can be blocked by applying antibodies specific to MHC of the stimulator. It is noteworthy that both MLR and CTL actually model the direct presentation of MHC antigens. *In-vitro* models of the indirect MHC presentation are being developed recently (see Section 15.9).

15.9 What is known about the actual *in-vivo* recognition of grafts?

Most of our knowledge about the recognition phase of anti-allograft immune reactions comes from the above-described MLR model. Yet, experiments on inbred and especially congenic mice (see Chapter 5) shed some additional light on the nature of transplant recognition. Studies of graft rejection in congenic mice where the donor and the recipient differed only in particular MHC loci underscored the notion that what is actually recognized during the grafting is the foreign MHC allele. In addition, *in-vivo* studies showed that allogeneic MHC molecules can be indirectly presented to the host's T lymphocytes by the host's professional APCs, most importantly **dendritic cells**. Also, the donor's APCs, again, most likely dendritic cells of the donor origin, can play a role in the presentation of the donor's MHC molecules to the recipient's T cells. These donor's APCs reside in the interstitium (intracellular spaces) of the graft, and were sometimes called in older literature "**passenger leukocytes**." Experiments showed that rigorous washing of a graft (e.g., a piece of skin), resulting in removal of most "passenger leukocytes," leads to a delay in the rejection of this graft. Studies using morphological, histochemical, and cellular immunology techniques showed that the majority of the so-called "passenger leukocytes" are in fact dendritic cells.

15.10 What is known about the activation of the host's immune system during transplantation, as judged from *in-vitro* and *in-vivo* studies?

Early studies that employed MLR and CTL assays, performed in the late 1970s to early 1980s, have clearly demonstrated that the recipient's **T lymphocytes** are being activated in response to grafts (in particular, to alloantigens). In the course of their activation, the T cells proliferate; their further fate depends on their belonging to either helper of cytolytic subset. Most of the T cells that express CD4 and recognize Class II MHC molecules of the donor differentiate into cytokine-producing helper T lymphocytes. Most of the T cells that express CD8 and recognize the donor's MHC Class I molecules differentiate into CTL. Both populations strongly incorporate ^{3}H-thymidine in MLR. The proliferative response of the first subset can be blocked by monoclonal antibodies to the donor's MHC Class II molecules, or to CD4; the proliferative response of the latter can be blocked by monoclonal antibodies against the donor's MHC Class I molecules, or to CD8.

The *in-vivo* studies of the host immune system's activation focused on the role of co-stimulators in activation of the host's T lymphocytes. It has been established that recipients' T cells can be strongly activated by solid tissues that essentially lack professional APCs. In addition, it was shown that tissues from mice knockout for molecules such as B7-1 and B7-2 (see Chapters 7–9) can stimulate the recipients' T lymphocytes. It was concluded from these experiments that donor's APCs, albeit of some importance for the activation of host's T cells, are not mandatory for such activation.

CHAPTER 15   Transplantation Immunology

15.11 Were similar studies performed addressing B-lymphocyte activation during transplantation?

Yes, but the conclusion from these experiments was, essentially, that the B-cell activation during transplantation does not differ from the B-cell activation during the invasion of any foreign protein antigen. B cells are strongly activated by antigens of the graft, and differentiate into plasma or memory cells.

15.12 How does the effector phase of immunity manifest itself during transplantation?

Traditionally, the patterns in which rejection manifests itself were classified as **hyperacute**, **acute**, and **chronic** rejection. This classification is based on histopathology and/or on the time-course of rejection rather than on molecular and cellular immune mechanisms. However, some data about the involvement of immune system in graft rejection are available, and will be included in the following discussion.

15.13 What is hyperacute rejection?

It is a rejection of a graft that occurs within **minutes to hours** (seldom a few days) after grafting. This type of rejection is **mediated by pre-existing antibodies** to graft antigens. Clinical features that accompany the hyperacute graft rejection include hemorrhage (bleeding), thrombotic occlusion of the graft vasculature, and irreversible ischemic damage (necrosis) of the graft. It is thought that the hyperacute rejection begins from binding of pre-existing antibodies to antigens expressed on graft endothelial cells. If the bound antibodies bind complement (and many of them do), the activated complement cascades stimulate the endothelial cells to secrete a number of products, including high-molecular weight isoforms of the so-called Willebrand factor. The latter mediate platelet adhesion and aggregation, leading to **intravascular coagulation** in the graft.

In previous years, hyperacute rejection was often mediated by pre-existing **IgM alloantibodies**. Such antibodies can be triggered, for example, by carbohydrate antigens expressed on bacteria that colonize the gastrointestinal tract early in life. The best known example of IgM alloantibodies are antibodies directed to ABO antigens expressed on red blood cells (and also endothelial cells). Nowadays, hyperacute rejection by anti-ABO antibodies is not a clinical problem because all donor–recipient pairs are selected to have identical ABO antigens of the same ABO type. However, other alloantibodies, in particular, **IgG alloantibodies**, are a problem even today. Such antibodies can be triggered by previous blood transfusions, transplantations, or multiple pregnancies. If the titer of these alloantibodies is low, the hyperacute rejection can develop for as long as several days, and is in this case referred to as **accelerated acute rejection**. To prevent hyperacute or accelerated acute rejection, potential graft recipients must be tested for pre-existing antibodies to graft tissues before actual grafting.

15.14 What is acute rejection?

It is a rejection of a graft that begins **after the first week** of transplantation. This type of rejection is **mediated by T cells**, **macrophages**, and **antibodies** of the recipient. In contrast to hyperacute rejection, the effectors of this reaction do not pre-exist, but have to be elicited by the graft recognition: T cells must go through activation and differentiation, and antibodies must be produced. This explains the delay in the onset of rejection. Both T_h and CTL are involved in acute rejection: CTL directly lyse the graft cells, and T_h produce cytokines (mostly "signature" T_h1 cytokines IFN-γ and IL-2) that activate macrophages and recruit inflammatory leukocytes. It is believed, however, that CTL play a more important role than T_h because the infiltrates present in grafts undergoing acute rejection are markedly enriched by $CD8^+$ CTL. In addition, alloreactive CTL can cause acute rejection after their adoptive transfer. In recent years, graft biopsy followed by reverse-transcriptase polymerase chain reaction (see Chapter 5) amplifying such CTL genes as perforin and granzyme B became a specific and sensitive indicator of clinical acute rejection. Nevertheless, T_h1 cells and DTH reactions that they cause are also instrumental in acute graft rejection. $CD4^+$ T_h cells can also trigger acute rejection upon adoptive transfer; in addition, acute rejection still occurs in knock-out mice lacking CD8 (and therefore $CD8^+$ cells) or perforin. Histologically, acute rejection manifests as vascular damage ("endothelialitis"), parenchymal cell damage, and interstitial inflammation but, unlike in the case of hyperacute rejection, not in thrombosis.

15.15 What is chronic rejection?

It is a rejection of a graft that begins **several months** after transplantation and develops over an extended period of time. The pathogenesis of this type of graft rejection is still not completely understood. Chronic rejection manifests in what is called **graft sclerosis**, which is often manifested histologically as **proliferation of smooth muscle cells** in the tunica intima of graft vessels. It is believed that this proliferation is the result of **a special kind of DTH**, where activated macrophages secrete smooth muscle growth factors. Experiments in rodents show that chronic rejection often follows the pharmacologically suppressed acute rejection. However, in human clinical subjects, there is no correlation between chronic graft rejection and episodes of previous acute rejection, or infections, or lipid abnormalities that are known to cause "regular" atherosclerosis.

Another manifestation of chronic rejection is **fibrosis**, which is believed to be caused by activated macrophages producing mesenchymal (extracellular matrix) growth factors such as platelet-derived growth factor (see Chapter 11).

CHAPTER 15 Transplantation Immunology

15.16 What are the main principles of medical intervention to stop or prevent graft rejection?

These include: (1) **general immunosuppression**; (2) **minimizing the strength of specific allogeneic reaction**; and (3) **attempts to induce specific immune tolerance to the graft.** Several methods of immunosuppression are commonly used:

- **Immunosuppressive drugs that inhibit or kill T lymphocytes.** This is, currently, the most popular treatment against graft rejection. The most widely used pharmacological agent from this group is **cyclosporine**. This drug, essentially, inhibits the transcriptional activation of certain genes, notably cytokine genes, including IL-2 (see Chapters 7 and 10). Recall from Chapter 7 that antigen-triggered signal transduction pathways in T lymphocytes include activation of transcription factors like NFAT. Such activation requires the formation of a complex between calcium and a cytoplasmic phosphatase called **calcineurin**. Cyclosporine binds to another abundant cytoplasmic protein, called **cyclophilin**. The complex of cyclosporine and cyclophilin is capable of high-affinity binding to calcineurin, thus preventing its enzymatic activity and shutting down NFAT. A second mehanism of action of cyclosporine is that it stimulates the synthesis of the inhibitory cytokine TGF-β (see Chapter 10). Thus, T cells under the influence of cyclosporine do not proliferate or differentiate because of the lack of IL-2 and the hyperproduction of TGF-β. Due to the use of cyclosporine, most heart and liver transplants are now able to survive for over five years. However, some cases of graft rejection are resistant to cyclosporine; in addition, the doses of cyclosporine needed to inhibit NFAT tend to damage the patient's kidneys. **FK-506** is a fungal metabolite that binds to its specific binding protein (called **FKBP**, from **FK**-506 **b**inding **p**rotein) and acts very similarly to cyclosporine. It is sometimes efficient in cases of graft rejection that are resistant to cyclosporine. **Rapamycin** is another drug that, like cyclosporine and FK-506, inhibits T-cell proliferation, albeit through a different mechanism. It is an antibiotic that binds to FKBP, but the rapamycin–FKBP complex does not inhibit calcineurin. Instead, it binds another protein called **MTOR** (from **m**ammalian **t**arget **o**f **r**apamycin), which may serve as a regulator of a protein kinase that participates in cell cycle control. Rapamycin-treated T cells show abnormalities in their cyclin/cyclin-dependent kinase complexes (see Chapter 10). In addition, rapamycin, as well as a drug with similar effects called **brequinar**, inhibits antibody synthesis and is useful in prevention of B-cell reactions against grafts. Combinations of cyclosporine and rapamycin and/or brequinar are known to be very efficient in many cases of graft rejection.
- **Metabolic toxins that kill proliferating T lymphocytes.** Formerly, the drugs belonging to this group and used in transplantation clinics to fight graft rejection included cyclophosphamide and azathioprine. These pharmaceutical agents do not act selectively on T cells but, rather, kill any cell that is actively synthesizing new DNA. It is no wonder that such drugs caused many side effects, including negative effects on bone marrow precursors of blood cells and enterocytes of the gut. At present, a new drug called **mycophenolate mofetil (MMF)** has appeared and is widely used to selectively kill the proliferating T

lymphocytes during graft rejection. MMF is metabolized to **mycophenolic acid**, which blocks a lymphocyte-specific isoform of an enzyme called inosine monophosphate dehydrogenase. This enzyme is required for the synthesis of guanine nucleotides, and the isoform of this enzyme that MMF blocks is not present in any other cell type. Thus, MMF selectively kills activated T lymphocytes; it is very effective in many cases of graft rejection (especially in combination with cyclosporine), and is not known to cause serious side effects.

- **Antibodies reactive with T lymphocyte surface antigens**. Two of these are especially useful: **OKT3**, a mouse monoclonal antibody against CD3 (Chapter 7), and **anti-CD25**, an antibody to the alpha chain of the human IL-2R (Chapter 10). It is somewhat paradoxical that OKT3 is so effective, because *in vitro* this antibody acts like a potent T-cell mitogen. However, *in vivo* it is toxic to T lymphocytes, perhaps because it activates complement and/or acts as a specific opsonin (Chapters 11 and 13). The anti-CD25 antibody is thought to either block the alpha subunit of the IL-2R from binding IL-2, or to act similarly to OKT3. In the recent years, attempts were made to "humanize" anti-CD25 antibody by replacing its constant region with that of a human immunoglobulin through the use of genetic engineering techniques (see Chapters 3 and 4). The "humanized" anti-CD25 antibody does not trigger strong species-specific response of the host's immune system, and is thus very promising.

- **Anti-inflammatory agents**. The most popular pharmaceuticals from this group are **glucocorticoid** (or **corticosteroid**) **hormones**. These agents are very efficient in combating graft rejection because they strongly inhibit the production of TNF and IL-1 by macrophages and the recruitment and functions of inflammatory leukocytes (see Chapters 10 and 11). In very large doses, corticosteroids are also directly toxic to T cells, but these doses are usually not prescribed because of exceedingly strong side effects. Recently, new anti-inflammatory agents were put into clinical trials; these include soluble cytokine receptors and antibodies to cytokines and adhesion molecules.

15.17 What methods are used to minimize the strength of specific allogeneic reactions?

These include testing the recipient's serum for **pre-existing antibodies** to graft tissues before grafting (see above), and **tissue typing** (see Chapter 5). Testing the recipient's serum for pre-existing antibodies to graft tissues is sometimes referred to as **cross-matching**. Usually, suspensions of the donor's lymphocytes are used as representatives of the donor tissues, but in fact any tissue can serve this purpose. Tissue typing is done with the help of Terasaki's microlymphocytotoxic test, or through PCR (see Chapter 5). A large body of evidence indicates that matching of at least five out of six alleles (two HLA-A, two HLA-B, and two HLA-DR) between the donor and the recipient is crucially important for graft survival. If less than five out of six of the above alleles match, poorer results are achieved; for example, less than 60% of renal allotransplants survive for more than 5 years. The HLA-C alleles seem to be less important, although the HLA-C antigen seems to be

CHAPTER 15 Transplantation Immunology

an activator of allogeneic T cells. Typing of HLA-DQ alleles is less important because the HLA-DQ locus is in strong linkage disequilibrium with HLA-DR (see genetics textbooks for details). The significance of HLA-DP allelic match is not yet appreciated because few HLA-DP-specific typing reagents exist.

15.18 What methods are used to induce specific tolerance to the graft?

As was discussed in Chapter 9, P. Medawar's experiment with neonatal tolerance induction has demonstrated that **it is principally possible** to create a state of life-long specific unresponsiveness of a recipient to various grafts. Unfortunately, Medawar's approach, while being excellent for mice, did not yield practical results in humans, presumably because human infants are born with a relatively mature immune system (at least compared to mice). Attempts to induce specific tolerance to alloantigens in humans were relatively successful in bone marrow transplantation (see later) when, after transplantation, the donor's and the recipient's lymphocytes co-existed in the recipient (the so-called "**mixed chimerism**"). Human mixed chimeras turned out to be tolerant to grafts derived from the same donor as the donor of bone marrow, presumably because HLA and other alloantigens of the donor induced specific immune tolerance. This method has now been abandoned, however, because of the risk of the graft-versus-host reaction (see later).

One interesting and potentially practical approach to tolerance induction in transplantation clinics is to **block T and B co-stimulatory molecules** with monoclonal antibodies or soluble antagonists. Thus, specific tolerance was induced to cardiac allografts in mice that had been injected with such inhibitors of co-stimulation. To this end, soluble CTLA-4, which blocked the recipients' B7, was used (see Chapter 7), and antibody to CD40 ligand was used to block CD40-CD40L interaction (see Chapter 8). Similar experiments have already been successfully performed on primates, and it is likely that this approach will be beneficial for humans. Yet another approach to induce tolerance in humans is to inject potential graft recipients with allogeneic MHC-derived **immunodominant peptides** in their tolerogenic form, e.g., in high doses intravenously (see Chapter 9). These methods of tolerance induction are currently being used in clinical trials.

15.19 What is known about xenogeneic transplantation, and can it be of practical use in clinics?

Xenogeneic transplantation attracts much attention because of the availability of tissues and organs derived from domesticated animals. In some clinical situations, xenogeneic transplantation has already been long practiced with success; for example, the skin from pigs is usually grafted with good results to human patients with burns. Pigs are anatomically rather compatible with humans, so they have been the object of much attention as donors of other organs, e.g., heart, liver, and pancreas. A major **obstacle** to xenogeneic transplantation is the presence of **natural antibodies** that can, and do, cross-react with xenogeneic tissue and cause hyperacute rejection of xenotransplants. The consequences of antibody binding and complement activation in xenogeneic transplantation are even more severe than in allogeneic transplantation, possibly because pig cells do not express inhibitors of human complement. These consequences, however, can be made milder if grafts are taken from transgenic pigs that express human complement inhibitors. Another problem of xenogeneic transplantation is the so-called **delayed xenograft rejection**, which resembles hyperacute rejection in that it is accompanied by intravascular thrombosis. However, delayed xenograft rejection does not require complement activation but, in fact, is mediated by NK and macrophages. Xenografts are also rejected by T-cell-mediated mechanisms similar to those that are involved in the rejection of allografts (see Sections 15.14 and 15.15). Currently, attempts to induce specific tolerance to xenografts are under trial; these attempts are principally similar to those that are being developed for allografts (see Section 15.18).

15.20 What is blood transfusion, and how does the immune system work in blood recipients?

The term "**transfusion**" refers to infusions of blood taken from one individual to another. In the past, attempts to transfuse blood always led to disastrous consequences because of the incompatibility between donor's and recipient's blood cell alloantigens. The first alloantigen system to be defined in humans, and the system that posed a major obstacle in blood transfusions, was a family of red blood cell surface antigens called **ABO**. If the alleles of ABO antigens of the donor and the recipient mismatch, blood transfusion results in **extremely strong antibody- and complement-dependent lysis** of the donor's red blood cells by alloantibodies present in the recipient's serum. This lysis is accompanied by massive release of free hemoglobin, which at high concentrations is toxic to liver and kidney cells; thus, it rapidly leads to liver and kidney failure. In addition, massive lysis of red blood cells triggers an extensive production of such cytokines as **TNF and IL-1** (Chapter 10), causing high fever, shock, and disseminated intravascular coagulation, which may be fatal. IgM alloantibodies to foreign ABO antigens pre-exist in the recipient's serum before transplantation; it is believed that they are triggered by cross-reactive microbial antigens.

Chemically, all ABO antigens are **glycosphingolipids**. All individuals express a common core glycan, called the **O antigen**; it is attached to a sphingolipid. An enzyme which functions as a **glycosyltransferase** modifies the O antigen in a way that depends on the exact allelic form of this enzyme. The single genetic locus that codes this enzyme can exist in one of three allelic forms, called A, B, and O. The enzyme form coded by the O allele is inactive, so no glycosyl modification of the O antigen occurs if the O allele is expressed. This allele is recessive; therefore, to be expressed, it needs to be present in two identical copies on the two homologous chromosomes. The enzyme coded by alleles A and B is active, but its activity is different depending on which of the two allelles codes it. The enzyme coded by the A allele transfers a terminal N-acetylgalactosamine moiety to the O antigen, and the enzyme coded by the B allele transfers a terminal galactose moiety. The A and B alleles are co-dominant, so, if a person inherits both of them, that person will

CHAPTER 15 Transplantation Immunology

have both forms of the enzyme active. Thus, recessive homozygotes express unmodified O antigen on their red blood cells; AO heterozygotes or AA homozygotes express *N*-acetylgalactosaminated O antigen (called "A antigen"); BO heterozygotes or BB homozygotes express galactose-coupled O antigen (called "B antigen"), and AB heterozygotes express both A and B antigens.

Complement-dependent lysis of donor's cells can be caused by binding of preexisting anti-A antibodies to the A antigen, or anti-B antibodies to the B antigen, or anti-A and anti-B antibodies to both A and B antigens. In individuals who express the A or the B antigen, anti-A or anti-B antibodies are usually not detected, presumably because these antigens cause specific immune tolerance early in fetal life. However, in individuals who do not express these antigens, the anti-A and/or anti-B antibodies can be present in high titers. Currently, all potential recipients of allogeneic blood are rigorously tested for expression of A and B antigens, which minimizes the risk of the complement-dependent lysis of donor's cells after transfusion. It should be noted, however, that **the ABO alloantigen system is only one of the many existing alloantigen systems represented in human blood cells**. Because of this, transfusion of whole allogeneic blood remains a risky procedure, and must be avoided whenever infusions of blood substitutes or other therapeutic measures are effective.

The so-called **ABH** and **Lewis** alloantigen systems are examples of the above "other" systems. Both of them are systems of alloantigens created by enzymatic modification of the same above-mentioned O antigen; however, the enzymes that modify it are different from the enzymes that create the A and the B antigens of the ABO system. Lewis antigens began to attract much attention of immunologists recently, because it is now believed that their carbohydrate groups serve as ligands for E-selectin and P-selectin (see Chapters 11 and 12). Incompatibility between the donor and the recipient in the allelic forms of ABH, Lewis, MNSs, Kell, and other red blood cell antigens does not lead to massive complement-dependent lysis of the donor's red blood cells; nevertheless, it does lead to progressive anemia, jaundice (because of chronic liver failure), nephritis, and other negative consequences of allogeneic immune response. The Rhesus antigen system is yet another example of antigens associated with the membrane of red blood cells and able to trigger the synthesis of alloantibodies that may be dangerous to the fetus (see applied immunology and pediatrics textbooks for details).

15.21 What is bone marrow transplantation, and how do immune mechanisms work in marrow recipients?

Bone marrow transplantation is a powerful new method of combating such serious, life-threatening diseases as hemopoietic tumors (leukemias and lymphomas), anemias, and immunodeficiencies. Essentially, **bone marrow transplantation is the transplantation of pluripotent bone marrow stem cells** (see Chapter 2). Until recently, the only way to transplant these cells was to transfer the aspirated bone marrow from a donor to a recipient. At the moment, however, alternative methods exist; for example, treatment of the potential donor with GM-CSF (Chapter 10) can mobilize the donor's bone marrow stem cells, and they can be isolated from the donor's peripheral blood. Because of these new approaches, the term "bone marrow transplantation" gradually gives way to a more accurate term "stem cell transplantation."

Allogeneic bone marrow cells are readily rejected even by MHC-compatible recipient if the recipient is immunocompetent. In fact, even a complete MHC match between the donor and the imunocompetent host would not prevent the rejection, because the latter is known to be at least partially mediated by **NK cells** rather than by alloreactive T cells. The involvement of NK in bone marrow graft rejection explains the phenomenon of **hybrid resistance** observed in experimental allogeneic bone marrow transplantation in mice. Recall from Chapter 5 that if the reaction to graft is triggered by MHC alleles (which are expressed in co-dominant fashion), F1 hybrids between two murine inbred strains will not reject grafts from either parent. However, F1 hybrids do reject bone marrow grafts from both parents. The exact mechanism of bone marrow graft rejection mediated by NK cells is not known.

Because of the above, **the only way to ensure engraftment is to ablate the immune system of the host**. This is accomplished by irradiation and extensive chemotherapy, both of which must precede the transplantation. Insufficient irradiation and/or chemotherapy may result in residual occupancy of the bone marrow "niches" by the host's own bone marrow stem cells, and may inhibit or even abrogate the engraftment. Needless to say, patients who receive a nearly lethal dose of radiation and chemotherapy need very special care, which makes it a must for bone marrow transplantation units to have sterile conditions and special equipment.

If the transplanted bone marrow survives in its new host, the stem cells begin to give rise to hemopoietic precursors and, eventually, to mature blood cells, including T and B lymphocytes. The "brand new" immune system of the recipient, however, is prone to two major abnormalities, which are the **graft-versus-host disease (GvHD)** and **immunodeficiency**.

15.22 What is GvHD?

It is, essentially, **an immune attack developed by the newly matured grafted immune system against tissues of the bone marrow recipient (host)**. T cells of the donor origin are major mediators of GvHD. The target antigens that they are believed to recognize and respond to (in MHC-matched recipients) are **minor histocompatibility antigens** mentioned in Section 15.7. GvHD is a severe clinical problem, and a major limitation to bone marrow transplantation in humans. It may also develop after transplantation of such solid organs as lung or bowel, because these organs contain many mature lymphocytes of the donor origin. Just like graft rejection, GvHD can be classified into clinical forms based on predominantly histologic patterns.

Two forms of GvHD are known: **acute** GvHD and **chronic** GvHD. Acute GvHD is characterized by necrosis of three principal target organs: the skin, liver, and gastrointestinal tract. In the liver, biliary epithelial cells but not hepatocytes are involved. In severe cases of acute GvHD, the skin and the epithelial lining of the gastrointestinal tract may slough off, which may be fatal. Chronic GvHD involves

CHAPTER 15 Transplantation Immunology

the same target organs, but manifests as fibrosis (growth of connective tissue) rather than necrosis. Although both acute and chronic forms of GvHD develop because T cells recognize target antigens and respond to them by activation, CTL responses are not the major effector mechanism in these conditions. **The principal effector mechanism of GvHD is the action of NK cells activated by high local concentrations of cytokines**. Such NK cells are sometimes referred to as **LAK** (from **l**ymphokine-**a**ctivated **k**iller cells). Usually, GvHD is treated with the same immunosuppressants that are used against T-cell-mediated graft rejection (especially cyclosporine). However, NK cells may not respond to treatment with these drugs, and, therefore, attempts to treat GvHD so far have failed. Success in the fight against GvHD perhaps lies more in its prophylactics, which will be greatly advanced when the nature of minor histocompatibility antigens are better understood.

15.23 What is the nature and the character of immunodeficiency in bone marrow transplantation?

In virtually all cases of "successful" bone marrow transplantation, at least some long-lasting immunodeficiency (see Chapter 18) can be observed. The origin of immunodeficiency is, most likely, in that the unique architectonics of secondary lymphoid organs is damaged more or less irreversibly during the "conditioning" of the patient before transplantation (radiation, chemotherapy). Because of this damage, even if T and B cells eventually mature, they show some **partial developmental defects**. For example, B cells show defects in their **class-switch** (especially to IgA) and in their selection in the germinal centers. Peripheral blood B lymphocytes of long-term survivors of bone marrow transplantation accumulate **few or no somatic mutations** in their antibody V-regions (see Chapter 4). Due to partial immunodeficiency, many bone marrow transplant recipients are abnormally susceptible to bacterial and viral infections. The bacterial infections include those of pneumococci and *H. influenzae*, while viral infections include EBV infection (sometimes aggravated by EBV-induced B-cell lymphoma), and cytomegalovirus infection.

Questions

REVIEW QUESTIONS

1. What features of the fate of grafts prompted Medawar and his co-workers that graft rejection is a form of immune reaction?
2. What is common and what is different between a T-cell reaction to foreign MHC and a T-cell reaction to microbial superantigens?
3. What is the role of cross-priming in graft recognition?

CHAPTER 15 Transplantation Immunology

4. Briefly summarize the information that transplantation immunologists and clinicians obtain when they analyze the results of MLR and CTL assays.
5. It has been mentioned that rigorous perfusion of solid grafts before transplantation increases the chances for the graft's survival. What does this prove?
6. Why are T cells not made tolerant when they recognize foreign graft antigens in B7-1 or B7-2 knockout mice?
7. What is the significance of Willebrand factor in transplantation?
8. If we assume that DTH and CTL responses are equally important for acute graft rejection, which double knockout mice will not develop it (or at least develop it in a mild form)?
9. Explain the importance of cyclosporine for transplantation.
10. Which feature of brequinar, not shared by other anti-rejection drugs, validates its use in transplantation clinics?
11. Antibodies to which human antigens have the greatest potential for use in transplantation clinics?
12. What did immunologists and clinicians learn from the phenomenon of mixed chimerism?
13. Individuals who belong to blood group O (i.e. those who express neither A nor B antigens) used to be called "universal donors," meaning that their blood can be safely transfused to recipients who belong to any blood group. What was the reason behind such confidence, and is it really justified?
14. A patient who survived for three months after bone marrow transplantation suddenly begins to complain of unbearable skin itching and the appearance of multiple skin rashes. What condition would you suspect, and what is the mechanism of this condition?
15. What could be the consequences of the scarcity of somatic mutations in antibody V-region genes of bone marrow transplant recipients?

MATCHING

Direction: Match each item in Column A with the one in Column B to which it is most closely associated. Each item in Column B can be used only once.

Column A

1. Second-set rejection
2. Host's APC process graft's MHC
3. Responders
4. Mediated by pre-existing antibodies
5. Sclerosis
6. MTOR
7. Patients with blood group A
8. Cells that actually give rise to recipient's new immune system
9. Target biliary epithelium during GvHD
10. Often complicates marrow transplantation

Column B

A. Cells that incorporate ^{3}H in MLR
B. Cytomegalovirus infection
C. Stem cells
D. NK cells
E. "Sensitized" leukocytes
F. Have anti-B IgM in serum
G. Chronic rejection
H. Bound by rapamycin-FKBR
I. Hyperacute rejection
J. Indirect presentation

CHAPTER 15   Transplantation Immunology

 ## Answers to Questions

REVIEW QUESTIONS

1. The secondary graft is rejected faster and stronger than the first; sensitized leukocytes could trigger a second-set rejection when adoptively transferred.
2. Common: unusually high frequency of responding T cells and the ability to recognize unprocessed MHC molecules: Different: the site of interaction with peptides on TCR (in case of MHC antigen, it is a "conventional" site that includes CDR; superantigen sites lie outside of the "conventional" antigen-binding TCR site).
3. It involves APCs of the host or of the graft, and allows the host's T cells to recognize the graft's MHC as conventional antigen. It can be diminished by eliminating APC (e.g., by graft perfusion).
4. The MLR indicates how strongly the T cells are activated; the CTL assay indicates how extensive is their potential to differentiate into effector cytolytic cells.
5. That indirect presentation of MHC molecules occurs *in vivo*.
6. Because many T cells recognize unprocessed MHC molecules, which, as in the case of superantigen recognition, probably leads to TCR cross-linking (and thus bypasses the necessity of "signal two").
7. High-molecular weight isoforms of this enzyme mediate platelet adhesion and activation during hyperacute graft rejection.
8. Examples include IFN-γ and perforin, or IL-12 and perforin, or IL-12 and granzyme B, etc.
9. Cyclosporine revolutionized transplantation because it inhibits transcriptional activation of the IL-2 gene and thus prevents T lymphocytes from responding to grafts. Due to the use of cyclosporine, most liver and heart allografts can now survive for more than five years.
10. Brequinar inhibits antibody synthesis and thus prevents B-cell reactions against grafts.
11. CD3 and CD25.
12. That it is principally possible to induce tolerance to grafts in human recipients.
13. The reason for confidence was that, since they have no expressed ABO alloantigens, they will not be lysed by anti-A or anti-B alloantibodies of recipients. However, ABO is only one of the many alloantigen systems expressed on red blood cells; thus, the absence of A and B does not guarantee the match on ABH, Lewis, MNSs, Kell, and other alloantigen systems.
14. Acute GvHD. NK cells of the donor origin have attacked the patient's epithelial cells in the skin.
15. No affinity maturation, hence perhaps poor protection against bacterial infections.

MATCHING

1, E; 2, J; 3, A; 4, I; 5, G; 6, H; 7, F; 8, C; 9, D; 10, B

CHAPTER 16

Immunity to Tumors

Introduction

Tumor is an uncontrolled, deregulated growth of tissue that can, and often does, severely affect the overall homeostasis of the tumor-bearing organism and eventually kill this organism. Tumors, especially malignant tumors or **cancers**, are a serious health problem and a major cause of death worldwide. Because of that, the importance of knowledge about mechanisms of antitumor resistance cannot be underestimated. In this regard, it is remarkable that as early as in the 1950s, Burnet envisioned immune cells as sentinels or watchdogs who constantly perform what he called an "**immune surveillance**." He reasoned that since tumor cells are "changed," their antigens might stop being self for the organism, and the immune system thus might recognize the tumor antigens and react to them. According to Burnet's vision, new tumors appear in organisms constantly, but are normally recognized and destroyed by the immune system very early after their onset. Further, the immune system responds also to tumors that escaped the "surveillance," and keeps their growth under control. Only when the immune system fails both of these tasks – surveillance and restriction of tumor growth – does the tumor begin to grow rapidly and becomes a threat to its host. The hypothesis of immune surveillance has been often seriously criticized because, as we will discuss, most tumors are only weakly immunogenic and, in addition, they tend to evade immune responses. Nevertheless, the basic notion of immune surveillance against tumors received a sound experimental support.

In this chapter, we will begin our discussion from early experiments that supported Burnet's immune surveillance hypothesis. We will then classify tumor antigens and illustrate the importance of these antigens in tumor immunity, as well as in diagnostics of certain tumors. Our discussion will continue with a dissection of the antitumor immune response. We will then proceed to the analysis of ways and means that tumors use to evade immunity, and of possible ways to strengthen the immune response to tumors. At the end of the chapter, we will show some examples of successful use of immunotherapeutic strategies in the fight with human malignancies.

CHAPTER 16 Immunity to Tumors

Discussion

16.1 How was it demonstrated experimentally that the immune system reacts to tumors?

In the late 1950s, experimenters learned to induce tumors with the help of a chemical carcinogen **methylcholanthrene (MCA)**. This MCA-induced tumor can be continuously grown in syngeneic mice, i.e., excised from a mouse and grafted on a mouse that belongs to the same inbred strain, to be grown further. Cells separated from the excised tumor can be injected into a new ("naive") mouse that belongs to the same strain, and a growth of the tumor will be seen after a certain period of time. However, if a number of cells separated from the excised tumor are injected into the same individual mouse where the tumor originally grew, no tumor growth will be seen. The mouse that rejected cells from its own ("autologous") tumor, however, accepted cells taken from MCA-induced tumors that had been growing in other syngeneic mice. In subsequent experiments, it was demonstrated that lymphocytes are the cells that mediate autologous MCA-induced tumor rejection. This was interpreted to mean that **individual tumors express antigens that are recognized by the host's immune system and that can be the target of an immune reaction**. It was difficult to apply the above data to spontaneous tumors in animals and humans. However, the finding was reproduced when tumors induced by other chemical carcinogens or by viruses were analyzed. The overall conclusion was that the immune system does respond to tumor antigens, but in many cases this response is not strong enough. However, it is strong enough to cause tumor rejection when antigens of the host are substantially modified by chemical carcinogens (which are also strong mutagens), or when the immune system of the host recognizes foreign (viral) antigens associated with the tumor.

16.2 What tumor antigens exist, and can they be classified?

A wide variety of existing tumor antigens can be classified in a variety of ways. The earliest, and still very popular classification of tumor antigens is based on the pattern of their expression. According to this classification, tumor antigens can belong to one of the two groups: (1) **tumor-specific antigens (TSA)**; and (2) **tumor-associated antigens (TAA)**. TSA are antigens that are unique to tumor cells; they cannot be found in normal, untransformed cells. TAA are antigens that can be found in normal cells as well as in tumor cells; however, their expression in tumor cells is aberrant or deregulated. Another classification of tumor antigens is based on the way they are recognized by the immune system. According to this classification, there are **tumor antigens recognized by antibodies** and **tumor antigens recognized by T cells**. This classification is practically important because, as we will show later, tumor antigens recognized by T cells are thought to be the major inducers of antitumor immunity, and the most promising candidates for tumor vaccines. Yet another (modern) classification of tumor antigens is based on their molecular structure and source. This latter approach will be used in our subsequent discussion.

16.3 What categories of tumor antigens exist, according to their molecular structure and source?

They are the following: (1) products of mutated oncogenes and tumor suppressor genes; (2) mutants of cellular genes not involved in cell cycle control; (3) overexpressed or aberrantly expressed normal cellular proteins; and (4) tumor antigens encoded by genomes of oncogenic viruses. For practical purposes, it is useful to segregate three additional categories of tumor antigens: oncofetal antigens; altered glycolipid and glycoprotein antigens; and tissue-specific differentiation antigens.

16.4 What are products of mutated oncogenes and tumor suppressor genes?

Mammalian cells contain genes that are involved in cellular activation and control of cell cycle, called **cellular proto-oncogenes**. Genes that negatively regulate the cell cycle and prevent cells from continuous division are called **tumor suppressor genes**. If these two categories of genes are affected by point mutation, deletion, chromosomal translocation, or viral gene insertion, they may acquire transforming activity. The products of these altered proto-oncogenes or tumor suppressor genes are synthesized in the cytoplasm of the tumor cells and enter the Class I antigen-processing pathway. Thus, the peptides generated from them are recognized by $CD8^+$ T cells in association with MHC Class I antigens (see Chapter 6). In addition, these proteins can be processed in the endosomal-lysosomal compartment (Chapter 6) if dead tumor cells are phagocytosed by the professional APCs of the host. In this case, the peptides generated from them will be recognized by $CD4^+$ cells in the context of MHC Class II molecules. Thus, peptides derived from these protein products can be presented to both $CD8^+$ T cells (mostly CTL) and $CD4^+$ T cells (mostly T_h). Since these altered genes are not present in normal cells, they do not induce self-tolerance (see Chapter 9); therefore, T cells that have the TCR able to recognize the peptides derived from these antigens are not deleted and are functionally active. Examples of proto-oncogenes known to be prone to having transforming mutations are **Ras**, **p53**, and **Bcr-Abl**. In animals, immunization with antigenic products of these genes elicits strong tumor-specific CTL responses. In humans, however, the majority of CTL responses to tumor antigens do not target these antigens.

16.5 What are mutants of cellular genes not involved in cell cycle control?

These are genes that code for an extremely diverse group of products, united by their appearance in particular tumors as results of random mutation. Antigenic products of such random mutations were in older literature referred to as **tumor-**

CHAPTER 16 Immunity to Tumors

specific transplantation antigens (TSTAs). The antigen(s) that elicited the antitumor immune response in the experiment described in Section 16.1 can serve as examples of TSTAs. Unmutated predecessors of the genes that code for TSTAs often have no known function. As was mentioned in Section 16.1, spontaneous, not experimental, cancers in animals and humans rarely express TSTAs and rarely induce tumor-specific adaptive immune responses. However, TSTAs are often expressed in radiation- or chemical carcinogen-induced experimental tumors. Whilst not particularly valuable as vaccine candidates, these antigens are important for studies on tumor immunity.

16.6 What are overexpressed or aberrantly expressed normal cellular proteins?

These proteins are target of CTL responses in tumor patients and not in healthy individuals, although they can be found in certain tissues of the latter, usually in small quantities. One example of such proteins is **tyrosinase**, an enzyme involved in melanin biosynthesis and expressed in melanomas. Small amounts of tyrosinase can be detected in normal melanocytes. It is presumed that unless melanomas begin to grow, tyrosinase is present in such a small quantity that it does not trigger immune responses, although it does not induce tolerance either (a phenomenon called clonal ignorance). Another example of tumor antigen that can also be found in normal cells is the so-called **MAGE** protein (the term MAGE is an abbreviation from "melanoma antigen gene"). This gene was first identified in melanomas, and later found to be also expressed in carcinomas of bladder, breast, skin, lung, and prostate and in some sarcomas. It is also weakly expressed in normal cells of the human testis and placenta. Interestingly, the nucleotide sequence of the MAGE gene in tumor cells is identical to that in normal cells, i.e., the gene is not mutated. The function of MAGE gene and protein in normal cells, or its role in tumorigenesis, are unknown. Currently, tyrosinase and MAGE vaccines are undergoing clinical trial, with the hope to boost CTL responses in melanoma and other cancer patients.

16.7 What are tumor antigens encoded by genomes of oncogenic viruses?

A variety of viruses can transform host cells and cause tumors. Classical examples of oncogenic viruses include **papovaviruses**, such as **polyomavirus**, **simean virus 40 (SV40)**, and **adenoviruses**, which induce malignant tumors in neonatal and adult immunodeficient rodents. In humans, the **EBV** (see Chapter 8) and human **papilloma virus (HPV)** are associated with lymphomas and cervical carcinomas, respectively. These viruses are DNA viruses; protein products of their genomes can be found in the host cell's nucleus, cytoplasm, and surface. These antigenic products are processed in the cytosol, and peptides generated from them induce strong CTL responses, which often seem to be protective to adult individuals with a healthy immune system. At least one RNA virus, called **human T-cell lymphotropic virus 1 (HTLV-1)**, belongs to the category of tumor viruses; it is the cause of adult T-cell leukemia/lymphoma (ATL). Anti-HTLV CTL responses have been demonstrated, but their protective role remains uncertain.

16.8 What are oncofetal antigens?

These are antigens that were originally thought to be expressed in malignant tumors and in fetal tissues, but not in normal adult tissues. Recently, more sensitive detection techniques showed that these antigens can be also expressed in nontransformed tissues during inflammation, and also (in small quantities) in some normal tissues. Examples of oncofetal antigens include **carcinoembryonic antigen (CEA)** and **alpha-fetoprotein (AFP)**. CEA (or CD66) is a member of the immunoglobulin superfamily (see Chapter 3) and functions as an adhesion molecule (Chapters 7 and 11). It is expressed at high levels on cells of the gut, pancreas, and liver during first two trimesters of gestation, and at low levels in normal adult colonic mucosa and lactating breast. CEA expression is increased in many **carcinomas of the colon**, **pancreas**, **stomach**, and **breast**. In patients with these tumors, CEA can also be detected in serum. Although the expression of CEA is also increased in patients with chronic inflammation of the bowel or liver, it is still useful as a diagnostic marker for gastrointestinal tumors. AFP is a circulating glycoprotein that is present in fetal serum at very high concentrations (2–3 mg/ml), but is replaced by albumin soon after birth. Serum levels of AFP are significantly elevated in patients with **primary cancer of the liver (hepatocellular carcinoma)**, germ cell tumors, and, occasionally, gastric and pancreatic cancers. AFP is used as a diagnostic marker for these tumors, but only with caution because its levels are also increased in patients with liver cirrhosis. Immune responses to oncofetal antigens have not been demonstrated.

16.9 What are altered glycolipid and glycoprotein antigens?

Elevated expression of glycolipid and some glycoprotein antigens is characteristic of many tumors, and these molecules are often structurally changed compared to those expressed in normal cells. These tumor antigens include a variety of **gangliosides**, **blood group antigens**, and **mucins**. Their common feature is their ability to induce strong antibody responses in experimental animals as well as in human patients, which makes them attractive for tumor diagnostics and immunotherapy. In particular, antibodies to tumor-associated gangliosides GM2, GD2, and GD3 (expressed in melanomas), and mucins CA-125 and CA-19-9 (expressed in ovarian carcinomas) are used for these purposes. A mucin called MUC-1 is thought to be expressed exclusively in breast carcinomas, and can serve as an early diagnostic marker of these tumors. It is also considered as a vaccine candidate.

16.10 What are tissue-specific differentiation antigens?

These are antigens normally expressed by cells at certain stage of their maturation or differentiation, and retained by tumors originating from these cells. For example, B-cell-derived tumors may retain **CD10**, which is a marker of committed B-lymphocyte precursors, and **CD20**, which is a mature pan-B-cell marker. (CD10 is also known as "common acute lymphoblastic leukemia antigen," **CALLA**.) Antibodies to these antigens can be used for cancer immunotherapy if properly targeted to tumors. One interesting example of tissue-specific differentiation antigens associated with tumors is **idiotype**. As was mentioned in Chapters 4, 7, and 9, idiotypes are unique markers of lymphocyte clones associated with antibody or TCR V-regions. Lymphocyte-derived tumors are, essentially, expanded lymphocyte clones; therefore, anti-idiotype antibodies can be used as diagnostic or therapeutic tools able to discriminate between the transformed and expanded clone and other (normal) clones of lymphocytes. Work aimed at creation of idiotype vaccines is a major focus of many laboratories.

16.11 What is the role of T lymphocytes in antitumor immunity?

T cells, and **especially CD8$^+$ CTL**, play a major role in tumor immunity. This is expected because tumor antigens are processed in the cytosol, and their peptides presented in the context of MHC Class I molecules. Moreover, the above statement is directly supported by two observations. First, tumor-specific CTL can be isolated from the peripheral blood of tumor patients, e.g., patients with melanoma. Second, the mononuclear infiltrate of many human solid tumors contains CTL with the capacity to lyse the tumor. The role of CD4$^+$ T$_h$ in antitumor responses is less clear, but it is believed that some T$_h$-derived cytokines (especially IFN-γ and TNF, Chapter 10) can boost anti-tumor CTL responses by enhancing the expression of MHC Class I antigens by tumor cells.

It is still not quite clear exactly how T cells respond to tumor cells, because most tumor cells are not derived from professional APCs and, therefore, express little or no co-stimulators. The discovery of **cross-priming** (see Chapters 12 and 15) shed some light on the mechanism of antitumor T-cell response and opened new avenues of the practical application of this knowledge. Recall that cross-priming means internalization of a cell, e.g., a tumor cell, by a professional APC and presentation of the internalized cell's peptides by the MHC molecules of the APC together with the APC's co-stimulators. It is believed that cross-priming, especially mediated by dendritic cells, plays an important role in antitumor T-cell responses. In recent years, this idea began to be used in clinical trials. It is thought that dendritic cells isolated from a cancer patient can be pulsed with tumor cells of the patient and then re-injected as **dendritic cell "vaccine"** to boost CTL responses.

16.12 What is the role of B cells and antibodies in antitumor responses?

As was mentioned in the above sections, many tumor-associated antigens elicit rather strong antibody responses. The ability of antibodies to kill tumor cells with the help of **complement** or through **antibody-dependent cytotoxicity (ADCC**, see

Chapter 13) has been demonstrated *in vitro*. However, no evidence so far supports the idea that antibodies are efficient in elimination of tumors *in vivo*.

16.13 What is the role of NK cells and macrophages in antitumor responses?

NK play a major role in antitumor immunity because, as was discussed in Chapter 11, these cells are efficient against targets that down-regulate MHC Class I antigens. Many tumor cells, indeed, down-regulate Class I antigens up to the point of not expressing them at all. This may be a result of oncogenic virus infection, or of a selection against MHC Class I-expressing tumor cells by CTL. Lymphokine-activated NK (**LAK**, see Chapter 15) are currently under clinical trial as potential antitumor therapeutic agents. The role of macrophages in antitumor immunity has been demonstrated largely *in vitro*. The mechanisms of macrophage-mediated killing of tumor cells are perhaps the same as mechanisms of macrophage-mediated microbial cell killing (see Chapter 14). The role of these mechanisms for antitumor immunity *in vivo* is unclear.

16.14 How do tumors evade immune responses?

The principal mechanisms that tumors use to evade immune responses are the following: (1) down-regulation of the expression of MHC molecules; (2) selection of cells that do not express tumor antigens; (3) production of immunosuppressive substances; (4) induction of immunologic tolerance to tumor antigens; and (5) shielding the surface of tumor cells.

Down-regulation of MHC expression is very common in tumors, and can be explained either by the effect of oncogenic viruses (see Chapter 14), or by negative selection against tumor cells that express MHC molecules. Such negative selection may be performed by CTL at the early stage of tumor development. Tumor viruses employ all known mechanisms of MHC down-regulation and functional impairment mentioned in Chapter 14 (e.g., inhibit MHC transcription, interfere with TAP function, etc.). It is paradoxical (and still not explained) that metastatic tumors which apparently have evaded immune attack do not, on average, express fewer MHC molecules than nonmetastatic tumors.

Tumors often lose the expression of antigens that elicit immune responses; the resulting cells are sometimes referred to as **antigen loss variants**. These are especially common in rapidly growing tumors, and are thought to originate in high-rate mutation peculiar to these tumors. The growth of antigen loss variants can be accelerated by co-culturing the tumor with a source of CTL (an example of selection against antigen-expressing cells).

The best example of immunosuppressive substance produced by tumors is **TGF-β** (see Chapters 10 and 12). This is secreted in large quantities by many tumors, and can strongly inhibit the proliferation and function of lymphocytes and macrophages. In addition, some tumors express death-inducing molecule FasL (see Chapters 8 and 9). Paradoxically, when tumor cells are transfected by the FasL gene *in vitro*, the expressed FasL is nonfunctional.

CHAPTER 16   Immunity to Tumors

Tumors can induce tolerance to their antigens either because these are self, or because these are foreign antigens expressed and presented to the host in a **tolerogenic form** (see Chapter 9). For example, murine mammary tumor virus (MTV) can be passed from nursing mothers to neonatal mice, and the latter become tolerant to its antigens for life, which leads to the development of tumors. The tumors are immunogenic, however, if they are transplanted to uninfected mice. Similar tolerance has been demonstrated to SV40 and other tumor viruses. Recently, it has been demonstrated that blocking CTLA-4 (see Chapters 7 and 9) enhances some antitumor immune responses, suggesting that antitumor tolerance is in part due to B7–CTLA interaction.

In many tumors, cell surface antigens are shielded from external factors (including antibodies or T cells) by a thick polysaccharide wall called **glycocalyx**. In addition, some tumors locally activate blood coagulation and coat themselves in a "**fibrin cocoon**." Interestingly, some antibodies promote the growth of tumors after their binding to tumor surface antigens, the phenomenon known for decades as enhancement effect. Presumably, these antibodies shield tumor antigens from T cells; by themselves, these antibodies are not dangerous to tumors because they fail to activate effector mechanisms of humoral immunity.

16.15   How can the immune system of tumor patients be stimulated?

For many years, attempts to stimulate the patients' defense against tumors were limited to boosting their innate immunity (see Section 16.16). Today, however, several approaches to stimulate adaptive immune responses to tumor antigens bring promising results. These approaches include the following:

- **Vaccination with killed tumor cells and purified tumor antigens.** In previous years, killed tumor cells or purified tumor antigens were used to boost specific immune responses, usually by administering them with adjuvants. This often led to only weak immune responses. At present, killed tumor cells or purified tumor antigens are first incubated with patients' dendritic cells. Such "pulsed" dendritic cells are then reintroduced into the patient and often cause strong CTL responses. The responses can be even stronger if patients are injected with plasmids containing cDNA encoding tumor antigens (DNA vaccines), or with their own dendritic cells transfected with such plasmids. Currently, the MAGE, tyrosinase, Ras, gp100, p53, and other cellular and/or DNA vaccines are on clinical trial for efficacy in antitumor immunotherapy.
- **Augmentation of host immunity to tumors with cytokines and co-stimulators.** Cell-mediated immunity to tumor antigens can be enhanced by expressing co-stimulators and cytokines in tumor cells, or by administering cytokines to tumor patients. Making tumor cells express co-stimulators bypasses the need for cross-priming, and strongly activates T_h. The latter effect can also be achieved by cytokines; in addition, many cytokines can greatly enhance non-specific inflammatory responses, which by themselves may have antitumor

activity. In experiments on laboratory animals, transfection of tumor cells with the B7 gene and the use of the resulting transfectant cells as tumor vaccines has greatly enhanced protective T-cell-mediated antitumor immunity. Similar results were obtained when rodent tumor cells were transfected with IL-2, IL-4, IFN-γ, and GM-CSF genes (see Chapter 10). Clinical trials of co-stimulator- and cytokine-transfected tumor vaccines for the use in humans are currently under way. Of cytokines administered systemically, IL-2 is the most widely used, not as much because of its effect on T cells as because of its effect on NK (see Chapters 10, 11, and 15). **The LAK induced by IL-2 can be very efficient** in eliminating various tumors, notably melanoma (even at advanced stages) and renal carcinoma. However, the systemic administration of IL-2 may result in fever, pulmonary edema, and vascular shock, so it must be used with caution. Of cytokines that boost the innate immunity, **IFN-α** attracts much attention because of its ability to up-regulate the expression of MHC Class I molecules (thus stimulating antitumor CTL) and activate NK cells. IFN-α has documented beneficial effects in patients with melanomas, renal carcinomas, Kaposi's sarcoma (a common complication of AIDS, see Chapter 18), and lymphomas. Also, **TNF** can be beneficial to patients with melanomas or sarcomas when administered by regional perfusion (e.g., of the affected limb).

- **Nonspecific stimulation of the immune system.** In addition to cytokine therapies described at the end of the previous section, clinicians have for many years used **stimulators of inflammation like bacillus Calmette-Guerin (BCG)**. When injected at the site of tumor growth, BCG activates macrophages and thus is somewhat helpful in containing the growth of tumor. More recently, polyclonal T-cell activation has attracted attention as having a potential for tumor therapy. It has been shown that low doses of OKT3 (an anti-CD3 antibody, see Chapter 7) can activate T cells in tumor patients and have a degree of antitumor effect.

16.16 What methods of passive immunotherapy are used in tumor patients?

Passive immunotherapy involves the transfer of immune **effectors**, including tumor-specific T cells and antibodies, or nonspecific effectors like LAK, into tumor-bearing patients. The transfer of antitumor cells into a patient is called **adoptive cellular immunotherapy**. One approach to adoptive cellular therapy is to culture a patient's peripheral blood leukocytes with high concentrations of IL-2 and then to return the generated LAK back into the patient. This approach turned out to be very efficient in mice with experimental tumors, but produced variable results in humans. Currently, clinicians are trying to use **tumor-infiltrating lymphocytes (TIL)** instead of peripheral blood leukocytes, obtaining CTL from TIL (which are enriched with tumor-specific lymphocytes).

Monoclonal **antibodies** to a number of tumor antigens have been used in patients, mostly in the form of immunotoxins (see Chapter 3). Examples of successful and/or promising antibodies of this kind are antibodies to a Her-2/Neu

CHAPTER 16 Immunity to Tumors

antigen (currently on trial for breast cancer), CD10, CD20, CEA, CA-125, GD3 ganglioside, and others. All these antibodies are, or are expected to be, "humanized" by gene engineering through joining their V-regions with human C-regions (see Chapters 3 and 4). In the therapy of B-cell tumors (lymphomas and myelomas), several anti-idiotype antibodies are on clinical trial.

Questions

REVIEW QUESTIONS

1. Briefly outline the concept of immune surveillance. What are the strong and the weak parts of this concept?
2. In one sentence, explain what was learned from the experiment with reimplantation of autologous tumor.
3. Antigen X was detected on the surface of a lymphoma clone that grew in patient A, and on no other cell types. Antigen Y was detected in large quantities on breast cancer cells, and in small quantities on normal epithelial cells. What would you call antigens X and Y?
4. How would you summarize the role of CTL responses to products of mutated cellular proto-oncogenes?
5. Where would you look for a difference between the MAGE in melanoma cells and the MAGE in cells of normal testis?
6. EBV is frequently used in cellular immunology to activate or immortalize human B lymphocytes. Biosafety rules absolutely forbid investigators who use EBV to work with their own cells. Is there a reason for such a prohibition?
7. How would you evaluate the diagnostic potential of CEA and AFP?
8. How is it possible to induce T-cell immunity to tumors using idiotype vaccines? What tumors would be subject to this immunity?
9. What are the two major mechanisms of B-cell-mediated killing of tumors?
10. A primary culture of tumor cells is incubated with a CTL clone specific to an antigen that the tumor expresses. What will you observe if you analyze the expression of MHC molecules by these tumor cells daily?
11. What is, in your opinion, more efficient – to transfect tumor cells with cytokine genes, or to infuse cytokines? Why?
12. Explain the terms "fibrin cocoon" and "enhancement effect."
13. Why is Bacillus Calmette-Guerin still used in oncology clinics?
14. Why is it necessary to perfuse affected limbs in sarcoma patients with TNF rather than simply inject it intravenously?
15. Why is using T cells from tumor infiltrates more efficient for adoptive cellular immunotherapy than using peripheral blood T cells?

CHAPTER 16 Immunity to Tumors

MATCHING

Direction: Match each item in Column A with the one in Column B to which it is most closely associated. Each item in Column B can be used only once.

Column A

1. TSA
2. Bcr-Abl
3. Belongs to papovaviruses
4. MUC-1
5. Is nonfunctional
6. Antibodies kill tumors *in vivo*
7. Nursing mice
8. LAK
9. Inflammatory response
10. GD3

Column B

A. No evidence
B. A proto-oncogene
C. Transmit SV40
D. A ganglioside
E. Specific only to tumors
F. Polyomavirus
G. Breast carcinomas
H. Tumor-expressed FasL
I. Efficient antitumor effect
J. Triggered by BCG

Answers to Questions

REVIEW QUESTIONS

1. This concept states that tumors express antigens that are recognized as nonself, and trigger immune responses that eliminate or contain the constantly emerging tumors. The strength of this concept is that antitumor immune responses have really been demonstrated. The weakness of this concept is that many, if not most, tumors are at best weakly immunogenic.
2. It became clear that immune surveillance works at least in the case of MCA-induced tumors, and that it is tumor antigen-specific.
3. Antigen X is a tumor-specific antigen (TSA), and antigen B is a tumor-associated antigen (TAA).
4. These responses can be demonstrated experimentally, but their protective role in patients is questionable.
5. In the adjacent regions (promoters, enhancers, neighboring genes, etc.), because this gene in tumor cells is not mutated.
6. Yes. If an investigator transforms his own cells and accidentally injects even a small number of them into his finger etc., he/she may die of lymphoma. Another individual's cells, although still hazardous, may be less dangerous because they will probably be rejected as allogeneic.
7. It is limited because both of these antigens are expressed in diseases other than tumors. However, their strong expression can be used as an additional diagnostic criterion.
8. By making APC present the idiotype. It is possible although not trivial, because idiotype is only a small part of antibody molecule. Targets will be lymphomas and myelomas.

CHAPTER 16 Immunity to Tumors

9. Complement-dependent lysis and ADCC.
10. It will steadily go down because of the negative selection against MHC "high expressors" mediated by the CTL.
11. To transfect, because it is less likely to cause systemic effects (which are often harmful).
12. Fibrin cocoon is a layer of fibrin produced around tumor cells due to local activation of blood coagulation. Enhancement effect means that bound antibodies promote the growth of tumor cells they bind to, because they fail to use effector mechanisms such as complement of phagocytes, but instead shield the tumor from T cells.
13. Because it strongly induces local inflammation, boosting macrophage-mediated killing of tumor cells and other antitumor effects.
14. Because of the danger of septic shock, which TNF may cause if administered systemically in high doses.
15. Because these cells are enriched by clones specific to tumor antigens.

MATCHING

1, E; 2, B; 3, F; 4, G; 5, H; 6, A; 7, C; 8, I; 9, J; 10, D

Autoimmunity and Autoimmune Diseases

Introduction

People's "common sense" views our body defense systems (including the immune system) as protective against potentially or actually harmful microbes and other foreign substances, and not reacting to substances of our own bodies. Yet, as it becomes apparent from learning about molecular mechanisms of antibody and TCR repertoire diversification (Chapters 4 and 7), molecules that are reactive with our own body antigens are constantly being made, because *the above mechanisms are "blind" as far as "foreign" or "self" is concerned*. There must exist some mechanisms that prevent these autoreactive molecules (antibodies and TCRs) from interacting with self antigens and causing the immune system to launch an attack against self tissues, or autoimmunity. Such mechanisms indeed exist (see Chapter 9), and if they are impaired, the autoimmune attack can be launched. A net result of such an autoimmune attack may be an autoimmune disease.

Autoimmune diseases are common, and some of them are severe and debilitating. Because these diseases are a common and serious medical problem, it is important to know the exact mechanisms of autoimmunity-mediated tissue damage during these diseases. We have already discussed basic concepts of tolerance, including tolerance to self, in Chapter 9. In this chapter, we will touch on the possible mechanisms that break the immunologic tolerance to self; it should be noted, however, that none of these mechanisms is known for certain. Further, we will analyze the effector phase of the autoimmune attack. At the end of the chapter, we will briefly discuss the issues of genetic susceptibility to autoimmune diseases.

CHAPTER 17 Autoimmune Diseases

Discussion

17.1 How does the breakdown of central or peripheral immunologic tolerance to self antigens contribute to the development of autoimmune diseases?

The answer to this question is more difficult than might have been expected. The notion that autoimmunity results from a failure of the selection processes that normally delete immature self antigen-specific lymphocytes seems logical based on what is known about these processes (Chapter 9). However, *there is very little if any experimental evidence that would support this hypothesis* in any human or experimental autoimmune disease. Somewhat more is known about failure of peripheral self-tolerance, but in this case, too, the available evidence is biased towards T lymphocytes. Several experimental models and genetic abnormalities (including those that are created artificially by targeted gene disruption) indicate that autoimmunity may result from a failure of peripheral T-cell tolerance. The tolerance in the form of either deletion or anergy of mature, peripheral T cells may fail. The exact molecular mechanism that accounts for such failure is not established. We also do not know enough about how, or whether, loss of peripheral tolerance in B lymphocytes may contribute to autoimmunity.

17.2 What experimental evidence supports the hypothesis that a failure of peripheral T-cell tolerance contributes to autoimmunity?

One very strong experimental evidence in favor of this hypothesis was obtained in laboratory rodents, including mice, rats, and guinea pigs, immunized with a protein called **myelin basic protein** (**MBP**), or with a protein called **proteolipid protein** (**PLP**), together with adjuvants. Both of these proteins are major components of the myelin "cover" of the brain and spinal cord. The immunized animals develop a disease called **experimental autoimmune encephalomyelitis** (**EAE**) (Fig. 17-1), the symptoms of which include accumulation of lymphocytes around the vessels of the brain and spinal cord, and signs of demyelination. A chronic form of EAE is reminiscent of the human disease **multiple sclerosis** (**MS**), although the exact cause of MS is not known.

It has been firmly established that in mice and in rats, **EAE is caused by $CD4^+$ T_h1 cells specific for MBR or PLP**. Several lines of evidence support this statement. First, $CD4^+$ T cells isolated from the diseased animals develop a strong proliferative reaction to the above antigens presented by APCs *in vitro*. The proliferating cells produce IFN-γ and IL-2, i.e., they belong to the T_h1 subset (Chapter 10). Second, the disease can be transferred to unimmunized mice by $CD4^+$ T cells of MBP- or PLP-immunized syngeneic animals, or by MBP- or PLP-specific cloned $CD4^+$ T_h cell lines. Third, development of the disease can be prevented by injecting antibodies specific for CD4 or MHC Class II molecules into immunized mice.

CHAPTER 17   Autoimmune Diseases

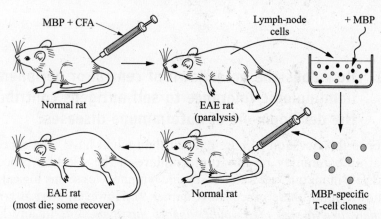

Fig. 17-1. Experimental encephalomyelitis in rats. Experimental encephalomyelitis (EAE) can be induced in rats by injecting them with myelin basic protein (MBP) or proteolipid protein (PLP) (not shown) in complete Freund's adjuvant (CFA). MBP-specific clones can be generated by culturing lymph node cells from EAE rats with MBP. When these T cells are injected into normal animals, most develop EAE and die (although a few recover).

The reason why immunization with myelin leads to such a pronounced T-cell-mediated autoimmune disease is not clear. However, it is reasonable to believe that myelin components belong to the relatively rare category of self antigens that are **not present in the thymus**. Because of that, T-cell clones with the appropriate TCRs are not deleted in the thymus, but are, rather, anergized when they encounter myelin in the periphery. Immunization with a myelin antigen together with an adjuvant leads to breakdown of anergy, and a T-cell response to this antigen develops. Perhaps adjuvants induce local inflammation and thus up-regulate the expression of co-stimulators like B7-1 and B7-2 in the APCs of the peripheral lymphoid organs where the antigen-specific T cells and the presented antigen interact. Interestingly, EAE can be induced in mice of a certain inbred strain ($H-2^u$) spontaneously, without immunization, by expressing an MBP-specific TCR as a transgene. In this case, the severity of EAE directly correlates with the degree to which the mice are exposed to infectious bacteria. Again, this supports the hypothesis that local inflammation helps to break clonal T-cell anergy in the periphery.

17.3   Is there a human autoimmune disease where a failure of peripheral T-cell tolerance can be traced?

Yes. This disease is **insulin-dependent diabetes mellitus (IDDM)**. Murine models of IDDM exist, and both the cause and the symptoms of the disease in these models is exactly the same as in human patients. In both humans and mice, IDDM manifests as a profound impairment of metabolism resulting from **destruction of pancreatic islets**, normally responsible for insulin production. A likely cause of the pancreatic islets destruction is an autoimmune attack on these structures, mediated by the host's own T cells. In some inbred mouse and rat strains, an IDDM-like disease develops spontaneously, and can be transferred to young animals with T cells of older syngeneic animals, in which the disease is already pronounced. In strains where IDDM does not develop spontaneously, it can be induced by co-expression

CHAPTER 17 Autoimmune Diseases

in pancreatic islet cells of a viral protein (with an insulin promoter) and co-stimulatory molecule B7-1 as transgenes. The likely **mechanism** of autoimmune attack in this case is **the break of anergy in T-cell clones that recognize the viral antigen, together with the aberrantly expressed B7-1**. The resulting autoreactive T_h cells presumably activate the production of insulin-specific antibodies and other autoantibodies with specificities for islet proteins, notably **GAD** (glutamine-adenosine deaminase). Together, these autoreactive T cells and antibodies contribute into the destruction of the islets, which is followed by severe insulin deficiency.

17.4 Can a failure of activation-induced cell death result in an autoimmune disease?

Yes, at least in an animal model system it can. A debilitating and life-threatening human disease called **systemic lupus erythematosus (SLE)** can be modeled in inbred mice that belong to one of the two strains – *lpr/lpr* or *gld/gld*. In mice that belong to these strains, as well as in human patients, the disease manifests in the appearance of autoreactive T and B cells and, most importantly, **high-affinity antibodies to double-stranded DNA and nucleoproteins**. These aggressive autoantibodies tend to form **immune complexes** with abundant double-stranded DNA of the host. Such complexes can obstruct small capillaries, especially in the kidney, causing glomerulonephritis and kidney failure. In both *lpr/lpr* and *gld/gld* mice, a clear failure of activation-induced cell death can be observed. In the *lpr/lpr* mice, lymphocytes do not die in the periphery because of the mutation in their **Fas antigen** (see Chapter 9), rendering this protein inactive and unable to transmit the death signal. In *gld/gld* mice, the activation-induced cell death is impaired because of the mutation in **Fas ligand**. Although Fas and Fas ligand are intact in most adult human SLE patients, some children with SLE-like symptoms do have mutations in Fas and/or Fas ligand.

17.5 Can an impairment of negative signaling through the CTLA-4 molecule result in an autoimmune disease?

Again, at least in an animal model it can. Knockout mice lacking CTLA-4 develop a fatal autoimmune disease with T-cell infiltrates and tissue destruction involving the heart, pancreas, and other organs. Although these results are quite impressive, their relevance to any known human autoimmune disease is not known.

17.6 Can an abnormal activity of immunoregulatory lymphocytes result in an autoimmune disease?

Yes, again in an animal model with no known correlate to human autoimmune diseases. The role of active immunosuppression was demonstrated in mice that lack both T and B lymphocytes because of a maturational defect. If these mice receive an injection of naïve syngeneic T lymphocytes, they develop a severe systemic autoimmunity. However, this autoimmune reaction can be prevented if *activated* syngeneic T cells are co-transferred with the naive T cells. This was interpreted to mean that autoreactive T-cell clones exist among naive T lymphocytes and are kept under check by some of the activated T cells of the host. In the absence of the latter, the naive autoreactive T cells can be activated and develop an autoimmune attack.

17.7 Can a failure in peripheral B-cell tolerance lead to an autoimmune disease?

As we have mentioned in earlier sections, there is little formal evidence that would directly link an impairment of B-cell tolerance to autoimmune diseases. Yet, it is suspected that a break in B-cell tolerance does play a role. Exposure of murine B lymphocytes to **LPS** (see Chapter 11) activates many different B-cell clones, some of which are autoreactive. However, these antibodies are usually of low affinity and cause no damage to self tissue. It is thought that in normal individuals, the absence of pathogenic self-reactive antibodies is primarily the result of the absence of T cell help and not of the break in B-cell tolerance.

17.8 What are examples of autoimmune diseases where not T cells, but autoantibodies, play a major role in the development of tissue injury/damage?

We have already mentioned SLE, where both autoreactive T cells and autoreactive antibodies are thought to contribute to the tissue damage. There are some autoimmune diseases, however, where antibodies and not T cells *per se* play a major role in the pathogenesis. One of such diseases is **myasthenia gravis**, in which high-affinity antibodies specific for the acetylcholine receptor in the motor end-plates of neuromuscular junctions are produced. Binding of the antibodies interferes with acetylcholine-mediated neuromuscular transmission, and may impair the ability of muscles to respond to neural impulses. A disease resembling myasthenia gravis can be produced in rats or mice by immunizing them with purified acetylcholine receptors. The experimental disease can be adoptively transferred to normal animals by antibodies against the acetylcholine receptor. However, in some animals, the disease can be also adoptively transferred by acetylcholine receptor-specific $CD4^+$ T cells, in agreement with the point made in Section 17.7.

Another example of antibody-mediated autoimmune disease is **Graves disease**, or autoimmune hyperthyroidism. In patients with this disease, antibodies are produced to the **receptor for thyroid-stimulating hormone** (**TSH**). Binding of the antibody to TSH receptor has the same effect as the binding of TSH itself, i.e., stimulates the production of thyroid hormones. Yet another example is **autoimmune hemolytic anemia**, where high-affinity antibodies are produced to self antigens expressed on red blood cells, especially the so-called I/i glycoproteins. In patients with autoimmune hemolytic anemia, autoreactive B cells may be oligo- or monoclonal in origin and use one individual V_H segment (V4-34) for expression of their antibody heavy chain, in conjunction with various light chains (see Chapter 4). It is noteworthy that anti-red blood cell antibodies in this disease tend to agglutinate (glue together) red blood cells and may react with their anti-

CHAPTER 17 Autoimmune Diseases

gens at temperatures about 22°C better than at 37°C. For that reason, these antibodies are sometimes referred to as **cold agglutinins**. The antibodies can bind complement and lyse red blood cells, often causing liver dysfunction and failure because of toxicity of the free hemoglobin. The exact cause for elicitation of high-affinity autoreactive antibodies in all these diseases is unknown.

17.9 What is rheumatoid arthritis?

Rheumatoid arthritis is one of the most prevalent autoimmune diseases, with almost 1% of adult Americans suffering some form of it. Its main feature is destruction of the joint cartilage and inflammation of the synovium. Small joints, most frequently joints of fingers and toes, are usually involved. The disease often begins during adolescence or at young adult age, and can be crippling. Many parameters of the disease indicate that a local immune reaction to self antigens is instrumental in the tissue damage. Both autoreactive T cells and autoantibodies contribute to the development of lesions. The most interesting and unusual feature of the involvement of the immune system in rheumatoid arthritis is the **formation of ectopic sites of lymphoid tissue** (see Chapter 2). CD4+ T cells, activated B lymphocytes, and plasma cells are found in the inflamed synovium, and in severe cases, one can find well-formed lymphoid follicles with germinal centers there.

The exact mechanism of tissue damage, as well as the exact antigen specificity of most activated T and B cells involved, remain unknown. Many different cytokines are detected in the inflamed synovial cavities, but this is more likely to be a consequence, rather than a cause, of the inflammation and tissue lesions. Many B cells in rheumatoid arthritis patients produce the so-called **rheumatoid factors** – antibodies to constant regions of self IgG antibodies. However, these antibodies are not involved in tissue damage, and their significance remains obscure. Another interesting characteristic of B cells in these patients is the increased frequency of somatic mutations in their V_H genes (see Chapter 4). It has been recently shown (by E.C.B. Milner and his associates) that many more resting B cells in the patients have heavily mutated V_H-regions, and accumulate more mutations per cell than B lymphocytes of healthy individuals. The significance of this discovery is still not appreciated, however.

17.10 Are there examples of autoimmune diseases caused predominantly or exclusively by autoreactive CTL?

In humans as well as in experimental models, $CD8^+$ CTLs reactive with self tissues can be detected. In the case of human or experimental murine autoimmune myocarditis, for example, CTLs lyse cardiac tissue, thus leading to severe heart failure. However, the "autoimmune" myocarditis and perhaps other CTL-mediated diseases are not really autoimmune because, in fact, these immune cells react to viruses that infect the host's cells. Anti-"cardiac" CTLs actually are specific to antigens coded by Coxsackie viruses that asymptomatically infect cardiac cells. Therefore, a break or failure of normal mechanisms of self-tolerance hardly plays a role in the development of such diseases.

17.11 Are there other examples of "confusion," when a disease that may be viewed as autoimmune is, essentially, not autoimmune?

Yes. These examples include diseases when an immune reaction is aimed at self antigens, but **these antigens are altered by an inflammatory or necrotic process**. In these cases, the cause of disease is not the abnormality of the immune system but the infectious agent that caused the inflammation, or an etiologic agent that caused necrosis. For example, rheumatic fever is often described in literature as an autoimmune disease. In fact, the immune attack on the patient's joints and cardiac tissue is secondary; it develops because the antigens in these anatomical locations are altered by inflammation caused by certain strains of streptococci. Also, myocardial infarction is often accompanied by production of antibodies reactive with necrotically changed heart muscle of the patient. Obviously, these antibodies are not the cause of the problem. Physicians must clearly understand the difference between autoimmune diseases, where the root of the problem is an abnormality in the patient's immune system, and diseases accompanied by autoimmune reactions, where the root of the problem is the disease that alters normal human tissue.

17.12 Is there a genetic predisposition to autoimmune diseases?

Yes, although the exact genetic factors that determine this predisposition are complex and not completely understood. Perhaps the best example that demonstrates this point is IDDM (see Section 17.3). Studies on identical twins revealed that **the concordance in IDDM in this group varies from 35% to 50%,** while nonidentical twins have a concordance of only 5% to 6%. This statistic proves that a gene or genes are definitely controlling the development of the disease. Similar (although not always that impressive) results were obtained for other autoimmune diseases. Many approaches including, most recently, approaches of molecular genetics (PCR, microsatellite analysis, footprinting, RFLP and other methods – see molecular biology and genetics textbooks for details) have shown that susceptibility to most autoimmune diseases is **a polygenic trait**, i.e., is controlled by a group of polymorphic genes.

It is quite remarkable that among all genes studied for possible association with autoimmune diseases, the **best candidate genes are MHC genes**. There is one presumably autoimmune disease (the exact mechanism of which is not known), called **ankylosing spondylitis**, where there is an enormously strong association with one allele of an MHC Class I gene (HLA-B27). Individuals who express HLA-B27 have a 90- to 100-fold stronger risk of developing this disease than individuals who do not express this allele. However, **most autoimmune diseases are associated with MHC Class II genes** rather than with MHC Class I genes. This makes sense, because MHC Class II genes are involved in the selection and activation of CD4+ T lymphocytes (see Chapters 2, and 5–7) and thus regulate both humoral and cell-mediated immune responses.

CHAPTER 17 Autoimmune Diseases

The following features of the above MHC association are noteworthy. First, in many cases, the MHC allele associated with an autoimmune disease is in **linkage disequilibrium** (inherited together) with another allele, and that other allele may be in the casual association with the disease. Therefore, it is more accurate to talk about "**extended haplotypes**" (or sets of genes) associated with autoimmunity, than about single alleles. For example, an extended haplotype including HLA-DR and HLA-DQ alleles is implicated in the susceptibility to IDDM. Second, structurally, the HLA molecules in patients with autoimmune diseases may differ from the HLA molecules in healthy subjects only in their peptide-binding clefts. This is logical, because the structure of these clefts is what actually determines peptide and TCR binding (see Chapters 5–7) and, therefore, the function of MHC molecules in normal or pathologic immune reaction. Third, serological MHC typing implies, and recent advances in molecular biology directly show, that **disease-associated HLA sequences are found in healthy individuals**. Therefore, expression of a particular HLA gene is not by itself the cause of any autoimmune disease, but it may be one of several factors that contribute to autoimmunity together.

17.13 Are non-MHC genes among the factors contributing to autoimmune diseases?

Yes, obviously they are. Again, IDDM is perhaps the best illustration. In human IDDM and in its experimental animal model (the so-called NOD mouse strain), about 20 genes appear to be associated with disease susceptibility. Among non-MHC genes that are associated with IDDM and other autoimmune diseases, **genes that code for immunologically significant molecules** seem to be of special importance. In mice, polymorphisms on the IL-2 gene correlate with the diabetes-like disease very strongly, and in both mice and humans a gene that is mapped close to CTLA-4 is apparently implicated. There is also some association between SLE and polymorphisms on complement proteins C2 and C4, as well as between SLE models in mice and mutations in Fas and Fas ligand genes (see Section 17.4).

Questions

REVIEW QUESTIONS

1. From the point of view of immunology, what is the key difference between diseases such as IDDM and diseases such as rheumatic fever?
2. One may occasionally read in the literature that the discovery of T-cell selection in the thymus, or Goodnow's experiments with HEL-anti-HEL double transgenic mice helped us to understand autoimmune diseases. How would you comment on such statements?

3. Why does the injection of MBP and/or PLP without adjuvants fail to induce EAE?
4. Explain the development of EAE in H-2^u mice transgenic for anti-MBP TCR. Is it possible to use this transgene in other mice strains? Why, or why not?
5. What is the significance of *lpr/lpr* and *gld/gld* mutations? Is it limited only to the pathogenesis of SLE?
6. The experiment described in Section 17.6 implicates T cells with suppressive properties in the development of systemic autoimmunity. How would you extend this observation to link these suppressor cells with the T_h1 or the T_h2 subset?
7. The so-called "natural" antibodies to conserved self antigens (components of cytoskeleton, single-stranded DNA, some polysaccharides, etc.) are constantly produced by B-1 cells, and their titers can be rather high. Can these antibodies cause autoimmune diseases? Why, or why not?
8. What is similar and what is different in the pathogenesis of EAE and myasthenia gravis, and what experiment can be performed on an animal model to provide proof?
9. Design an experiment showing that a patient's hyperthyroidism has an autoimmune nature.
10. Patients with autoimmune hemolytic anemia often have blue fingertips and earlobes, and dark spots of the wings of their noses. Can you explain these symptoms?
11. V4-34 is used among CD40-activated human peripheral blood B lymphocytes less frequently than among freshly isolated human peripheral blood B lymphocytes. Explain these data (name two possibilities based on the knowledge of peripheral B-cell tolerance from Chapter 9).
12. Summarize the differences between rheumatoid arthritis and rheumatic fever.
13. The so-called Wallenberg–Roose reaction is, essentially, a test that determines the titer of rheumatoid factors. How would you characterize the significance of this test in the diagnosis of rheumatoid arthritis?
14. How did recent advances in molecular biology, and in particular the establishment of nucleotide sequences of HLA genes, contribute to our understanding of associations between genes and autoimmune diseases?
15. Is there any reason behind the association of IL-2 alleles and autoimmunity in mice?

MATCHING

Direction: Match each item in Column A with the one in Column B to which it is most closely associated. Each item in Column B can be used only once.

Column A

1. MBR
2. H-2^u
3. GAD
4. Target for autoantibodies in SLE
5. *gld/gld*
6. Acetylcholine receptor
7. All cold agglutinins use
8. Rheumatoid arthritis lesion site
9. Ankylosing spondylitis
10. Susceptibility to autoimmune diseases

Column B

A. V4-34
B. Polygenic trait
C. Finger joints
D. TCR transgenics develop EAE
E. HLA-B27
F. Spinal cord myelin
G. Myasthenia gravis
H. Autoantibodies in IDDM
I. dsDNA
J. Fas ligand

CHAPTER 17 Autoimmune Diseases

Answers to Questions

REVIEW QUESTIONS

1. IDDM is truly an autoimmune disease, because its pathogenesis includes a failure of tolerance to self; rheumatic fever manifests as a reaction to the patient's own antigens, but these antigens are altered by inflammation caused by a factor that is irrelevant to the patient's immune system (*Streptococcus*).

2. Yes and no. These experiments are indeed crucial in our understanding of tolerance, and as such, help us to understand autoimmune diseases better. On the other hand, the significance of central tolerance (which these experiments address) in the development of the known autoimmune diseases is questionable.

3. EAE is mediated by T_h1 cells that cannot be activated without co-stimulators. If they recognize myelin proteins without co-stimulators, they become anergic and do not attack self tissue. Adjuvants, however, up-regulate local co-stimulators.

6. The TCR that binds MBP-derived peptides was genetically cloned from the $H-2^u$ strain. To develop an immune attack against myelin, T cells that carry this transgene must recognize their specific peptide in association with self MHC. "Self" for them is exclusively the $H-2^u$ allele; therefore, no other murine strain will develop EAE with this particular transgene.

5. This significance is broader than that, because these mutations point to the role of activation-induced cell death as a factor that prevents autoimmunity.

6. Activate the T cells to be co-transferred in the presence of either IFN-γ or IL-4, and see which of the populations suppresses autoimmunity generated by naive T cells.

7. No, because the T cells that potentially could deliver them help during cognate interaction are usually deleted in the thymus or peripheral lymphoid organs.

8. Similar: both can be caused by $CD4^+$ T_h cells. Different: myasthenia gravis, but not EAE, is mediated by antibodies. Adoptive transfer is the approach to show this.

9. Detect anti-TSH antibodies in the patient's blood, and then show in a tissue culture-based system that these antibodies stimulate the production of thyroid hormone.

10. Because of red blood cell agglutination at these sites, where the temperature is lower than in the rest of the body.

11. Possibility one – V4-34 expressors are more susceptible to activation-induced cell death because they bind some abundant ligand when repeatedly stimulated; possibility two – receptor editing.

12. In rheumatoid arthritis, patients first present as young adults; small joints (fingers, toes) are usually involved; there is an ectopic growth of lymphoid tissue; mutations in patient's antibody V_H-regions are increased. In rheumatic fever, the disease can start at any age; big joints (usually knees) are involved; there is no ectopic growth of lymphoid tissue; the frequency of mutation in V_H-regions is normal.

13. It is limited, because antibodies to Fc fragments of self IgG can be detected in a wide variety of diseases, as well as in healthy patients, and these antibodies do not damage self tissues.

14. They showed that the so-called "disease-associated" allelic HLA sequences can be detected in healthy patients, thus indicating that the mere expression of a certain HLA allele cannot be the sole cause of an autoimmune disease.

15. Perhaps some allelic forms of IL-2 make this cytokine nonfunctional (possibly, by abrogating its ability to bind its receptor). This leads to an impairment of activation-induced cell death (see Chapter 10). T-cell responses are not impaired, though, because other cytokines may take over IL-2.

MATCHING

1, F; 2, D; 3, H; 4, I; 5, J; 6, G; 7, A; 8, C; 9, E; 10, B

CHAPTER 18

Immunodeficiencies

Introduction

Immunodeficiencies is a collective name for a large group of disorders when the function of the immune system is not strong enough to protect the organism against foreign invaders, especially microbes and their toxic products, or against the organism's own malignantly transformed cells. It is remarkable, and not quite appreciated by nonspecialists that immunodeficiencies are rather common: approximately 1 in 500 Americans have some form of immunodeficiency. Yet, the severity of these disorders varies very widely, and only a small fraction of immunodeficiencies are serious and even life-threatening diseases.

Immunodeficiencies are usually classified into two large groups: **congenital** and **acquired**. We will begin our discussion with a description of congenital immunodeficiencies where, because of a genetic defect, the humoral and/or the cell-mediated immunity in the patient is inadequately weak, or even absent. We will show that congenital immunodeficiencies include some cases with a rather complex pattern of inheritance and unclear pathogenesis, and emphasize on the notion that further research is needed to design new, efficient therapeutic procedures against these disorders.

Our analysis will proceed to acquired immunodeficiencies. We will briefly describe some common acquired immunodeficiencies related to such environmental factors as malnutrition, ionizing radiation, and infection. Further, we will concentrate on the most threatening, pandemically widespread of all immunodeficiencies: the acquired immunodeficiency syndrome (AIDS), highlighting some important aspects of its etiology, pathogenesis, prevention, and treatment.

Discussion

18.1 What are congenital immunodeficiencies, and how can they be further classified?

Congenital (or primary) immunodeficiencies are genetic, hereditary abnormalities of the immune system. They can be classified into congenital deficiencies of adaptive immunity and congenital immunodeficiencies of innate immunity. The congenital deficiencies of adaptive immunity can be further classified into abnormalities in B-lymphocyte maturation, T-lymphocyte maturation, combined B- and T-lymphocyte maturation, and abnormalities in lymphocyte activation.

18.2 What congenital immunodeficiencies can be placed into the first group (i.e., abnormalities in B-lymphocyte maturation)?

The best known congenital immunodeficiency that belongs to this group is called **X-linked agammaglobulinemia**, or Bruton's agammaglobulinemia. It is also perhaps the very first congenital immunodeficiency to be described as such in medical literature (in 1952). It is one of the most common congenital immunodeficiencies. The disease is characterized by the absence of immunoglobulins in the blood and by the absence of B lymphocytes, including mature and immature B cells. There are no germinal centers in peripheral lymphoid organs, and no plasma cells in patients' tissues. The number and functions of T lymphocytes are close to normal, and the generation of B-cell precursors in the bone marrow, including pro-B cells and pre-B cells, is normal. The disease is inherited as an X-linked recessive trait (similarly to hemophilia and other X-linked recessive diseases; see genetics textbooks for details). Because of this pattern of inheritance, the X-linked agammaglobulinemia patients are mostly infant boys, who suffer from numerous infections (especially pyogenic bacterial infections). Women may be clinically healthy carriers of the disease. In these female carriers, only those B cells that inactivate ("Lionize") the X chromosome carrying the mutant allele mature. Interestingly, in some patients the numbers of activated T cells are somewhat reduced, probably due to nonexistent B-cell antigen presentation (see Chapter 8). Also, in as many as 20% of the patients, autoimmune disorders develop, for no known reasons.

X-linked agammaglobulinemia develops because of a recessive mutation in the gene that codes for an enzyme called B-cell tyrosine kinase, or **Bruton's tyrosine kinase (Btk)**. The exact function of this enzyme is not known, but it is clear that it is somehow involved in transducing signals from pre-B lymphocyte receptors that are required for continued maturation of the cells. Knockout mice lacking Btk show less severe immunodeficiency than humans, indicating that other, compensatory enzymes may exist in mice but not in humans. An inbred murine strain called **CBA/N** has a point mutation in the Btk gene and an X-linked defect in B-cell maturation. However, the numbers of B cells and the total concentration of serum immunoglobulins in CBA/N mice are normal. The immunodeficiency in

CHAPTER 18 Immunodeficiencies

these mice manifests in selective unresponsiveness to certain polysaccharide antigens. The therapy of the disease in humans is periodical (monthly or weekly) intravenous infusions of pooled human gamma globulin preparations.

18.3 What congenital immunodeficiencies can be placed into the second group (i.e., abnormalities in T-lymphocyte maturation)?

A prototypical congenital T-cell maturation defect is called **DiGeorge syndrome**. The origin of this disease lies in the hereditary malformation of the third and fourth pharyngeal pouches during fetal life. Such malformation leads to the absence of the thymus, and also to defective development of parathyroid glands and to facial deformities. Peripheral blood T lymphocytes in DiGeorge patients are absent or greatly reduced in numbers, and even if some T cells are detected, these cells do not respond to polyclonal activators or allogeneic lymphocytes *in vitro*. Antibody levels may be normal, but all antibodies are IgM with unmutated V-regions, and there is no affinity maturation. In patients with severe DiGeorge syndrome, antibody concentrations are reduced.

DiGeorge syndrome patients suffer from numerous infections, and especially from viral, fungal, and mycobacterial infections. Because of the absence of parathyroid glands, the patients also tend to suffer from the low concentration of calcium in their blood, which manifests as tetany (muscle twitching). The immunodeficiency can be corrected by thymic transplantation or bone marrow transplantation. However, paradoxically, **the symptoms of DiGeorge syndrome tend to reverse spontaneously** after the first year of life, and the affected children seem to be almost normal at 5 years of age. This phenomenon can be explained either by the presence of some remnants of thymus, or by the existence of some yet unknown extrathymic sites of T-lymphocyte maturation. It may be that these sites take over the thymus in these patients later in their lives. A murine analogue of the human DiGeorge syndrome is the "nude" (nu/nu) mutation mentioned in Chapter 8.

18.4 What congenital immunodeficiencies can be placed into the group that includes abnormalities in both B- and T-lymphocyte maturation?

These abnormalities are collectively called **severe combined immunodeficiency (SCID)**. Recent studies of the cellular and molecular basis of SCID, as well as of its genetics, allowed immunologists to classify it into three large categories.

1. **X-linked SCID caused by mutations in the cytokine receptor common γ chain**. Recall from Chapter 10 that the common γ chain is the signal-transducing transmembrane protein shared by the receptors for IL-2, IL-4, IL-7, IL-9, and IL-15. The single gene that codes for the common gamma chain is mapped to the X chromosome; the immunodeficiency is caused by a

recessive mutation in this gene and manifests in homozygotes or hemizygotes as a greatly impaired lymphocyte maturation (because of the lack of signaling through lymphopoietic cytokine IL-7, see Chapter 10). In mice that express this mutation, both T- and B-cell numbers are greatly reduced. In human recessive homozygotes or hemizygotes (male patients), T-cell numbers are severely reduced, but the numbers of B cells may be normal or even increased. Yet, antibody production is severely impaired, especially in responses to protein antigens (because of the absence of T cell help). In female carriers of the disease, lymphocytes that inactivate ("lionize") the normal X chromosome do not mature; therefore, in nonlymphoid cells the "lionization" is random, while all lymphoid cells that survive inactivate the same diseased X chromosome. This feature may be used in the differential diagnosis of this immunodeficiency.

2. **Autosomal recessive SCID caused by adenosine deaminase (ADA) deficiency**. ADA is an enzyme that catalyzes the deamination of adenosine and 2′-deoxyadenosine to inosine and 2′-deoxyinosine, respectively. If the gene that codes for this enzyme is mutated, deoxyadenosine and its precursors (especially deoxyadenosine triphosphate, dATP) tend to accumulate in cells and cause toxic effects. Lymphocytes are more sensitive to ADA deficiency than other cells because they lack the capacity to degrade the toxic dATP. Therefore, many lymphocytes in patients with ADA deficiency do not mature. The numbers of peripheral B and T lymphocytes in these patients are usually normal at birth, but decrease precipitously during the first year of life. A small fraction of patients with similar symptoms and the same autosomal recessive pattern of inheritance have the PNP (purine nucleoside phosphorylase) rather than the ADA defect. PNP is an enzyme that converts inosine into hypoxanthine and guanosine to guanine; its deficiency leads to the accumulation of dATP, hence it mimics the ADA deficiency.

3. **SCID caused by other factors.** This group includes cases when maturation of both B and T cells is impaired, but its cause is unknown. In some cases of SCID, it is suspected that a mutation affects **RAG-1 and/or RAG-2 genes** (see Chapter 4). Recall that these genes code for enzymes that catalyze DNA looping and excising intervening sequences during antibody and T-cell receptor V-region gene rearrangement. In other cases of SCID, known under a somewhat "fuzzy" name **reticular dysgenesis**, the suspected defect is in enzymes or other proteins that function very early in maturation, perhaps at the level of hemopoietic stem cells. It is characterized by a complete absence of not only lymphocytes but also myeloid cells (granulocytes and red blood cells), and is very rare and lethal.

4. **"SCID mouse."** This is an experimental murine model of SCID that develops in mice with mutations in the enzyme called DNA-dependent protein kinase. This enzyme catalyzes the repair in double-stranded DNA breaks, and its impairment results in inability of lymphocytes to rearrange V-region genes and therefore to mature. There are no known human correlates of DNA-dependent protein kinase deficiency, but this murine model is widely

CHAPTER 18 Immunodeficiencies

used in experimental immunology where the diseased mice serve for "housing" various foreign lymphocyte subsets.

18.5 What congenital immunodeficiencies can be placed into the group that includes abnormalities in both B- and T-lymphocyte activation?

This relatively large and diverse group of disorders includes the following categories: (1) selective immunoglobulin isotype deficiencies; (2) common variable immunodeficiency; (3) defects in T-cell-dependent B-cell activation; (4) defects in T-lymphocyte activation; and (5) defective MHC expression.

- **Selective immunoglobulin isotype deficiencies.** This category includes two congenital immunodeficiencies that are rather prevalent – IgA deficiency and IgG subclass deficiency. The cause of either of these abnormalities is unknown. Patients with selective IgA deficiency often show few or even no clinical symptoms, and the abnormality is diagnosed in the laboratory quite accidentally. Those patients that do show symptoms suffer from occasional respiratory infections and diarrhea. Rarely, patients have severe and recurrent infections in airways and the gastrointestinal tract. Remarkably, the expression of Cα gene in these patients is normal, and many B cells have membrane-bound IgA; however, the secretion of IgA is profoundly impaired. The total numbers of B and T cells in these patients are normal. Among selective IgG subclass deficiencies, the IgG3 deficiency is most common. In most patients with this abnormality, IgG3 can be expressed on the membrane, but cannot be secreted. In other patients, the Cγ3 gene is deleted. A combination of IgA deficiency and IgG3 deficiency is sometimes observed in children, suggesting that these abnormalities have some common denominator in their etiology and/or pathogenesis.
- **Common variable immunodeficiency (CVI)** is a clinically heterogeneous group of abnormalities where the common factor is a defect in B-cell differentiation. Most patients with CVI show little or no plasma cells and/or impaired B-cell differentiation in response to antigens. It is noteworthy that CVI patients often have manifestations of autoimmunity (especially autoimmune hemolytic anemia and rheumatoid arthritis), and tumors. CVI may be sporadic or familial; the latter can be inherited in either autosomal-dominant or autosomal-recessive fashion. The cause or the exact pathogenesis of CVI remain unknown.
- **Defects in T-cell-dependent B-lymphocyte activation.** The best known of these abnormalities is the so-called X-linked hyper-IgM syndrome – a relatively rare disorder that manifests in a profound decrease of serum IgG and IgA levels and compensatory increase of serum IgM levels. This disease is associated with mutations in CD40 ligand (see Chapter 8). The mutant forms of CD40L either do not bind CD40, or do not transduce the signal upon binding. In addition to the lack of downstream antibody isotypes, patients show multiple defects in

cell-mediated immunity, probably because the CD40L-mediated activation of macrophages (Chapter 12) in these patients is deficient.

- **Defects in T-lymphocyte activation** are rare and became known only recently, after the extensive molecular dissection of T-lymphocyte activation and signal transduction-associated molecules (see Chapters 7–9). An interesting group of patients that demonstrate this congenital immunodeficiency has a mutant and functionally impaired form of **Zap-70 kinase** (see Chapter 7). T-lymphocyte activation in these patients is very weak, although numbers of T and B cells are normal or sometimes elevated. These patients often show an increased proportion of $CD8^+$ T lymphocytes and a decreased proportion of $CD4^+$ T lymphocytes, the reason for this abnormal ratio being unclear. Another abnormality that falls into this category is the so-called X-linked lymphoproliferative disease. This genetic disorder manifests in weak T-cell activation and, remarkably, an inability of the patients to contain EBV infection. This leads to severe infectious mononucleosis and often B-cell lymphomas. The pathogenesis of this immunodeficiency is poorly understood; it is thought to be related to mutations in the gene coding for an adaptor molecule used in T- as well as B-lymphocyte signal transduction.

- **Defective MHC expression** can manifest as either defective MHC Class II expression (also called "**bare lymphocyte syndrome**"), or defective MHC Class I expression. Bare lymphocyte syndrome is a rare disease that is inherited in an autosomal recessive fashion. The patients express little or no MHC Class II antigens, and normal or only slightly reduced numbers of MHC Class I antigens and β-2 microglobulin. Several MHC Class II transcription factors, most notably RFX5 and CIITA, were shown to have mutations in many of these patients. Both humoral immunity (especially responses to protein antigens) and cell-mediated immunity is profoundly impaired in these patients, and they usually die within one year of life unless treated by bone marrow transplantation. Defective MHC Class I expression is a rare abnormality that manifests in the absence of MHC Class molecules, reduction in $CD8^+$ T cell numbers, and impaired responses to respiratory tract bacterial infections. Surprisingly, responses to viral antigens in these patients are intact, presumably because of the "back-up" by NK cells. In some patients with this immunodeficiency, mutations in TAP1 or TAP2 proteins (see Chapter 6) were found, indicating that the pathogenesis of this disease may involve abnormalities in MHC Class I peptide loading and final assembly.

18.6 Are there genetic diseases where immunodeficiency is an important part of a disease, but is associated with other hereditary abnormalities?

Yes. Two of such diseases are worth mentioning. In **Wiskott–Aldrich syndrome**, two principal symptoms – eczema (skin rashes) and thrombocytopenia (reduced number of platelets) – are seemingly unrelated to the immune system. However, patients also demonstrate an increased susceptibility to infections, especially infections with encapsulated bacteria like pneumococci or *H. influenzae*. Recently, it

CHAPTER 18   Immunodeficiencies

has been shown that the principal defect in Wiskott–Aldrich syndrome is an X-linked recessive mutation causing an abnormality in a cytosolic protein expressed in bone marrow-derived cells, including precursors of platelets and white blood cells (granulocytes, monocytes, and lymphocytes). This protein participates in signal transduction by interacting with adapter molecules like **Grb2** and other molecules related to GTP pathway (see Chapter 7). The main defect in the adaptive immunity of these patients is their unresponsiveness to T-cell-independent antigens (hence their susceptibility to encapsulated bacteria). These patients also show inadequate recruitment of leukocytes to the sites of inflammation (hence skin rashes), probably because of inadequate expression of adhesion molecules. Interestingly, the size of lymphocytes and platelets in these patients is smaller than in healthy individuals.

Ataxia-telangiectasia (AT) is another example of hereditary disease where immunodeficiency is associated with other genetic defects. AT is an autosomal recessive disorder. Clinically, the disease manifests as abnormal gait (ataxia), peripheral blood vessel dilations (telangiectasia), neurological disorders, high incidence of tumors, and immunodeficiency, especially impairment in IgA and IgG2 class switch and thymic hypoplasia. The gene associated with AT codes for an enzyme that is structurally related to phosphatidylinositol-3-kinase, which is involved in TCR-triggered T-cell activation (Chapter 7). The mutated form of this enzyme might cause abnormalities in T-cell signaling. The high incidence of tumors in TA patients may be related to the immunodeficiency, although another explanation is an abnormal sensitivity of the patient's cells to ionizing radiation.

18.7   What congenital immunodeficiencies concern primarily innate immunity?

Several hereditary disorders involve primarily nonspecific, innate immunity, in particular, phagocytes, inflammatory leukocytes, NK cells, and complement. **Chronic granulomatous disease** (**CGD**) is a rare X-linked recessive disorder that manifests in chronic infections with intracellular bacteria and develops because of a mutation or a transcriptional defect in the enzyme called **phox-91** (from "**ph**agocyte **ox**idase, mol.wt. **91** kDa"). This enzyme is a part of the phagocyte oxidative system, and its deficiency leads to inability to produce ROI (see Chapter 11). Since phagocytosis without ROI is incomplete, T_h1-mediated immune response tends to result in chronic granulomatous inflammation (see Chapter 12), which is often fatal to the patients. Interestingly, IFN-γ can up-regulate the deficient transcription of phox-91 and thus be a life-saving drug to these patients.

Inflammatory leukocytes are the primary target of two rare autosomal recessive disorders, called leukocyte adhesion deficiency-1 (LAD-1) and LAD-2. In LAD-1, the so-called β-2 integrins or CD11CD18 family of molecules (see Chapters 7 and 11) are abnormal, and in LAD-2, there are mutations in the gene that codes for an enzyme controlling the sialyl Lewis X component of E- and P-selectin ligands (Chapters 11 and 12 and Section 15.19).

Granulocytes and NK cells are impaired in patients with the **Chediak–Higashi syndrome** – a rare autosomal recessive disease in which leukocyte granules tend to fuse because of a mutation in a cytosolic protein participating in granule formation. Symptoms of the disease include the appearance of giant granules and giant lysosomes in various cells, albinism (because of a dysfunction in melanocytes), neurological, and blood clotting disorders. In some patients, dendritic cells are also affected, leading to the impairment of both innate and adaptive immunity. There is a mouse model for this disease, called "**beige mouse**." In mice with "beige" disorder, innate immunity is severely impaired and giant granules are detected in granulocytes and NK cells; the affected gene is homologous to the mutated gene in the Chediak–Higashi patients.

Complement is the target of various mutations that can lead to abnormalities in both innate and adaptive immunity. The most common genetic defect in complement system, however – **the defect in C2 and C4 complement proteins** – is **not** associated with an increased susceptibility to infections, perhaps due to the fact that the alternative complement pathway and Fc-mediated immune functions (that do not involve C2 and C4) are sufficient to contain most infections. The mutations in C2 and C4, rather, lead to an autoimmune disease that somewhat resembles SLE (see Chapter 17). Mutations in C3, factor D, and properdin (Chapter 13), however, lead to severe immunodeficiency and rampant bacterial infections.

18.8 What are the main environmental factors that lead to weakening of normal immune responses?

Among these factors, the four most common must be mentioned: malnutrition, ionizing radiation, tumor growth, and infection. **Malnutrition** is the most common and very underappreciated cause of immunodeficiency, especially in Third-World countries. It has been estimated that worldwide, a person dies of hunger every 15–20 seconds; many of these deaths are actually caused by infections that could be contained should it not be for the immunodeficiency. **Ionizing radiation** became a serious medical problem recently, when large contingencies of people were exposed to it because of nuclear reactor breakdowns. Small doses of radiation tend to stimulate immune responses slightly (a phenomenon known as "radiohermesis"), but they also negatively affect the fine tuning of the immune system and may lead to exacerbation of infections, autoimmune reactions, and tumor growth. Larger doses of radiation are detrimental to lymphocytes, and lead to severe immunodeficiency. Patients with advanced **tumors** are always immunodeficient, likely because TGF-β and many other products of tumors are either inhibitory or toxic to lymphocytes. Among **infections** that lead to acquired immunodeficiencies, the infection with human immunodeficiency virus (HIV) requires special attention, and will be discussed in subsequent sections.

18.9 What is the nature of immunodeficiency caused by HIV?

The virus known as **human immunodeficiency virus (HIV)** causes AIDS, a disease that leads to a profound impairment of immunity and, eventually, to the ablation of the patient's immune system and death. There is no known cure or prophylactic treatment against AIDS. So far, HIV has caused the death of almost 14 million people.

CHAPTER 18 Immunodeficiencies

Almost 50 million people are infected worldwide, making the search for an effective treatment against AIDS one of the highest medical and humanitarian priorities.

18.10 What is HIV?

HIV is a species of RNA-containing viruses that belongs to the family Lentiviridae. Other viruses that belong to this family are visna virus, and the bovine, feline, and simian immunodeficiency viruses. All these viruses share the ability to cause slowly progressing, fatal diseases that manifest as immunodeficiency, wasting syndrome, and central nervous system (CNS) degeneration. Two closely related types of HIV, called HIV-1 and HIV-2, have been identified. These types differ in genomic structure and antigenicity. HIV, as well as other Lentiviridae, has a tropism to lymphoid cells and some other cell types (in particular, macrophages and dendritic cells), and uses specialized receptors expressed on host cells in order to enter them.

18.11 What is the genetic structure of HIV?

An infectious HIV particle (Fig. 18-1) contains **two identical strands of RNA**, each approximately 9.2 kb long. These strands are packaged into a **core** of viral proteins, and surrounded by a phospholipid bilayer **envelope** derived from the host cell membrane but including virally encoded membrane proteins. HIV genes are stretches of RNA and, to be transcribed, need a special enzyme called **reverse transcriptase (RT)**. The RT, together with a number of other viral enzymes, is encoded by one of the HIV RNA genetic regions called **pol**. Viruses that use RNA as their genome and RT as the means for transcription are called **retroviruses**.

The genetic structure of HIV is typical for all retroviruses and is, essentially, as follows. There are two **long terminal repeats (LTRs)** at each end of the RNA strand of the viral genome; these are responsible for integration of the reverse-transcribed viral DNA into host cell genome, and serve as binding sites for host transcription factors. The **gag** sequences code for core structural proteins. The **env** sequences code for the envelope glycoproteins gp120 and gp41, which are required for infection of cells. The above-mentioned pol sequences encode, besides RT, the enzymes protease, integrase, and ribonuclease, which are required for viral replication. In addition, an HIV particle contains regulatory genes such as tat, rev, vif, nef, vpr, and vpu. We will briefly discuss their role in later sections.

18.12 What is the HIV life cycle?

Briefly, it can be summarized as follows. The HIV life cycle begins when the viral particle binds to its receptor and co-receptor expressed on target cells (see later sections). The particle then enters the cell via membrane fusion. Once this happens, the enzymes within the viral nucleoprotein complex become active and begin the viral reproductive cycle. The nucleoprotein core of the virus becomes disrupted, the RNA genome is reverse-transcribed into a double-stranded DNA by

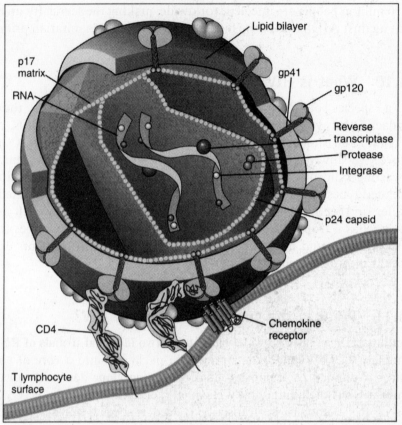

Fig. 18-1. HIV. An HIV-1 virion is shown next to a T-cell surface. Shown are the RNA genome, the core, and the envelope. The virion is shown to bind CD4 and chemokine receptors on the T cell. From Abbas, K., A.H. Lichtman, and J.S. Pober, *Cellular and Molecular Immunology*, fourth edition, W.B. Saunders, 2000, p. 455 (their Fig. 20-3).

the viral RT, and the viral DNA enters the nucleus. The viral integrase also enters the nucleus and catalyzes the integration of the viral DNA into the host cell genome. The resulting **provirus** (the integrated HIV DNA) may remain transcriptionally inactive for months or even years, but may be actively transcribed at once; this is a crucial factor that makes the infection latent or productive (i.e., resulting in the production of new virions and death of the infected cells).

Transcription of the provirus is determined by the above-mentioned LTR, and these, in turn, can be activated by host cell transcription factors such as NFκB and SP1. No wonder, **physiologic activation of target cells (T lymphocytes, dendritic cells and macrophages) greatly enhances HIV gene expression and viral replication**, since it involves signal transduction pathways that activate the above transcription factors (see Chapters 7 and 11). At first, regulatory genes are expressed; structural genes are then expressed, followed by processing and splicing of the transcripts and synthesis of viral proteins by the host cell. When the proteins are assembled into new viral particles, the host cell is completely destroyed (**lysed**), and new particles are released and ready to infect new host cells.

CHAPTER 18 Immunodeficiencies

18.13 What is important to know about the HIV binding to host cells?

The binding of the viral particles to host cells is mediated by two viral envelope proteins, **gp120** and **gp41**, which together form a complex called **Env**. The Env complex is expressed as a trimeric structure of three gp120/gp41 pairs. For a successful viral entry, **the Env must bind to the receptor of HIV, which is CD4, and a co-receptor of HIV, which is a chemokine receptor** (see Chapter 10). There is a certain sequence of events during this binding. First, the gp120 binds to CD4, and the resulting conformational change in gp120 promotes secondary binding of the same gp120 to the chemokine receptor. This, in turn, causes a conformational change in gp41, and the "activated" gp41 exposes a hydrophobic peptide and inserts into the cell membrane, promoting fusion between the viral particle and the cellular membrane. It is also remarkable that newly produced HIV particles do not need to be released and bind to new host cell; because HIV-infected cells tend to fuse, these particles (of HIV genomes) can be passed from one cell to another. The fusion of infected cells is facilitated by the expression of gp120 and gp41 on the infected cell's membrane.

18.14 What chemokine receptors serve as co-receptors for HIV?

Recent studies with the use of recombinant DNA transfection showed that at least seven different chemokine receptors can be such co-receptors. Among them are the receptors that are classified as CXCR4 (which bind chemokines called SDF-1α and SDF-1β), and CCR5 (which binds chemokines called RANTES, MIP-1α and MIP-1β, and other chemokines). According to the preference in cytokine receptors, all HIV isolates are now classified as **X4** (bind to CXCR4), **R5** (bind to CCR5), or **R5X4** (bind to both receptor families). Individuals who are deficient in chemokine receptor expression because of mutation may be resistant to infection with particular HIV isolates.

18.15 How exactly does HIV destroy the immune system?

HIV infection initially occurs when virions in the blood, semen, or other body fluids from one individual enter the cells of another individual as described in Section 18.13. The first cell type infected may be $CD4^+$ T cells or monocytes in the blood, or $CD4^+$ T cells and macrophages in mucosal tissues. Dendritic cells are very likely "disseminators" of this early infection because they are able to pick up various antigens (including virions) in epithelia and then migrate into the lymph nodes. There, the dendritic cells further pass HIV to T cells by direct cell–cell contact. Within several days, there is extensive viral replication in the lymph nodes, accompanied by **viremia** (presence of virions in the blood) and an **acute HIV syndrome** (a variety of nonspecific symptoms such as fever, headache, sore throat, swelling of regional lymph nodes, and skin rashes). The viremia helps the virus to disseminate throughout the body and infect target cells in all peripheral lymphoid organs. The mounted immune response to HIV at this stage may partially contain the infection, however. By approximately 12 weeks postinfection, the viremia usually drops to a low (but detectable) point. The immune response does not, however, eradicate the virus, which remains hidden (and often replicating) in the peripheral lymphoid organs.

A second phase of the disease is often called the **clinical latency period** because little or no clinical symptoms can be detected during it. However, this is actually a period of extensive viral replication in the peripheral lymphoid organs, and many $CD4^+$ cells (especially $CD4^+$ T cells) in these organs are destroyed at this time. At the beginning of the clinical latency period, only a few of the peripheral blood $CD4^+$ T cells harbor the virus, and the $CD4^+$ T-cell numbers in the blood are normal. By the end of this phase, however, these numbers begin to decline.

The next phase of the disease is the phase of **chronic progression**, during which the numbers of $CD4^+$ T cells in the peripheral lymphoid organs and the blood steadily decline. This might be accompanied by a rapid increase in the production of new viruses and the destruction of lymphoid tissue. Paradoxically, as the immune system attempts to mount anti-HIV responses as well as responses to other infectious agents, it exacerbates the infection, because the activation of immune cells greatly enhances HIV gene expression and replication (see Section 18.12). By the end of this phase, patients suffer from a wide variety of opportunistic infections including those with *Pneumocystis carinii*, *Candida albicans*, mycoplasms, and other infections.

The **terminal phase** of HIV inection (sometimes referred to as "full-blown" AIDS) manifests in a profound drop in the $CD4^+$ T cell count (below 200 cells per cubic millimeter), a dramatic climb of viremia, and rampant opportunistic infections. At this stage, patients often suffer from extensive weight loss, fever, fatigue, diarrhea, kidney failure, and dementia (because of the direct toxic effect of shed gp120 and other viral proteins on brain cells, as well as because of neurotoxic effect of cytokines triggered by the virus). Also, patients at this stage often develop malignant tumors, such as Kaposi's sarcoma, B-cell lymphomas, and other malignancies.

18.16 Are there any effects of HIV on the immune system other than direct killing of $CD4^+$ T cells?

Yes. During the clinical latency and the early stages of chronic progression, not very many T cells are infected by HIV, and yet the immune system is already deregulated. Many abnormalities in the immune function at these stages can be explained by the binding of shed gp120 to uninfected $CD4^+$ cells. The bound gp120 can impair the ability of CD4 to interact with the MHC Class II molecules and thus inhibit T-cell responses to soluble protein antigens. It has been reported that gp120 can also deliver an inhibitory signal upon binding to CD4. It is thought that gp120 can interact with some B lymphocytes, because it is able to bind to certain V_H regions of antibodies. Thus, it has been recently shown that gp120 directly binds to the V_H region coded by some V_H genes. The interaction of gp120 with certain antibody sequences can explain the hyperactivation and overall deregulation of B lymphocytes during the early stages of HIV infection. Other

CHAPTER 18 Immunodeficiencies

HIV proteins, notably Tat, can bind to some T-cell proteins that regulate transcriptional activation of cytokines. This partially explains the imbalance in cytokine production (e.g., inhibition of T_h1 cytokines IFN-γ and augmentation of T_h2 cytokines) observed in some HIV-infected individuals.

18.17 What are the effects of HIV on macrophages, dendritic cells, and other cells?

Macrophages can be bound and infected with HIV. Although they express only low levels of CD4, they express large amounts of cytokine receptors (especially of the CCR5 family), which may compensate for low CD4 expression. In addition, macrophages can be infected after they phagocytose dead or dying HIV-infected cells. Usually, macrophages are not killed by the virus, but instead become a **reservoir** for HIV. Much more of the virus is stored in macrophages than in T cells; examples of tissues where macrophages are the number one "storage room" for HIV include the brain and lungs. HIV-infected macrophages may be impaired in antigen presentation and cytokine production. **Dendritic cells** can be infected by HIV and, like macrophages, are usually not killed by it; it is believed that dendritic cells play a major role in transmitting the virus to naive T cells. It has been recently demonstrated that the transmission of HIV from dendritic cells to naive T cells is greatly enhanced by triggering of the CD40 molecule expressed on dendritic cells. Among other cells that feature prominently in HIV infection, **follicular dendritic cells** deserve mention because they trap large amounts of HIV bound by antibodies, through their Fc receptors. This may impair the physiologic function of follicular dendritic cells in the selection of high-affinity antibodies in germinal centers (see Chapter 4).

18.18 How does the immune system respond to HIV infection?

Both humoral and cell-mediated immune responses specific to HIV gene products occur in patients infected with HIV. The initial adaptive immune response to HIV infection is, essentially, **a massive expansion of $CD8^+$ CTLs**. Up to 10% of all peripheral blood $CD8^+$ cells at early stages of HIV infection are specific to HIV products (see Chapter 12). This strong CTL response largely accounts for clearance of many infected cells, and the fall in viremia after 12 weeks post-infection (see Section 18.15). However, this response fails to eradicate all HIV and, moreover, it can be harmful for patients because it kills their T cells. **Antibody responses** to a number of HIV proteins are detectable within a few weeks after infection, but there is no evidence that these antibodies are protective or beneficial. In fact, evidence suggests that the most immunogenic epitopes on gp120 are the least important for the function of this molecule, so most of the anti-gp120 antibodies are not neutralizing. Nonetheless, antibodies to several HIV proteins serve for diagnostic purposes, and standard kits helping to detect these antibodies in ELISA (see Chapter 3) are widely used in clinical laboratories. Some anti-gp120 antibodies may be neutralizing, and some can act in ADCC, but there is little evidence that these functions of antibodies are of any benefit to patients.

18.19 How does HIV evade immune responses?

The most important factor that allows HIV to escape immune attack is an extremely high rate of **mutation** in its genes, mostly due to error-prone reverse transcription. Also, like most viruses, HIV down-regulates the expression of MHC Class I molecules in the infected cells and thus lowers the efficiency of CTL responses. A regulatory gene called Nef (see Section 18.14) is thought to play a major role in the MHC Class I down-regulation. Interestingly, Nef product down-regulates HLA-A and HLA-B molecules, but leaves HLA-C intact, which may be an adaptation that does not allow NK cells to become activated upon binding to HIV-infected cells (see Chapter 11). In addition, the above-mentioned decrease in the production of T_h1 cytokines leads to an inhibition of cell-mediated immunity and further increases the patients' susceptibility to HIV infection.

18.20 How is HIV transmitted?

It is well known that AIDS is a sexually transmitted disease; in fact, **sexual contact** is the most frequent mode of transmission. Although in the early 1980s most new AIDS cases were observed among homosexual male partners, within recent years AIDS became more prevalent among heterosexuals; in any case, heterosexuals are as prone to the disease as homosexuals. While male homosexuals and bisexuals remain a group at risk in the USA, in Africa the vast majority of HIV infections occur by transmission from one heterosexual partner to another. The virus is present in the semen of infected males, and gains access to previously uninfected partners through the vaginal or rectal mucosa. Females can also transmit the virus to previously uninfected males. The second most frequent mode of transmission is via **infected blood**, either during needle-sharing or during blood transfusion. Finally, **mother-to-child** transmission is a major mode of transmission in pediatric AIDS cases.

18.21 What anti-HIV drugs are currently used, and how do they work?

Two classes of drugs are used, usually in combination.

1. **Inhibitors of RT** are nucleotide analogues; they include 3′-azido-3′-deoxythymidine (AZT), 2′,3′-dideoxyinosine, and 2′,3′-dideoxycytidine. RT inhibitors are usually efective at early stages of HIV infection and can delay its progression, but they do not halt the disease when its symptoms are pronounced. The major drawback for RT inhibitors is mutations in the pol gene that lead to appearance of drug-resistant HIV strains.
2. Viral **protease inhibitors** are efficient anti-HIV drugs when used in combination with two RT inhibitors; such combination of drugs can delay the

CHAPTER 18   Immunodeficiencies

progression of HIV infection and reduce viremia to undetectable levels for up to three years.

Both RT and protease inhibitors, are extremely costly, and have complicated administration schedules. Currently, inhibitors of chemokine receptors attract much attention as potential anti-AIDS drugs and perhaps will soon be investigated in clinical trials.

18.22   How can AIDS be prevented?

Perhaps the most efficient measure aimed at the prevention of AIDS is **public awareness**. Also, much effort is currently being invested in search for an **HIV vaccine**. Several candidate SIV vaccines have been formulated and are on trial (it should be noted that SIV, i.e., simian, nonhuman primate, immunodeficiency virus, is closely related to HIV genetically and antigenically). These include live, attenuated SIV vaccines and recombinant vaccines with deletions of parts of viral genome, e.g. nef. Another approach is to use live recombinant non-HIV viral vectors carrying HIV genes. This approach seems less hazardous (for obvious reasons), but it retains the possibility of eliciting efficient anti-HIV CTL responses. One of such vectors, called the canarypox vector, containing some HIV genes has already been tried on human volunteers and was shown to elicit strong CTL responses to the expressed HIV antigens.

Questions

REVIEW QUESTIONS

1. Why do X-linked agammaglobulinemia patients lack germinal centers?
2. Why does the condition of patients with DiGeorge syndrome improve after the first year of life?
3. In having an ability to detect genetic polymorphisms linked to X chromosomes, how would you tell that a woman is a carrier of an X-linked SCID mutation?
4. Can flow cytometry be used for the diagnosis of selective IgA deficiency?
5. A patient presented with chronic infections and a complete absence of IgG3 and IgA in the serum. How would you diagnose the patient? (Name two possibilities and tests to differentiate between them.)
6. What is the cause of the bare lymphocyte syndrome, and what impairment of the immune system features most prominently in these patients?
7. A kindergarten personnel worries about a three-year-old boy who cannot learn to walk, has dilated blood vessels seen on his hands, feet, and ears, and is extremely prone to "flu." What hereditary disease would you suspect, and why? How would you confirm the diagnosis?

8. Why do mutations in C2 and C4 lead to a disease that resembles SLE?
9. What facts argue against the notion that AIDS is a "gay plague?"
10. How many LTRs does any one HIV particle have? Where are they located? What is their importance for the course of HIV infection?
11. What is the difference between HIV virus and HIV provirus?
12. If a murine cell that does not express the human CD4 is successfully transfected with the human CD4 DNA, will it make it susceptible to HIV infection? Why, or why not?
13. A patient presented with the acute HIV syndrome and high viremia. After four months of extensive anti-HIV therapy, the level of viremia dropped to barely detectable. The patient's relatives say that he is "cured," and insist on termination of the therapy. How would you explain to them that the therapy must be continued? What tests, besides the test for viremia, must be done to be better oriented in the patient's status?
14. HIV is sometimes referred to as a "B-cell superantigen." What justifies this definition?
15. In volunteers treated with the canarypox vector with inserted HIV genes, what positive and what negative consequences should be expected, given that they show strong anti-HIV CTL responses?

MATCHING

Direction: Match each item in Column A with the one in Column B to which it is most closely associated. Each item in Column B can be used only once.

Column A

1. CBA/N
2. Cannot degrade dATP
3. Differentiation
4. CIITA
5. No ROI produced
6. Malnutrition
7. Transmit HIV to naive T cells
8. Some HIV isolates
9. Toxic to neurons
10. Nef

Column B

A. Does not affect HLA-C
B. Lymphocytes
C. Impaired in CVI
D. gp120
E. No response to polysaccharides
F. Bind to CCR5
G. Mutation in phox-91
H. Acquired immunodeficiency
I. Dendritic cells
J. Bare lymphocyte syndrome

Answers to Questions

REVIEW QUESTIONS

1. Because they have no mature B cells, which are required for the formation of germinal centers.
2. It is not known, but likely reasons are that either a small part of thymic tissue still develops and functions, or that extrathymic sites of T-cell maturation exist and begin to function by that time.

CHAPTER 18 Immunodeficiencies

3. Use an assay that can detect a parameter (e.g., size) of two homologous X-linked alleles simultaneously. In normal control females, both alleles must be present in a DNA preparation from B cells because homologous X chromosomes are inactivated at random. In the female carriers, only the diseased X chromosome is inactivated in surviving B cells; therefore, only one of two homologous alleles will be detected.
4. No, because the membrane expression of IgA in these patients is normal.
5. Possibility one: selective antibody isotype deficiency. Possibility two: X-linked hyper-IgM syndrome. Additional tests: concentration of IgG1 and IgG4; macrophage activation *in vitro* in the presence of the patient's polyclonally activated T cells; pedigree.
6. The cause of this disease is a mutation in transcription factors that regulate MHC Class II expression, notably RFX5 and CIITA. The most characteristic feature of this disease is the absence of antigen presentation to $CD4^+$ T cells.
7. Ataxia-telangiectasia, because the symptoms point to neurological disorders and immunodeficiency. To confirm the diagnosis, the boy's cells (e.g., skin fibroblasts) should be tested for radiation sensitivity), and a number of laboratory tests should be used to evaluate his immune system.
8. Probably because these complement components are important for the clearance of immune complexes.
9. The incidence of AIDS among heterosexuals is currently the same as among homosexuals. In Africa, the vast majority of HIV-infected individuals are known to have been infected after heterosexual contact.
10. Four (two on each RNA strand). They are located at the ends of each RNA strand.
11. The virus exists independently of cell, and has its own protein core and envelope. The provirus is the genetic information of the virus incorporated into host cell's genome; it does not exist independently of the host's cell.
12. No, because it lacks the human chemokine co-receptor.
13. They must be told that viremia means the presence of virus in the blood; viremia can go down, and yet the virus can live and replicate in the patient's lymph nodes, tonsils, spleen, etc. Additional tests: $CD4^+$ T cell count and microbiology laboratory tests for opportunistic pathogens.
14. Because its protein, gp120, can interact with some, although not with all, surface antibody V-regions and thus activate some, although not all, B-cell clones.
15. Positive: eradication of the virus. Negative: killing of the patients' own T cells.

MATCHING

1, E; 2, B; 3, C; 4, J; 5, G; 6, H; 7, I; 8, F; 9, D; 10, A

INDEX

ABO antigens, 259, 264
ABO blood groups, 264–265
Abscess, 192
Accessory cells, 3, 17–21
   (*See also* Antigen, processing and presentation)
Accessory molecules, 16, 115–116, 122–123
   (*See also* Tolerance, immunologic)
Activation-induced cell death, 17, 150–151
Adapter proteins, 118
Adenosine deaminase (ADA), 296
Adhesion molecules, 135–136, 202
Adjuvants, 45, 152, 283–284
Adoptive transfer, 129, 288
AIDS, 300–301, 306–307
   (*See also* HIV)
Alexin (*see* Complement)
Allelic exclusion:
   in B cells, 62
   in T cells, 115
Allergies, allergic reactions (immune mechanisms of), 227–233
Alloantisera, 81
Allotype, 56–57
Alpha-fetoprotein, 274
Alpha-1-antitrypsin, 192
Altered self model of T cell recognition, 96
Alternative complement pathway, 217–219
Anemia, autoimmune hemolytic, 286–287
Anergy, clonal 67, 147–148
Ankylosing spondilitis, 288
Antibodies:
   affinity and avidity, 41–42
   affinity maturation, 69–70
   antigen binding site of, 37–38
   definition, 3, 5
   detection of (*see* Immunoassays)
   classes, 38–39
      (*See also* Isotypes)
   complementarity-determining regions (CDRs), 38

Antibodies (*Cont.*):
   effector functions of, 32–33, 38, 213–216, 225–227
   framework regions (FRs), 38
   heavy and light chains of, 35, 62, 66–67
      (*See also* Isotypes; Kappa light chain; Lambda light chain)
   hinge, 37
   isotype, 38–40, 62, 68
   membrane-bound and secreted, 40
   monoclonal, 43–47
   polyclonal, 43
   structure of, 34–38
   variable and constant regions, 32, 34–40
Antibody-dependent cell-mediated cytotoxicity (ADCC), 215–216, 233
Antibody diversity, 6, 9, 43, 60–62
   genetic mechanisms of, 60–62
Antibody feedback, 137
Antigen(s), 2, 3, 40–41
   carcinoembryonic, 274
   processing and presentation, 94–107
   T-cell-dependent and -independent, 129–130
   tumor-specific and tumor-associated, 271–275
   (*See also* Bacteria; Parasitic infections; Transplantation; Tumor antigens; Viral infections)
Antigenic determinant(s) (*see* Epitope(s))
Antigenic variation, 240, 247
AP-1, 121
Apaf-1, 151
Apoptosis, 6, 16–17, 150–151
Ataxia-teleangiectasia, 299
Atopy, atopic individuals, 229
Autoimmunity, 282–288
AZT, 306
Azurophilic granules, 185

B7-1, B7-2, 123
   (*See also* CD28; Co-stimulation; CTLA-4)

INDEX

Bacteria, immune responses to, 238–242
Bacterial superantigens, 239–240
   (*See also* Superantigens)
Bare lymphocyte syndrome, 298
Basophils, 22, 229–233
B cells, 3–6, 14–15, 59–60, 66–73, 128–139
   activation of, 128–139
   antigen presentation by, 22, 132
   B-1, 15, 138, 184
   B-2, 15
   CD40, role in, 133–135, 138–139
   development of, 14, 23, 26, 66–73
Bcl-2, 151
Bcl-X, 151
Beige mouse model, 300
β-2 microglobulin, 83, 103
Blk, 117
Blood group antigens, 264–265
Bone marrow, 14, 23, 177–178
   transplantation, 265–267
Bruton's tyrosine kinase (Btk), 294
Bursa of Fabricius, 14

C1, 219–221
C1 inhibitor, 223–224
C3, 216–218
C3b, 217–218
C3 convertase, 217
C5, 217, 222
C5 convertase, 218, 221
Cachexia, 165
Calcineurin, 116, 120, 261
Calmodulin, 120
Calnexin, 100
Cancer, 270
   antigens associated with, 272–275
   immunotherapy of, 277–279
Carrier, 3
Caspases, 134, 151
CBA/N mice, 294
CC and CXC subgroup chemokines, 170
CCR5 and CXCR4 receptors, 170
   role in HIV binding, 303
CD1, 124
CD2, 193
CD4, 15, 84, 118, 146
$CD4^+$ cells, 15, 94, 200
   (*See also* T cells, helper T_h, T_h1, T_h2)
CD5, 15, 138, 184
CD8, 15, 84
$CD8^+$ cells, 15, 84, 94, 205, 206–207
   (*See also* Cytotoxic T lymphocytes)
CD10 (CALLA), 274–275
CD11aCD18, 136
CD11bCD18 (*see* Complement receptors)

CD11cCD18 (*see* Complement receptors)
CD14, 184–188
   (*See also* Lipopolysaccharide; Toll-like receptor protein)
CD19, 131
CD20, 274–275
CD21 (*see* Complement receptors)
CD28, 122–123
   (*See also* B7-1, B7-2; Co-stimulation)
CD30, 135
CD40, 133, 134
   (*See also* B cells, CD40, role in)
CD40 ligand, 133–135, 203, 206, 263, 297
CD44, 203
CD45, 118, 123
CD80, 123
   (*See also* B7-1)
CD86, 123
   (*See also* B7-2)
Cell-mediated immunity, 4, 199–209
Centroblast(s), 70
Centrocyte(s), 70
Chediak–Higashi syndrome, 299–300
Chemokine(s), 160, 169–170, 190, 303
   receptor(s), 161, 169–170, 303
Chronic granulomatous disease (CGD), 299
c-Kit ligand, 177
Class switch, 71–73, 114, 139, 172–175, 239, 299
Classical complement pathway, 217, 220–221
CLIP, 101
Clonal selection hypothesis, 8
Cognate interaction, 26
Cold agglutinin(s), 286–287
Collectin family, 195
Colony-stimulating factor(s) (CSFs), 178
Common variable immunodeficiency (CMI), 297
Complement, 33, 39, 194–195, 216–225
   receptors, 187, 222–223
Conformational epitopes, 41
Congenic mice, 78–80
Co-stimulation, 122–123, 134, 148
   (*See also* B7-1, B7-2; CD28; CTLA-4)
C-reactive protein (CRP), 195
Cross-priming, 206, 275
Cryptocidins, 184
CTLA-4, 123, 153, 263
CXCR4, 170, 303
Cyclooxygenase, 231–232
Cyclosporine, 261
Cytokine(s), 6, 15, 71, 158–178
   classification, 160
   general properties, 159–160
   receptors, 160–161
   signal transduction, 161–163
   (*See also* under names of particular cytokines)

INDEX

Cytotoxic (cytolytic) T lymphocytes (CTL), 15, 199, 205–209
   granules of, 205, 207–208

Dark zone, 70
Decay-accelerating factor (DAF), 139, 225
Defensins, 184
Degranulation:
   in CTL, 207
   in immediate hypersensitivity, 228–232
Delayed-type hypersensitivity (DTH), 201–205
Dendritic cells (DC), 17, 20–21, 70, 133
   in cancer therapy, 277
   development, 20–21
   subsets, 21
D gene segments:
   in B cells, 60, 63, 66
   in T cells, 114
Diacylglycerol (DAG), 120, 131
Dialysis, equilibrium, 42
DiGeorge syndrome, 295
Domains, immunoglobulin, 33, 36–38
Dryer–Bennett two gene hypothesis, 56–57
Dual recognition model of T cell receptor, 96

Effector caspases (*see* Caspases)
Effector lymphocytes, 6
Elastase, 191
ELISA, 48–50
   (*See also* Immunoassays)
Encephalitis, experimental autoimmune, 283–284
Endosome–lysosome compartment, 98
Endothelial cells, 22, 202–203
Eosinophil(s), 22, 233
Epitope(s), 6, 40–41
   conformational, 41
   cryptic (subdominant), 104
   dominant, 104
   linear, 41
   recognition by T cells, 96, 104, 115
Epstein–Barr virus, 10, 46, 244, 279
ERK, 119
E-selectin, 202–203
Extracellular proteins, processing and presentation of, 97–101

Fab fragment(s), 34
F(ab′)₂ fragment(s), 34
FACS (*see* Fluorescence-activated cell sorter)
Factor B, 218
Factor D, 218, 300
Factor H, 223
FADD, 151
Fas–Fas ligand system, 133–134, 150–151, 208, 285
   in autoimmunity, 285

Fc fragment(s), 34
FcγRI, 187, 214
FcγRII, 137, 187, 214
FcεRI, 219–230
Fc receptor(s), 137, 187, 214–215, 229–230
Fever, role of cytokines, 18, 165
Fibrin cocoon, 277
FK-506, 121, 261
FLICE, 156
FLIP, 151
Flow cytometry, 50–51
Fluorescence-activated cell sorter (FACS), 51
Follicles, lymphoid, 25–27
Follicular dendritic cells (FDC), 21, 70, 305
Follicular exclusion, 152
Fos, 119
Framework regions (FRs), 38
Fyn, 117

Gamma globulin, 33
γδ-T cell receptor, 123–124
γδ-T cells, 123–124
GDP–DTP exchange system (*see* GTP-binding proteins)
Germinal centers, 26, 70
   selection of high-affinity mutant B cells in, 70
Gld/gld mice, 285
Glycocalyx, 277
Graft(s), 77–78, 253–255
   allogeneic graft, 254
   autologous graft, 254
   syngeneic graft, 254
   xenogenic graft, 254, 263–264
Graft rejection, 259–260
   acute, 260
   chronic, 260
   hyperacute, 259
   prevention of, 261–263
Graft-versus-host disease (GvHD), 266–267
Granulocyte(s), 22
   (*See also* Innate immunity; Neutrophils)
Granulocyte colony-stimulating factor (G-CSF), 178
Granulocyte-macrophage colony-stimulating factor (GM-CF), 178
Graves' disease, 286
Grb2, 164
GTP-binding proteins, 161, 186, 207, 225, 230

H-2 genetic region (complex), 78
   (*See also* Major histocompatibility complex)
H-2M, 101
Haemophilus influenzae, 185, 248–249, 267
   vaccine against, 248–249
Haplotype(s) (in major histocompatibility complex), 82

Hapten(s), 3, 229
Heavy (H-) chain(s), in antibodies, 35, 61–62, 66–68
  gene rearrangements in, 61–62
Helminthic infections, 233, 245
Helper T cells, 15, 94–95, 128–129, 133–135
  subsets of, T_h1 and T_h2, 154, 169, 174–175, 196, 200
Hemolytic anemia, autoimmune, 286–287
Hemopoiesis, stimulation by cytokines, 160, 177–178
Hen egg lysozyme (HEL), 97, 152
  Goodnow's HEL-anti-HEL double transgenic system, 152
Hepatitis B virus, immunity to, 243
Herpes simplex virus, evasion of immunity, 244
High endothelial venules, 26
Histamine, 230
HIV (Human Immunodeficiency Virus), 301–307
  antibodies to, 305
  antigenic variation in, 306
  structure of, 301
  transmission of, 306
  vaccines against, 307
HLA complex, 81, 88–90
  (*See also* Major histocompatibility complex)
HLA-DM, 101
HLA typing, 86–88, 262–263
  (*See also* Tissue typing)
Homing receptors, 190
Homologous restriction factor (HRF), 225
Humoral immunity, 4, 19, 199, 212
  (*See also* Immunity)
Hybridomas, 10, 43–47, 109
  discovery of, 10, 43
  selection, 44
  T-cell hybridomas, 109
  use for monoclonal antibody production, 44–46
Hyper-IgM syndrome, 297–298
Hypermutation, somatic, in V-genes (*see* Somatic mutation in antibody genes)

ICAM-1, 136, 190
Idiotope, 154–155
  antibodies to, in cancer immunotherapy, 279
IFN-γ (*see* Interferon(s))
Igα and Igβ, 130
IgA, 28, 38–39, 226–227
IgD, 38–39, 67–68
IgE, 38–39, 227–230
IgG, 38–39, 72, 175, 214, 227
  subclasses of, 39, 175, 214
IgM, 38–39
  affinity and avidity of, 42
  complement-binding properties of, 221
IκB, 122
Immature B cell, 67
Immediate hypersensitivity, 205, 227–228

Immediate hypersensitivity (*Cont.*):
  mechanisms of, 228–233
Immune complexes, 47
  in disease, 285
Immune response, 1–2, 4, 6–7
  general features, 6–7
  phases, 6
  primary and secondary, 2, 6, 68–69
  regulation of, 154–155
  (*See also* Immunity)
Immune surveillance, 270–271
Immunity, 1–10, 182–197
  adaptive, 2
  cellular and humoral theories of, 19
  innate, 1, 182–197
  instructionist theory of, 7
  selectionist theory of, 7
Immunization, 4, 247–250
  active, 4
  passive, 4
  (*See also* Vaccination)
Immunoassays, 47–52
  solid-phase, 47–50
Immunodeficiencies, 293–307
  acquired, 293, 300–307
  congenital, 293–300
Immunodiffusion, 47
Immunofluorescent assay, 50–51
Immunogenicity vs. tolerogenicity, 152–153
Immunoglobulin(s) (*see* Antibodies)
Immunoglobulin gene(s), 44–73
  germline content of loci, 60–61, 68, 72
  rearrangements, 58–60, 62–67
  switch recombination, 71–72
  V, D, and J segments, 60–61
  (*See also* Variable regions)
Immunologic memory, 2, 6–7, 68–70
  (*See also* Memory)
Immunoprecipitation, 51
Immunoreceptor tyrosine-based activation motif(s) (ITAMs), 113, 130
Immunoreceptor tyrosine-based inhibition motif(s) (ITIMs), 193
Immunotherapy of cancer, 277–279
Immunotoxin(s), 46, 279
Inflammation, 22, 164, 185, 201–205
Innate immunity, 1–2, 182–196
Inositol-1,4,5-triphosphate (IP3), 120
Inside-out signaling, 135
Insulin-dependent diabetes mellitus (IDDM), 284–285
Integrin(s), 135–136, 187, 190, 202–203
Interdigitating dendritic cells, 21
Interferon(s), 91, 166–167, 174–175, 200, 203
  interferon α (IFN-α), 166–167
  interferon β (IFN-β), 166–167

INDEX

Interferons(s) (*Cont.*):
  interferon γ (IFN-γ), 174–175, 200, 203
    T_h subsets and, 154, 175, 200, 203
Interleukin-1 (IL-1), 151, 160, 165–166
  receptor, 166
Interleukin-2 (IL-2), 119, 120–121, 160, 170–172
  functions of, 172
  receptor, 171
Interleukin-3 (IL-3), 160, 177
Interleukin-4 (IL-4), 160, 173
  functions of, 173
  receptor, 173
  T_h subsets and, 154, 173, 228
Interleukin-5 (IL-5), 160, 174, 233
Interleukin-6 (IL-6), 160, 168
Interleukin-7 (IL-7), 160, 177–178
Interleukin-8 (IL-8), 170
Interleukin-9 (IL-9), 160, 178
Interleukin-10 (IL-10), 160, 168
Interleukin-11 (IL-11), 160, 178
Interleukin-12 (IL-12), 160, 168–169, 194, 202
  functions of, 168–169, 194, 202
  T_h subsets and, 168–169, 194, 202
Interleukin-13 (IL-13), 160, 176
Interleukin-15 (IL-15), 160, 169
Interleukin-16 (IL-16), 160, 177
Interleukin-17 (IL-17), 160, 177
Interleukin-18 (IL-18), 160, 169
Intraepithelial lymphocytes, 184–185
Isotypes, antibody, 38–40
  heavy chain, 38–39
  light chain, 39–40
  (*See also* Antibodies; Class switch)

Janus kinases (JAKs), 117, 161–162
J- (joining-) chain, 39
JNK, 119
J segments, 58–63, 114–115
  in antibody genes, 60–61, 63
  in TCR genes, 114–115
Jun, 119, 121–122

Kappa (κ) light chain, 39
  genes encoding, 61–63, 67
  receptor editing in, 148
Killer cell inhibitory receptor(s) (KIRs), 193
Kupffer cells, 20

Lambda (λ) light chain, 39
  genes encoding, 61–63, 67
Langerhans cells, 21
Large granular lymphocytes, 16
Late-phase reaction, 233
LBP protein, 189
Lck, 118

Lectin pathway of complement activation, 221
Leu-13, 131
Leukocyte adhesion deficiency (LAD), 299
Leukocytes:
  inflammatory, 22, 183, 190, 202–203
    recruitment of, 190, 202
  "passenger," 255, 258
Leukotrienes, 186, 231, 232
Light (L-) chains, 35, 37, 39, 61–63, 67
  gene rearrangements in, 61–63, 67
  isotype exclusion, 62
Light zone, 70
Linkage disequilibrium, in MHC typing, 263
Lipid mediators:
  in immediate hypersensitivity, 231–232
  in inflammation, 186
Lipopolysaccharide (LPS), 184, 187–189, 286
  receptor, 187–189
  signal transduction, 189
Lipopolysaccharide-binding protein, (LBP), 189
Lipoxygenase, 231
LMP2 and LMP7 genes, 102
Lpr/lpr mice, 255
L-selectin, 202
Lymph, 24–25
Lymphatic system, 24–25
Lymph nodes, 24–26
Lymphoblasts, 16
Lymphocytes, 3–8, 14–17
  activated, 16
  heterogeneity, 15
  clonal selection of, 7–8
  (*See also* B cells, T cells)
Lymphoid follicles (*see* Follicles, lymphoid)
Lymphoid organs, 13–15, 23–29
  generative (primary), 14–15, 23–24
  peripheral (secondary), 24–29
Lymphoid tissue, 13
  mucosal, 28, 226–227
Lymphokine-activated killer cells (LAK), 172, 267
Lymphokines, 159
Lymphotoxin (LT), 160, 176
Lysosomes, role in antigen processing, 98
Lysozyme, 183

Macrophage(s), 17–19, 185–193, 201–204
  activation of, 19, 189–193
  alveolar, 20
  characteristics of, 18
  in delayed-type hypersensitivity, 201–204
  phagocytosis, 190
  ROI formation, 190–192
Macrophage colony-stimulating factor (M-CSF), 160, 178
MAGE, 273

Major histocompatibility complex (MHC), 4, 32, 77–104, 115, 255–263
  alleles, 78–82, 262–263
  autoimmunity and, 288–289
  Class I genes/molecules, 81–83
    antigen presentation by, 97, 101–103
  Class II genes/molecules, 81, 83–84
    antigen presentation by, 97–101, 103–104
  expression, 90–91, 132
  mapping of, 78–81, 88–89
  organization of, 78–81, 88–90
  peptide binding, 84–85
Mannose-binding lectin (MBL), 195, 221
MAP kinases, 119
Marginal zone, 27–28
Mast cells, 185, 228–233
  in immediate hypersensitivity, 228–233
Mature B cell, 67
MBP protein, 283
Medulla:
  of lymph node), 25
  of thymus), 23–24
Membrane attack complex (MAC), 195, 216, 222
Memory, immune, 2, 6–7, 16, 68–73
  (*See also* Immunologic memory)
Memory cells, 6–7, 16
Microlymphocytotoxic test, 86
Migration inhibition factor (MIF), 160, 177
Minor histocompatibility antigens, 266
Mixed lymphocyte reaction (MLR), 86–87, 257
Monoclonal antibodies (*see* Antibodies, monoclonal; Hybridomas)
Monocytes, 17
Monokines, 159
Multiple sclerosis, 283
  EAE as an animal model of, 283–284
Mutations, somatic, in antibody genes, 69–70, 73
Myastenia gravis, 286
Mycobacterium tuberculosis, 201, 203–205
Myelomas, 43–45
  as a fusion partner in hybridoma technology, 43–45
  as a source of homogenous antibodies, 43

Natural killer cells (NK), 16, 163, 169, 172, 174, 175, 192–194
  mechanism of killing, 192–193
  receptors of, 192–193
Negative selection, 146–147
Neglect, death by, 145
Neonatal Fc receptor (FcRn), 227
Network hypothesis, 155
Neutralization by antibodies, 213–214
Neutrophils, 22, 164, 185–187, 192, 202–203
  in inflammation, 185–187, 192, 202–203
NFAT, 121–122

NFκB, 121–122
Nude mice, 129
Nurse cells, 24

Oligonucleotide primers (in PCR), 88
Oncogene(s), cellular proto-, 272
Opportunistic infections, in AIDS, 304
Opsonins, 19, 214
Opsonization, 19, 214, 224–225
Oral tolerance, 153

Parasitic infections, 245–247
Passenger leukocytes (*see* Leukocytes, "passenger")
PECAM-1 (CD31), 150
Peptide(s), 4, 82–86, 97–104
  antagonists (altered ligands), 150
  binding to MHC molecules, 84–86
  generation in the acidic vesicular compartment, 98–99
  generation in the cytosol, 101–102
  trafficking to MHC molecules, 100–104
    to MHC Class I molecules, 102–104
    to MHC Class II molecules, 100–101
Peptide-binding domain, of MHC molecules, 82–84
Perforin, 208, 222
Periarteriolar lymphoid sheath (PALS), 27
Peyer's patches, 14, 28
pH, role in antigen processing, 101
Phagocytosis, 18, 187, 190–191
  mechanisms of, 190–191
Phagolysosome, 190
Phospholipase C, 119–120, 186
Phox-91, 299
Pilin, 240
Plasma cells, 16, 23, 70–71, 168
  generation of, 16, 70–71
  IL-6 and, 168
Platelet-activating factor (PAF), 231
Platelet-derived growth factor (PDGF), 186, 192
Pleiotropy, cytokine, 159
Polyclonal antibodies (*see* Antibodies, polyclonal)
Poly-Ig receptor, 226
Polymerase chain reaction (PCR), 88
Polymorphism (in MHC), 77, 80
Polysaccharide antigens, 3, 28, 129–130
Positive selection, 145
Praustnitz–Kustner phenomenon, 228
Pre-B cell, 66
Precipitation, 47
Pre-Tα chain, 115
Primary follicles, 26
Pro-B cell, 66
Probes, labeled (in immunoassays), 51, 57–60
Programmed cell death (*see* Apoptosis)
Properdin, 218, 219

INDEX

Prostaglandins, 186, 231–232
Protease inhibitors, in AIDS therapy, 306–307
Proteasome, 101–102
Protein kinase C (PKC), 120, 186, 230
Protein tyrosine kinase(s) (PTK), 117–118, 130–131, 230
P-selectin, 202
Pseudogene(s), 89–90

Rac, 119–120
Raf, 118–119
RAG-1 and RAG-2, 65–66, 148
Ras·GTP system, 118–119
Ras proteins, 118–119
Reactive oxygen intermediates (ROI), 190–192
Reagins, 228
Recognition sequences, in antibody gene rearrangement, 64
Red pulp, 27–28
Redundancy, cytokine, 159–160
Regulators of complement activation (RCA), 223–225
Respiratory burst, 191, 203
Restriction enzymes, use in immunology, 57–58
"Reticuloendothelial system," 20
Retroviruses, 301
Rheumatoid arthritis, 29, 287
Rheumatoid factor(s), 287
RIP, 134
RNA splicing, role in B cell maturation, 67–68, 71

Sandwich solid-phase immunoassays, 50
Secondary immune responses, 7, 68–73
Secondary lymphoid organs, 24–29
Secretory component, 227
Self-MHC recognition by T cells, 96
Self tolerance, 7, 143, 145
Septic shock, 165–166
Serpentine(s), 161, 170
Severe combined immunodeficiency (SCID), 295–297
SHIP, 137
SHP-1, 137
Signal transduction (in the immune system), 16, 116–122, 133–134, 137, 161–163
Skin, as anatomic barrier, 1, 183
Slow-reacting substance of anaphylaxis, 231
SOCS, 163
Somatic mutation in antibody genes (see Mutations, somatic, in antibody genes)
SOS, 164
Southern blotting, 57–58,
Specific immunity (see Immunity, adaptive)
Spleen, 26–28
STAT(s), 161–163
Stem cells, hemopoietic, 14, 265–266
Sterile (germline) transcripts, 71

Steroids, use in transplantation, 262
Stromal cells, 160, 177–178
Subunit vaccines, 248
Superantigens, 239–240
 (See also Bacterial superantigens)
Suppressor cells, 154
Surrogate light chains, 63
SV40 virus, 277
Switch recombination, 71–72
Syk, 131
Synthetic vaccines, 249
Systemic lupus erythematosus (SLE), 285

Tail pieces, 39
TAP proteins, 102–103
TAPA-1, 131
Target cells, killing of, 205–208
Tetraspanin, 131
T_c (see T cells, cytotoxic)
T cells, 3, 4, 15, 23–24, 27, 70, 94–97, 108–124, 128–130, 133–135, 149–150, 199–209, 255
 activation, 115–124
 antigen recognition by, 4, 94, 96, 103, 104
 antigen specificity of, 4, 94–95
 cytotoxic (T_c, CTL), 15, 94–95, 199, 205–209
 $\gamma\delta$, 123–124
 helper (T_h), 151
  T_h0, 202
  T_h1, 154, 174–175, 202, 204
  T_h2, 154, 173, 175, 202, 205
 maturation of, 15, 23–24
 signal transduction in, 116–124
 subsets (see Cytotoxic T cells; Helper T cells)
 thymic selection of, 146–147
 (See also Lymphocytes)
Thoracic duct, 24–25
Thymocytes, 23–24
Thymus, 23–24, 146–147
Tissue typing, 86–88, 262–263
 (See also HLA typing)
Tolerance, immunologic, 7, 143–155
Toll, 166, 189
Toll-like receptor protein 4, 189
Tonsils, 28–29
Toxoids, 248
TRAF proteins, 134, 139
Transcription factors, 121–122
Transfection, in tumor immunotherapy, 277
Transforming growth factor beta (TGF-β), 160, 175–176
Transfusion, 253, 264–265
Transplantation, 253–267
 bone marrow, 265–297
Trypanosomes, immunity to, 247
Tuberculosis (see *Mycobacterium tuberculosis*)

Tumor, 270
Tumor antigens, 271–275
　tumor-associated, 271
　tumor-specific, 271
Tumor-infiltrating lymphocytes (TILs), 278
Tumor necrosis factor (TNF), 160, 163–166
　pathologic effects, 165
　physiologic effects, 164
　receptor, 164

Ubiquitin, 102

Vaccination, 32, 277
　(*See also* Immunization)
Vaccine(s), 247–250
　conjugate, 248
　killed, 248
　live attenuated, 247–248
　plasmid, 249–250
　subunit, 248
　synthetic, 249
　viral vector, 249
Variable regions, 32–38, 112, 114–116, 123
　in antibodies, 32–38, 56–67, 69–70

Variable regions (*Cont.*):
　genetic encoding of, 56–67, 112, 114–116, 123
　in T cells, 112, 114–116, 123
Variant surface glycoproteins (VSGs), 247
Viral infections, immunity in, 241–245
Viral neutralization, by antibodies, 213–214
V regions (*see* Variable regions)

Western blotting, 51
Wheal and flare reaction, 230–231
White blood cells (*see* Leukocytes)
White pulp, 27
Wiscott–Aldrich syndrome, 298–299

Xenogeneic transplantation, 254, 263–264
X-linked agammaglobulinemia, 294
X-linked hyper-IgM syndrome, 297–298
X-linked SCID (*see* Severe combined
　immunodeficiency)

Zap-70, 116–118, 298
Zeta (ζ) chain, in T-cell receptor complex, 112–117
　role in signal transduction, 116–117
Zymogene(s), 194